THE EARTH FROM THE AIR 366 DAYS

YANN ARTHUS-BERTRAND

ptions edited by Isabelle Delannoy

ntributors: Christian Balmes, Dominique Bourg, Hosny El-Lakany, Peter H. Gleick, Brian Groombridge,
an-Marc Jancovici, Alain Liébard, René Passet, Maximilien Rouer, Pierre Sané,
an Simcock, Laurence Tubiana, John Whitelegg and Gary Haq

Through working on this project for more than ten years, I have understood and wanted to stress that, more than ever, the present levels and methods of consumption, production, and exploitation of resources are not viable in the long term. I wanted to illustrate this conclusive stage where the alternative that sustainability offers should help bring about changes that allow us "to respond to the present needs without compromising the capacity of future generations to respond to theirs." The essays and images that follow, inseparable from the texts that accompany and complement them, invite everyone to reflect on the evolution of the planet and on the future of its inhabitants. We all can and should act every day for the future of our children.

YANN ARTHUS-BERTRAND

Contents

THE EARTH IN FIGURES

TAKING STOCK

The world's population has more than doubled in the last fifty years, and a further 50 percent increase is expected by 2050.

YEAR	WORLD POPULATION (IN BILLIONS)
1950	2.5
2002	6.2
2050	8 to 9[1]

If we imagine the Earth as a village with 100 inhabitants, sixty of these people live in Asia. Fourteen of them live in Africa. There are nine each in South America and Europe, five in North America, two in Russia, and one in Oceania.[2]

A large proportion of the population has no access to adequate health care or education.

- 815 million people (one in seven) are malnourished.[3]
- 1.1 billion people (one in six) have no access to clean drinking water.[4]
- 133 million children (one child in five) do not go to school;[5] 97 percent of these children live in developing countries.
- 860 million adults (one adult in five) cannot read or write;[6] of these, 544 million are women.
- 19 percent of children aged five to fourteen go to work.[7]

Rising demand for energy and a massive reliance on fossil fuels, which are not renewable, mean that these resources will not last long.

ENERGY SOURCE	SHARE OF WORLD OUTPUT
Oil	34.9 percent
Gas	21.1 percent
Coal	23.5 percent
Nuclear	6.8 percent
Other (renewable energy sources)	13.8 percent (10.7 percent of which is biomass, mainly wood)

- Between now and 2020, our energy use has been projected to increase at the rate of 1.5 percent per year.[8]
- In 2000, 79.5 percent of primary energy production came from fossil fuels.[9]

- Today, the world burns as much oil in six weeks as, in 1950, it burned in one year.[10]

IMPORTANT INEQUALITIES

For many, good health is out of reach.

LEAST DEVELOPED COUNTRIES	DEVELOPING COUNTRIES	INDUSTRIALIZED COUNTRIES	WHOLE WORLD
Mortality rate in children under the age of five:[11]			
157	89	7	82
Risk of mother's death as a result of childbirth:[12]			
1 in 16	1 in 61	1 in 4,085	1 in 75

- In 2002, average life expectancy worldwide was sixty-seven. In Africa, average life expectancy was fifty-three, while in North America it was seventy-seven—and in Japan, eighty-one.[13]
- HIV/AIDS affected 42 million people[14] in 2002. Of those, 90 percent lived in developing countries and 75 percent lived in sub-Saharan Africa.[15]

A minority of the population is using up most of the world's resources.

- 20 percent of people live in developed countries:
  —They consume 53 percent of the world's energy.[16]
  —They eat 44 percent of the world's meat.[17]
  —They own about 80 percent of the world's motor vehicles.

Inequality in access to clean water produces huge differences in consumption.

COUNTRY	DRINKING WATER CONSUMPTION PER PERSON PER DAY[18]
France	255.2 quarts (290 liters)
United States	519.2 quarts (590 liters)
China	77.44 quarts (88 liters)
Mali	10.56 quarts (12 liters)

HUMANITY FACES COMMON ISSUES

Human activity is enhancing the greenhouse effect, leading to climate change.

- For the last 150 years, industry has been releasing carbon dioxide (CO_2) into the atmosphere at a rate millions of times greater than the rate at which it was originally accumulated underground.
- If nothing is done, global temperatures could rise by up to 42.8 °F (6 °C) by 2100, with economic, social, and environmental consequences.

- To limit the disastrous effects of global warming, the world's CO_2 emissions would have to be cut by 50 percent; this would require an 80-percent reduction by developed countries.

Measures must be taken to ensure that our level of energy consumption allows development in a sustainable way.

COUNTRY	EMISSIONS OF CARBON EQUIVALENT IN KILOGRAMS PER PERSON PER YEAR (BASE 2000[19])	FACTOR BY WHICH PRESENT EMISSIONS EXCEED SUSTAINABLE LEVEL (OF 500 KG/PERSON/YEAR)
United States	6,718	13.5
Germany	3,292	6.5
France	2,545	5
Mexico	1,000	2
Mozambique	416	Less than

Reduction in biodiversity does not just damage the fertility of our fields—it reduces our chances of finding new medicines.

- In 2002, 24 percent of mammals, 12 percent of birds, and 30 percent of fish were in danger of extinction.
- Half of the world's mangrove forests, essential to the survival of 75 percent of the world's commercially exploited marine species, have disappeared.
- Primary tropical forests, which are reserves for the world's biodiversity, are disappearing fast—at a rate of about 15 million hectares per year. (The nation of Ireland occupies about 7.5 million hectares, half the area of the primary tropical forests lost every year).[20]

SIGNS OF REALIZATION

- Since the United Nations held the 1992 Earth Summit in Rio de Janeiro, the serious risks, and consequences, of the destruction of natural resources have been increasingly taken into account by politicians all over the world.
- Since 1990, more than 7,000 international non-governmental organizations (NGOs) have been formed[21] with a focus on environmental and social issues. The number of international NGOs and the actions they are taking are both growing inexorably.
- The growing success of fair-trade products allows everyone to take action as a consumer. An example is the Max Havelaar label, whose sales outlets grew in number from 250 to 3,500 in just three years.

THERE IS STILL TIME TO ACT

- Our apparent wealth is founded on economic growth, whose indicators ignore the exhaustion of natural resources. Prices take no account of environmental and social costs. For example, the price of a pound of wheat does not include the cost that the public pays to clean up the polluted water supply used to grow the crop. According to Amartya Sen, 1998 Nobel Laureate in Economics, there is now an urgent need to reassess the basic workings of the market.
- The ball is in our court. We can all take action by reducing unnecessary consumption and putting pressure on governments and industry to embrace sustainable development. Today, the choice is still ours between regulating ourselves or having regulations forced upon us. There is more room to maneuver if we have twenty years to implement those regulations than if we have only one week.

MAXIMILIEN ROUER

President–General Director of BeCitizen

Engineer at the National Agronomy Institute in Paris and M.S.

BeCitizen was formed in 2000 with the mission of promoting sustainable development. In France, the company is a leader in sustainable development expertise and consultancy. BeCitizen is a member of France's national sustainable development council.

NOTES

[1] *Total Midyear Population for the World: 1950–2050.* U.S. Bureau of the Census, International Data Base (*www.census.gov/ipc/www/worldpop.html*).
[2] *World Population Data Sheet 2002.* Population Reference Bureau (*www.prb.org*).
[3] U.N. Food and Agriculture Organization—1997/1999 figures.
[4] U.N. Joint Monitoring Programme, September 2002.
[5] *Human Development Report, 2002.* U.N. Development Programme (*http://hdr.undp.org/reports/global/2002/en*).
[6] UNESCO—2000 figures.
[7] UNICEF.
[8] Energy Information Agency (*www.eia.doe.gov*).
[9] International Energy Agency (*www.iea.org*).
[10] International Energy Agency (*www.iea.org*).
[11] *Vital Signs 2001.* Worldwatch Institute (*www.worldwatch.org*).
[12] UNICEF—2001 figures.
[13] UNICEF.
[14] *2002 World Population Data Sheet,* Population Reference Bureau (*www.prb.org*).
[15] UNAIDS: Joint United Nations Programme on HIV/AIDS (*www.unaids.org*).
[16] *Human Development Report, 2002.* U.N. Development Programme (*http://hdr.undp.org/reports/global/2002/en*).
[17] *Energy Statistics of OECD Countries, 1999–2000.* International Energy Agency (*www.iea.org*).
[18] Source: UN Food and Agriculture Organisation—2002 figures (http://www.fao.org/WAICENT/faoinfo/economic/giews/english/fo/fo0205/Y0000e13.htm).
[19] *Energy Statistics of OECD Countries, 1999–2000.* International Energy Agency (*www.iea.org*).
[20] *FAO/GIEWS—Food Outlook No. 2—May 2002.* U.N. Food and Agriculture Organization (*www.fao.org*).
[21] AQUASTAT: FAO's Information System on Water and Agriculture (*www.fao.org*).
[22] U.N. Framework Convention on Climate Change (*www.unfccc.int*); for emissions figures for Mexico and Mozambique, refer to Intergovernmental Panel on Climate Change (*www.ipcc.ch*).
[23] This is also the size of the state of Florida.
[24] *Human Development Report, 2002.* U.N. Development Programme (*http://hdr.undp.org/reports/global/2002/en*).
[25] *Ibid.*

SUSTAINABLE DEVELOPMENT: A PROJECT FOR CIVILIZATION

JANUARY

01	02	03	04	05

Sustainable development is like God in negative theology. Beyond slogans and a few sacred formulae, we cannot say, tangibly and for certain, what it is. On the other hand, we can see much more clearly what it is not, and could never be. It is not, for example, the indefinite continuation of our society's present habits. Sustainable development involves nothing less than building a new civilization, one that makes a partial break with the habits that characterize the way we live now.

The time-honored definition of sustainable development is that suggested by the World Commission on the Environment and Development, set up by the United Nations in 1983. Its president was the former Norwegian prime minister Gro Harlem Brundtland, and its conclusions were published in the 1987 report entitled *Our Common Future*. According to the commission, sustainable development is "development that meets the needs of the present without compromising the ability of future generations to meet their own needs." This means it "must not endanger the natural systems that support life on Earth: the atmosphere, the waters, the soils and living beings."

The contradiction with our present mode of development is glaring. We have disrupted all the main biogeochemical cycles in the biosphere, for example: the carbon cycle to the point that we are courting climatic disaster, the nitrogen cycle to the point of saturating soil and water, and the sulfur cycle to the point of destabilizing entire forest systems, such as that of the Vosges. Our obsession with short-term wealth thus leads to long-term disturbance of the system Earth. By the end of the century, average temperatures could rise by more than 10°C in the highest latitudes. The oceans' capacity to absorb carbon could be altered for thousands of years to come. The destruction of biodiversity threatens the evolution of species for millennia, and so on. Thus, what does not constitute sustainable development is blindingly clear: this negative definition includes ignoring

06 07 08 09 10 11 12 13 14 15

or denying the long-term effects of our actions, cynically relying on the supposed technological expertise of our descendants to resolve the problems we create, and leaving market forces in sole charge of managing our relations with the environment.

In contrast, what might be the features of a sustainable society? Its production would no longer rely on using ever greater amounts of energy and raw materials. Consumption would be based on provision of lasting services rather than the increasing obsolescence of goods; trade would not be wasteful of energy; and research would be driven by social need rather than by market forces. Important decisions in the social sphere would be taken collectively, based on widely available information and the participation of the greatest possible number of people.

Sustainable development is also a new way of interpreting the common good, which includes our future interests—that is, those of future generations—as well as those of other living things. Thus, the common good also includes preservation of the natural systems on which our existence depends—such as air, water, soil, biodiversity, and climate—and the ecological services they provide, such as rain, pollination, regulating temperature, keeping soil fertile, and purifying the air and water. This universal conception of the common good does not mean that a multitude of different types of sustainable development could not flourish, based on natural and cultural heritage and fueled by the products of society's creativity.

Let us come back to the big decisions and collective choices that form the bedrock of society. Everything is run as if we had handed over control to two automatic mechanisms, the market and progress, both of which are inadequate. Apart from exceptional circumstances, such as in a referendum, the popular will is hardly ever expressed in practice. The individual is supposed to choose his or her own goals directly—with happiness first among these—and, indirectly, a host of other things via the market. But is happiness possible in this threatening situation, when the choices made by others are damaging to

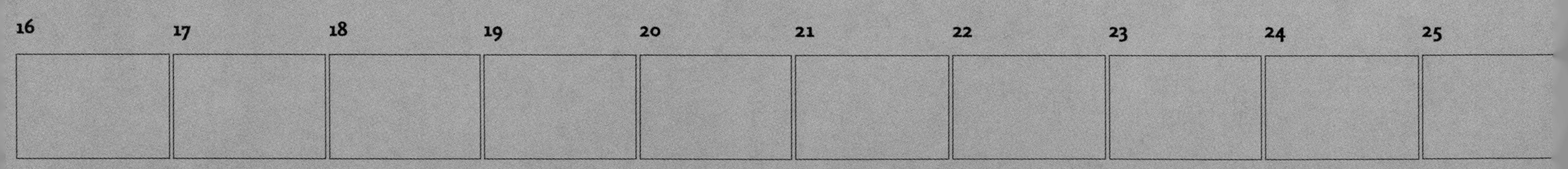

health, even to human nature itself? As for technological progress, it increasingly takes the form of a headlong race toward speed and power, with little sign that it is improving the condition of the majority. Some of the visions of progress touted today have nothing to do with Francis Bacon's hope of returning to the blissful state of Adam and Eve before the Fall. Take the prospect of humanity, devastating the universe in its quest for energy supplies, or that of people whose lives go on and on for centuries. Are those not terrifying rather than appealing?

A sustainable society should, on the contrary, tackle the problems and especially the contradictions between our different interests. It should allow us both to decide collectively what sort of future we want and to reject the future we do not want at any price. It should apply itself to promoting true individual choice, without endangering the ability to make choices that can only be made collectively and that are partly responsible for safeguarding our individual freedoms. The environment has always been a sphere in which individual and collective freedoms must perforce come together.

DOMINIQUE BOURG
Professor, University of Troyes, France
Director of the Centre de recherches et d'études interdisciplinaires sur le développement durable

JANUARY

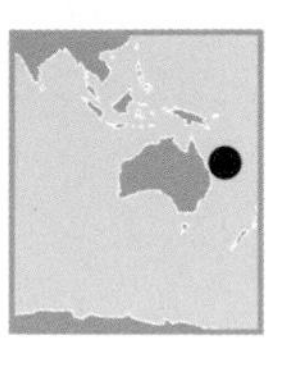

Heart in Voh in 2002, New Caledonia, France (20°57' S, 164°41' E).

No, this clearing wasn't carved by man. Nature is the originator of this traced heart in the mangrove, near Voh, on the west coast of Grande Terre island. The mangroves were formed by trees adapted to the brackish-water tides. In these forests, surfaces of bare ground ("tannes") appear, in which the forms spring up by chance. It's in the more elevated areas, hence less often flooded, where salt is concentrated by evaporation, causing the death of the mangroves. It's this phenomenon that is at the beginning of the heart of Voh. Flying over the heart in 2002 reveals its evolution since the 1990 shot. The vegetation grew back into the inside of the heart, where the salt got rid of almost 10 acres (4 ha) of it after a drop in the salinity caused by a change in the tidal flood conditions. The light patch in the foliage is a result of the blast of air from the second helicopter's blades. If the salinity continues to drop, the mangrove will close up completely within the heart. If the salinization comes back, the heart will rebuild itself. Nature will decide. But perhaps it must return?

01

JANUARY

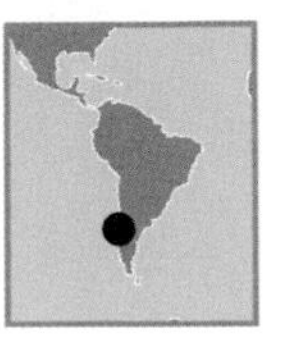

Snow-covered summit of Villarrica volcano, Chile (39°25' S, 71°57' W).

Villarrica is one of the most active volcanoes on the planet, and the sulfurous gases it gives off are a constant reminder that its crater contains a lake of boiling lava. Each of its most recent eruptions—in 1964, 1971, and 1984—killed about thirty people. Those coming may be more deadly, for more and more tourists visit the region to ski. However, the regional emergency bureau keeps a close watch on the volcano's activity, and the slightest explosion—accompanied by a plume of smoke—triggers an alert. Skiers are then advised to leave the *pistes*, while local residents begin carrying out their prepared evacuation plans. Over the last 30 years, Chile, which has 2,085 volcanoes and frequent earthquakes, has perfected an efficient means of dealing with natural disasters, thanks to the advice of international experts from bodies such as the World Bank and United Nations Development Programme, and to close collaboration with other South American countries. Together, the latter have succeeded in reducing deaths from such events by a third.

02

JANUARY

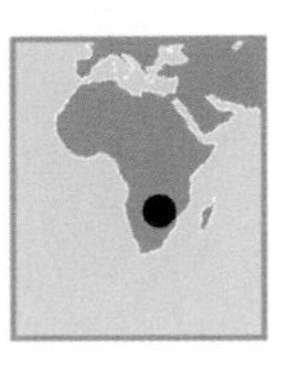

Boat in the marshes of the Okavango Delta, Botswana (18°45' S, 22°45' E).

The *mokoro,* a delicate traditional craft made from a hollowed-out tree trunk, is the only means of transport local people use to negotiate the marshy labyrinth where southern Africa's third-longest river meets its end. After an 800-mile (1,300-kilometer) journey that began in Angola, the Okavango ends here, north of Botswana, in a vast inland delta covering some 5,790 square miles (15,000 square kilometers). It will never reach the sea, for the 12 billion cubic meters of water it carries each year are gradually soaked up by the Kalahari Desert, or evaporate in the dry air. Before disappearing, the river forms a large wetland, inhabited by a prodigious number of wild animals. But the annual invasion of some 45,000 tourists and a plan to drain rivers are threatening the marsh and its wildlife. The rapid shrinkage of wetlands and estuaries is a worldwide problem: half the world's wetlands have disappeared since 1990.Yet they play a central part in human communities, notably by controlling floods and preserving drinking water supplies.

03

Roof terrace, Congress Center Auditorium, Monte Carlo district, Monaco (43°42' N, 07°23' E).

This is the remarkable story of the blocks of lava from Volvic. They were hewn from the quarries of the Massif Central, cut into slabs, and enameled in Provence; then they were cut into 24,000 lozenges in fourteen dazzling colors, and they now adorn the roughly 16,100 square feet (1,500 square meters) of the roof of the Congress Center Auditorium in Monaco. Built on rock-solid concrete piles, this hexagonal building perches above the Mediterranean in front of the famous casino, like a new growth on Monte Carlo's very urban seafront. The geometrically tiled terrace that serves as its roof, across which these two people are strolling, is the work of Victor Vasarely, who created it in 1979 with the architect Jean Ginsberg. Entitled *Hexagrace*, it evokes a recurring theme in Vasarely's work and gives concrete form to the artist's central ideas of integrating plastic beauty with the everyday urban environment. The avant-garde building of the Vasarely Foundation, in Aix-en-Provence, completed in 1976, is also a remarkable illustration of this. It contains an exhibition of the work of Vasarely (who died in 1997, aged 91), a master of abstract geometric and kinetic art.

JANUARY

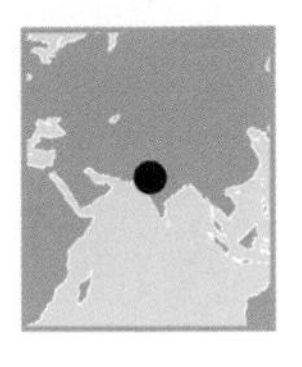

Cloth for saris drying, Jaipur, Rajasthan, India (26°24' N, 75°48' E).

Textiles are an ancestral craft in Rajasthan, northwest India, dominating the Chipa community of painters and dyers. Curcuma and pomegranate bark are used along with a knotting technique to dye the cloth yellow. The saris are then laid out in the sun to dry, soaked in a solution to fix the colors and, after being washed two or three more times and dried, they are ready to be sold. This traditional feminine garment hides an equally "traditional" role for women in society. Although customs have become less strict, once they are married, many Indian women often live "in *purdah*," that is, closeted at home, for reasons of expediency. Almost 90 percent of marriages in India are still arranged, and advertisements in Sunday newspapers are still classified by caste. Divorced women and single mothers are so frowned upon that they do not even figure in official statistics.

05

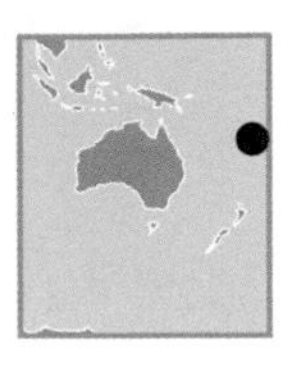

Catamaran in the Society Islands, French Polynesia, France (17°00' S, 150°00' W).

The waters around Indonesia contain 18 percent of the planet's coral. Australia has 17 percent, and the Philippines 9 percent. French overseas departments have 5 percent—that is, 5,512 square miles (14,280 square kilometers) of coral formations like those carpeting the seabed in the clear waters of the Polynesian archipelago in the Pacific Ocean, above which this catamaran appears to be flying. Tiny algae live in symbiosis with coral and stimulate its calcification. However, coral remains shrouded in mystery, for although the organisms that produce it are known, the mechanisms involved are not. There are local solutions to most of the threats that face coral, such as pollution, damage, and silting; the challenge is to put them into practice. However, the growing concentration of carbon dioxide (CO_2) in the atmosphere may lead to alteration of the chemical balance of seawater, which at present allows the coral skeleton to form. Having adapted to a 393-foot (120-meter) rise in sea level with the end of the Ice Age 15,000 years ago, will coral be able to cope with these new changes?

JANUARY

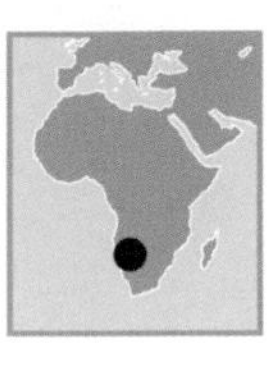

Spitzkoppe Massif at sunset, Damaraland region, Namibia (22°03' S, 17°02' E).

The peaks of the Grosse Spitzkoppe (5,666 feet, or 1,728 meters) and the Kleine Spitzkoppe (5,193 feet, or 1,584 meters) rise above the wild, desert region of Damaraland in northwestern Namibia. Thanks to the combined action of wind, rain, and lava, these granite domes emerged from the sedimentary strata that once covered them and which eroded 120 million years ago. At that time, the coast of Brazil was pressed against that of Namibia to form a single ancient continent, Gondwana. Much more recent are the many prehistoric sites in this region, which include rock carvings 27,000 years old. Damaraland bears the name of one of Namibia's oldest ethnic groups, the Damaras. This people still accounts for 7 percent of the country's present population, a quarter of whom live in the region. Damaraland is now very sparsely populated, and it is one of the few parts of Africa where antelope, giraffes, elephants, and even black rhinoceros can roam free outside national parks or protected areas.

07

JANUARY

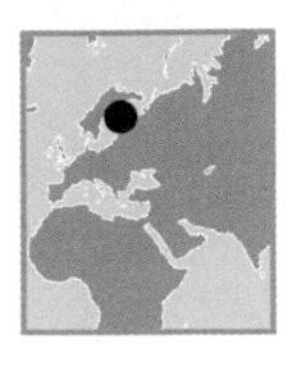

Formation of sea ice in the Turkü archipelago, Finland (60°27' N, 22°00' E).

Like a broken mirror, the thin ice of the Baltic breaks into sharp fragments that reflect the feeble light of a Finnish winter. This fragile sheet breaks only to reform more solidly. Sea ice floats on the current and can break up, but the fragments then meet again and pile together, until they reach their maximum thickness (an average of 20 to 25 inches, or 50 to 65 centimeters) in early April. The ice then disappears rapidly in spring, leaving the islands of the Turkü archipelago ice-free. Finland has no permanent ice sheet. Although the country is at the same latitude as Alaska and Greenland, it has a relatively temperate climate thanks to the Gulf Stream. This great warm sea current, which crosses the Atlantic and warms western Europe, may disappear as the climate changes and the Arctic ice cap melts. If this came about, Europe's climate would resemble that of Canada, which is at the same latitude.

08

JANUARY

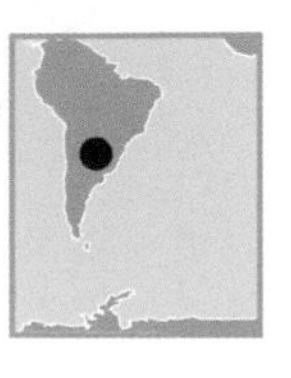

Herd of cattle on the plains bordering the Chimehuin River, Neuquén Province, Argentina (40°03' S, 71°04' W).

Patrolled by gauchos, this herd of Hereford cows crossing the Chimehuin River is returning to its home in the fields after seasonal migration to the high-lying pasturelands of the Andes cordillera. Partly covered by a thorny steppe, Neuquén is better suited for raising sheep than cattle, which remain a minority in the Patagonian region. Most of the country's bovine livestock live farther north, in the vast grassy plains of the pampas. These nearly 55 million cows consist primarily of breeds that originated in Great Britain or France. The world's fourth-largest producer, Argentina exports its beef products, famed for their fine flavor, throughout the world. Argentinians are the world's leading consumers of beef: nearly 145 pounds (65 kg) per person per year. Yearly per capita consumption is 100 pounds in the United States, 84 pounds in Australia, 14.3 pounds in the Philippines, and 9.3 pounds in China (double the consumption five years ago).

09

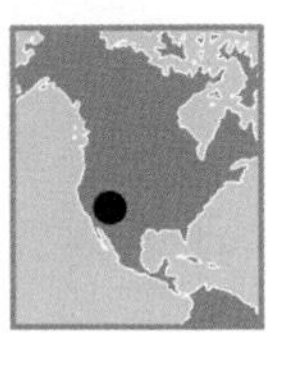

Prescott National Forest near Williams, Arizona, United States (35°14' N, 112°11' W).

There is still snow on the ground in Prescott National Forest. And there are still trees there too. The latter would be absent if this had not been designated a protected area in 1898. Gold prospectors, who flooded to Arizona in the mid-nineteenth century, did not skimp on their wood consumption. To save the forests, the government was forced to designate nature reserves. The first were declared in 1891. Since then, more than 185 million acres (75 million ha) of the 558 million acres (226 million ha) of forests in the United States have been saved from illegal logging. They are not only attractive and an asset for tourism, they are also of ecological value. The preservation of species depends, first of all, upon the preservation of their habitats. Worldwide, 12 percent of the 9,553 million acres (3,866 million ha) of forest are protected. But deforestation continues: 494 million acres (200 million ha) have been lost over the last 15 years—equivalent to twice the area of South Africa or four times that of Spain.

10

JANUARY

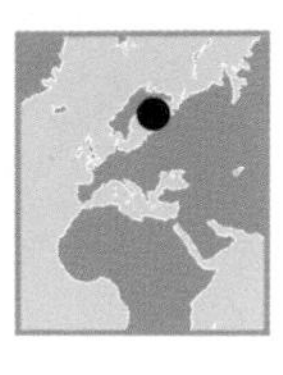

Illuminated Greenhouse near Sauvo, Varsinais-Suomi region, Finland (60°18' N, 22°36' E).

Finland occupies the most northern position in Europe, with a quarter of its territory located above the Arctic Circle. At such high latitudes agriculture faces natural challenges; in winter, night lasts uninterrupted for nearly two months in the north, while in the south the the sun does not appear for more than six hours daily. In this premature twilight the snow is scattered with gleams from greenhouses, where the daily duration of photosynthesis is extended by artificial lighting, as here near Sauvo, in the southwestern part of the country. Finland manages to thus produce 35,500 tons of tomatoes per year; but its greatest field of exploitation remains timber. It exploits pine and birch forests that cover 70 percent of its land and provide more than one-third of its export revenues. Residue from the timber industry and waste from cutting trees serve as fuel, an important source of renewable energy that has covered 20 percent of the country's energy consumption and 10 percent of its electricity consumption in the year 2000.

11

JANUARY

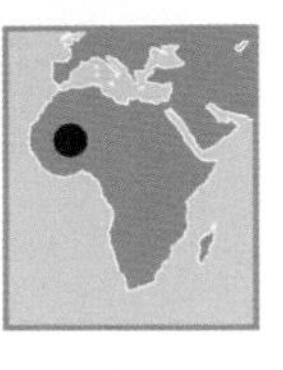

Market Gardens on the Senegal River near Kayes, Mali (14°34' N, 11°46' W).

In western Mali, near the frontiers of Senegal and Mauritania, the city of Kayes is a major ethnic and commercial crossroad. The Senegal River passes through the entire region, and many market gardens sit along its banks. The river is an important resource in this zone of the Sahel, and the waters are transported in containers by local women to provide for the manual watering of tiny parcels of land, which produce fruit and vegetables intended for the local market. The Senegal River, which begins at the confluence of the Bafing ("black river") and the Bakoy ("white river") rivers slightly upstream from Kayes, runs for 1,000 miles (1,600 km) across four countries. The hydraulic stations along its course allow the irrigation of only 234 square miles (600 km^2) of farms, but its basin of 136,500 square miles (350,000 km^2) provides water for nearly 10 million people.

12

JANUARY

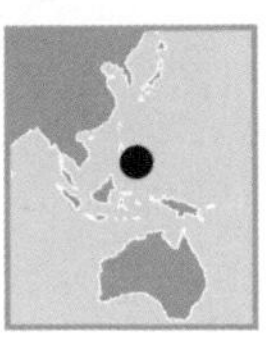

Boat near Bohol island, Visayan Islands, Philippines (9°50' N, 124°10' E).

This small vessel sailing through the Visayas archipelago provides an evocative picture of local resources. First, because the Filipinos are a sea-going people: almost a tenth of the world's sailors come from the Philippines. Second, because textiles, as represented by this colorful sail, are one of the region's main industries. Salago—whose fibers are used to make paper money but also rope, fishing lines and nets, bags, scarves, and hats—originally comes from Southeast Asia and notably from the islands of Bohol and Cebu. Wood, seen here transported by these sailors, is one of the country's chief natural resources. Once a big exporter, the Philippines has stopped producing wood owing to deforestation. Finally, the crystal-clear waters contain aquatic treasure: shoals of tuna, mackerel, and grouper, as well as shark, hide among the coral reefs, while dolphins and whales swim about farther out to sea.

13

JANUARY

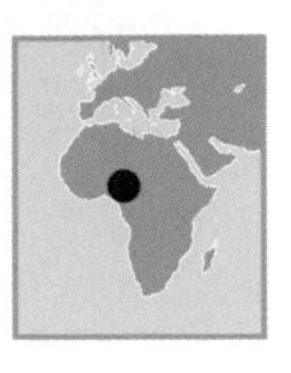

Kirdi village, Mandara Mountains, Cameroon (10°55' N, 14°00' E).

Each family has its *saré*, a group of huts huddled together that vary in number according to the number of residents. These dwellings are built according to precise rules: six rooms for the head of the family (living room, bedroom, and four grain stores), four for each of his wives (bedroom, kitchen, grain store, and hen coop), and several for the children. The whole is surrounded by tiers that follow land's contours, a system that combats erosion and maintains the few cultivable areas on the slopes of the Mandara Mountains. The animist Kirdis, who were driven into the mountains by the Muslim Peuls at the beginning of the nineteenth century, have retained their traditions and way of life. Now, however, they are threatened by mass tourism, with many people coming to look at them the same way they look at the animals in the nearby Waza National Park. Tourists often display attitudes that disrespect and devalue Kirdi beliefs. The small presents they give children (coins, pens, and other trinkets) teach them that it is easier to gather this "easy money" than to go to school.

14

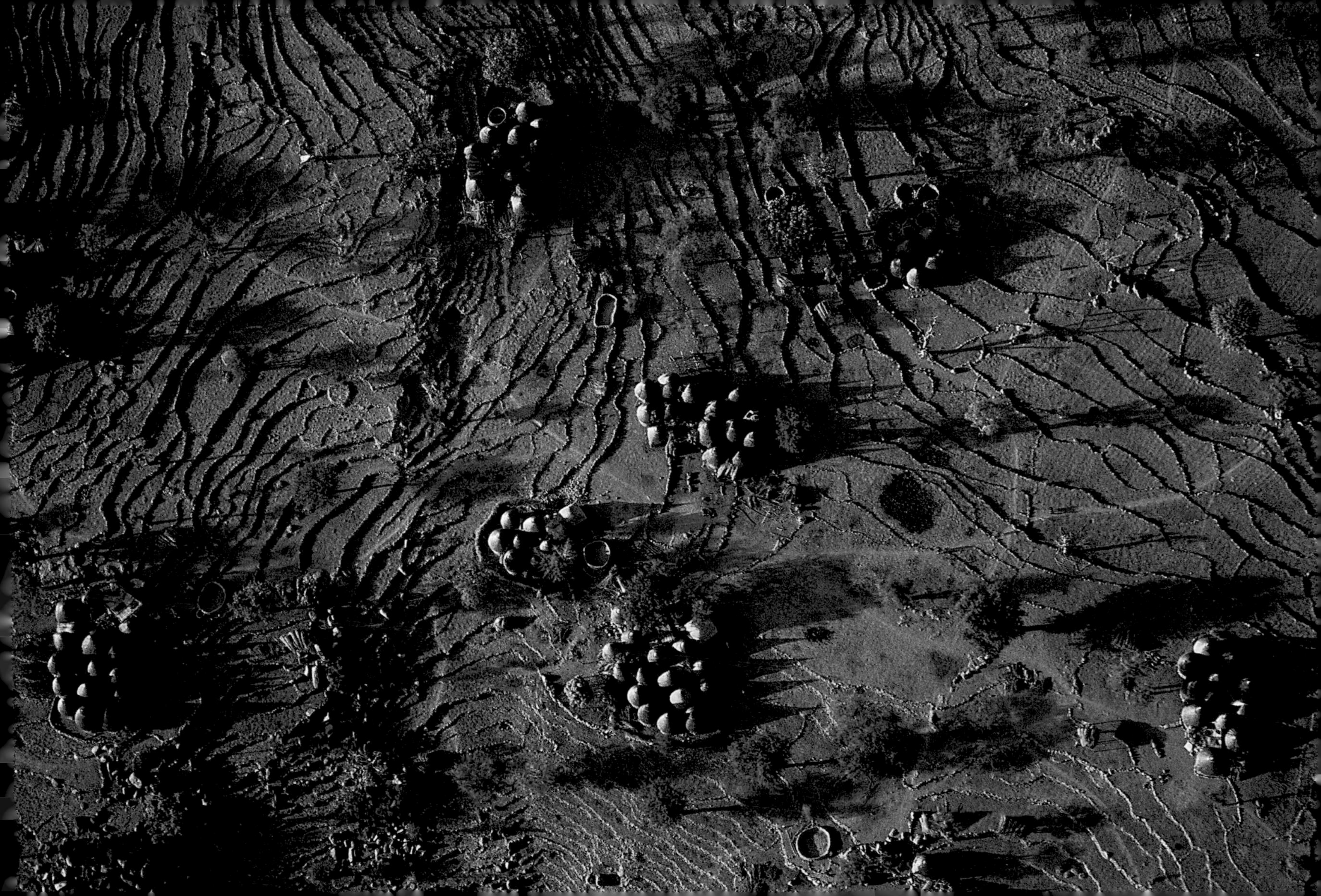

JANUARY

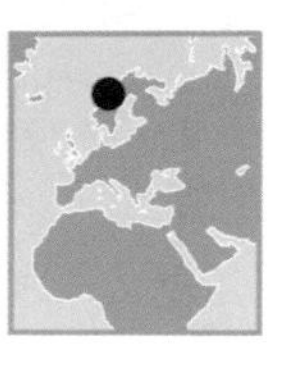

Fish farm, Boknafjorden, Norway (59°10' N, 5°35' E).

Boknafjorden, which lies between Stavanger and Bergen in Norway's south, is the first of the country's great fjords. These deep gashes in the coastline were carved by glaciers and, as Scandinavia's rocky foundations slowly sank, the sea gradually flooded them. In these enclosed bays, the sea is much calmer than it is offshore. As rainwater trickles down into them, it brings with it mineral salts that encourage algae and plankton to thrive, producing one of the planet's richest ecosystems. An added bonus for fish farming is the warm current of the Gulf Stream. Norway produces more than 450,000 metric tons of farmed salmon and trout per year. The world's seas yield about 100 million metric tons of fish, and a further 33 million metric tons come from farming. In other words, one in every four fish we consume a year is farmed. Fish farming is the fastest growing sector of the food industry, increasing at a rate of 11 percent per year since 1984.

15

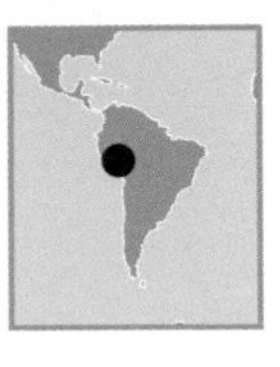

The Nazca astronaut, Ica, Peru (14°50' S, 74°60' W).

This is one of the mysterious, gigantic figures drawn on the pampas of San José and on some of the hillsides of the Rio Grande valley, between Palpa and Nazca. More than 386 square miles (1,000 square kilometers) of the rocky desert still bear the signs of Nazca culture, dating from the first to the sixth centuries AD. The animal figures, parallel or intersecting lines, and geometric shapes of these Nazca geoglyphs are formed by the contrast between the land surface, which is brown because it contains iron oxide, and the yellowish, sandy subsoil. They have survived thanks to the dry climate and low rainfall. Together, these drawings make up a calendar that allows the observation of certain astronomical phenomena, such as the solstices, but the meaning of the figures themselves, taken in isolation, remains a mystery. They certainly have some ritual or ceremonial significance. They might be a great prayer, an invocation to the gods made at the time of the severe drought of AD 550. For this ancient people of astronomers and hydraulic engineers, as for us today, water was more precious than gold.

16

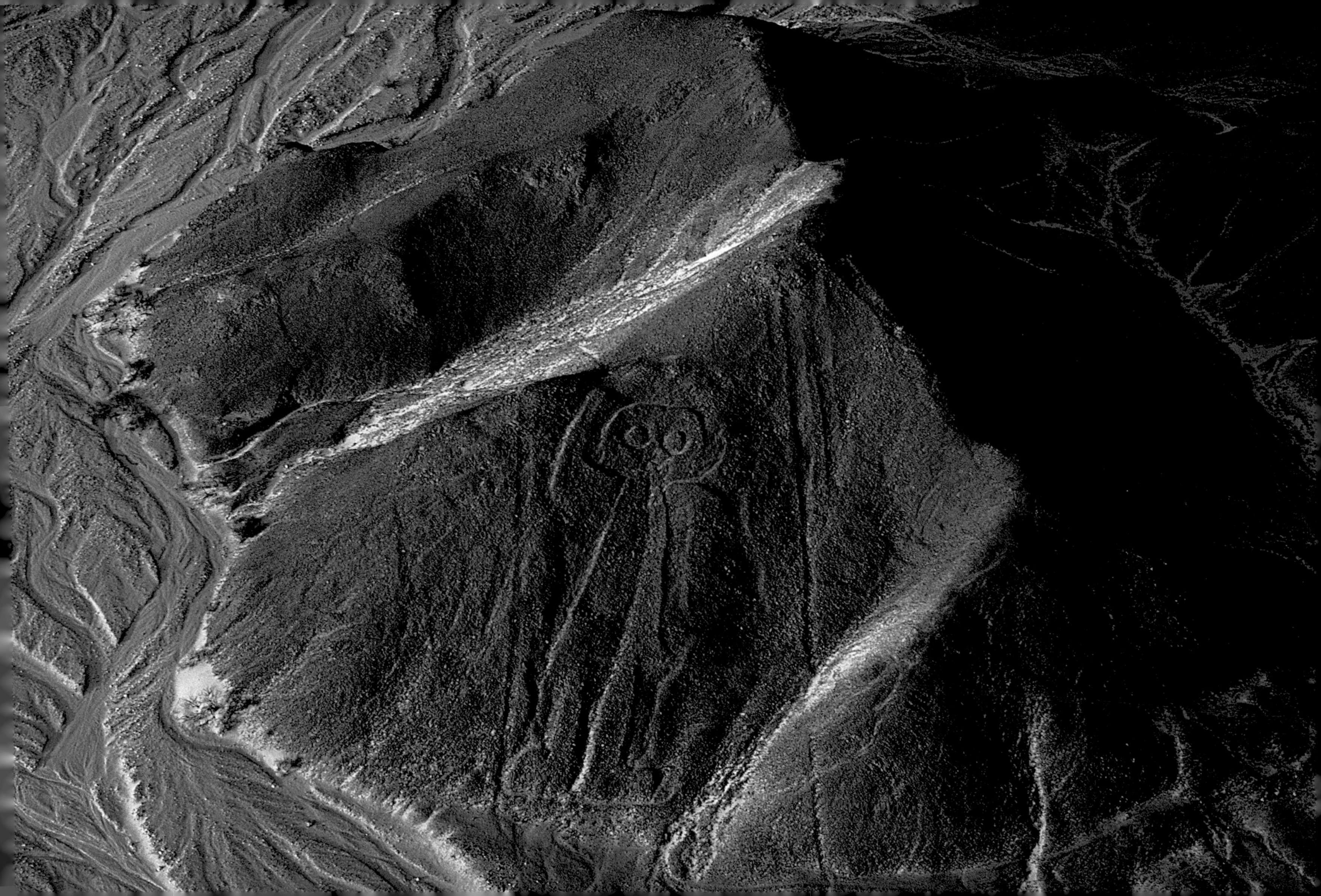

JANUARY

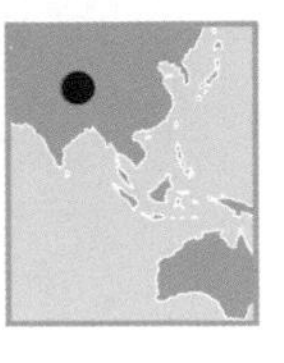

Terraced fields north of Katmandu, Bagmati region, Nepal (28°20' N, 85°55' E).

Nepal has the world's highest mountains and one of its highest rates of population growth, at 2.6 percent per year. Of Nepal's inhabitants, 80 percent are farmers who are obliged to cultivate hillside after hillside to grow enough food. However, not even ingenious irrigation networks, control of erosion, and intensive methods that allow up to four harvests per year—at a risk of exhausting the soil's fertility—are not enough to feed this growing population. Feeding humans and preserving the land is a problem that stretches well beyond Nepal's borders. Today, 23 percent of the planet's usable land is suffering degradation. Overgrazing, deforestation, poor irrigation, pollution by chemicals, and urban expansion are reducing its productivity. And yet it will have to feed 8 billion people by 2025.

17

JANUARY

***Barrios*, Caracas, Venezuela (10°30' N, 66°56' W).**

Founded in 1567 by the conquistador Diego de Lozada, Caracas was named after Los Caracas, the ferocious Caribbean Indians who lived in the region. The city has grown enormously in the last 40 years, attracting people from all over South America, filling its narrow valley and climbing up the steep sides of the surrounding hills. These new districts, known as *barrios* or *ranchos*—slums—are home to more than 50 percent of Caracas's 3.8 million inhabitants. As in São Paulo or Bogotá, whole streets are privatized and controlled by private militias. The gap between rich and poor is constantly widening in the cities of developing countries, but this is happening also in some industrialized nations such as the United States and United Kingdom. In 1979, the income of the richest 1 percent of U.S. households was ten times that of the average household; by 1997, it was twenty-three times greater.

18

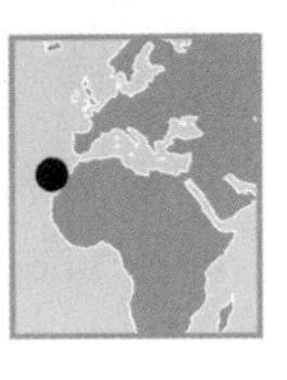

Hotel development near Arrecife, Lanzarote, Canary Islands, Spain (28°56' N, 13°35' W).

From the eighteenth and nineteenth centuries onward, naturalists flocked to Lanzarote, fascinated by the flora and geology of this volcanic island off Morocco. Later, the Canaries' landscapes and plentiful sunshine led to development for tourism, which started its boom in 1960. Today, 10 million tourists, mostly Germans (29 percent) and British (37 percent) visit the archipelago every year. Its coasts host people on both short summer trips and longer-term visits, especially retired people who stay all year. As a result the islands have been heavily built up. Acutely aware of the landscape and ecological impact of such infrastructure, Lanzarote has hosted two world conventions on sustainable tourism. Dealing with this industry's environmental impact is all the more urgent because over the next 20 years the sector is expected to grow at an average of 4.3 percent per year, and the number of tourists to increase by 300 percent.

JANUARY

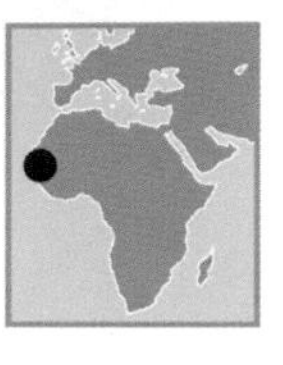

Fish market in Saint-Louis, Senegal (16°02' N, 16°30' W).

Fish is Senegal's biggest export, generating 66 percent of its export revenue, and is vital to its people's diet. Senegalese fishermen are still content to use their traditional dugout canoes, but they face competition from foreign trawlers. Since the late 1980s, the country has sold fishing permits to fleets from other countries, especially from the European Union. As a result of rampant overfishing, stocks have collapsed to the point that catches are now 75 percent lower than they were at the end of the 1970s. Poverty also leads to waste: 30 percent of the fish caught end up rotting on the dockside at Saint-Louis because there is no way of preserving them. Some fishermen also resort to drastic measures, using explosives, which damage habitats. To try to tackle this widespread problem, in 2001 the United Nations Food and Agriculture Organization (FAO) produced the Reykjavik Declaration on responsible fisheries in the marine ecosystem.

20

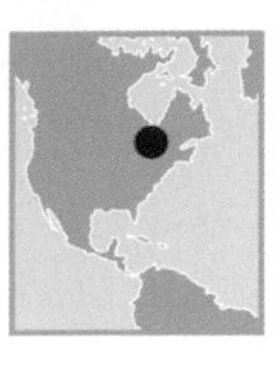

Polar bears being moved by helicopter, near Churchill, Manitoba, Canada (58°47' N, 94°12' W).

The town of Churchill on the shores of Hudson Bay is proud of its nickname "Polar Bear Capital of the World." The west side of the bay is home to between 1,000 and 1,200 specimens of *Ursus maritimus*—or 5 percent of a world population estimated at 25,000 individuals. Attracted by the waste produced by Churchill, the polar bears could pose a danger to residents, who inform the authorities responsible for removing the intruders. The number of these encounters is increasing, and scientists are blaming this on climate change. The ice around the bay breaks up two weeks earlier, and forms a week and half later, than it did 20 years ago. The bears, which depend on ice floes in order to hunt seals, are deprived of this food source for a longer time, and are forced to seek nourishment elsewhere. According to the Canadian Wildlife Service, they weigh 214 to 227 pounds (80 to 85 kilograms) less than they did in 1985. In 2001, it was estimated that summer sea ice in the Northern Hemisphere had shrunk by 10 to 15 percent since 1950.

21

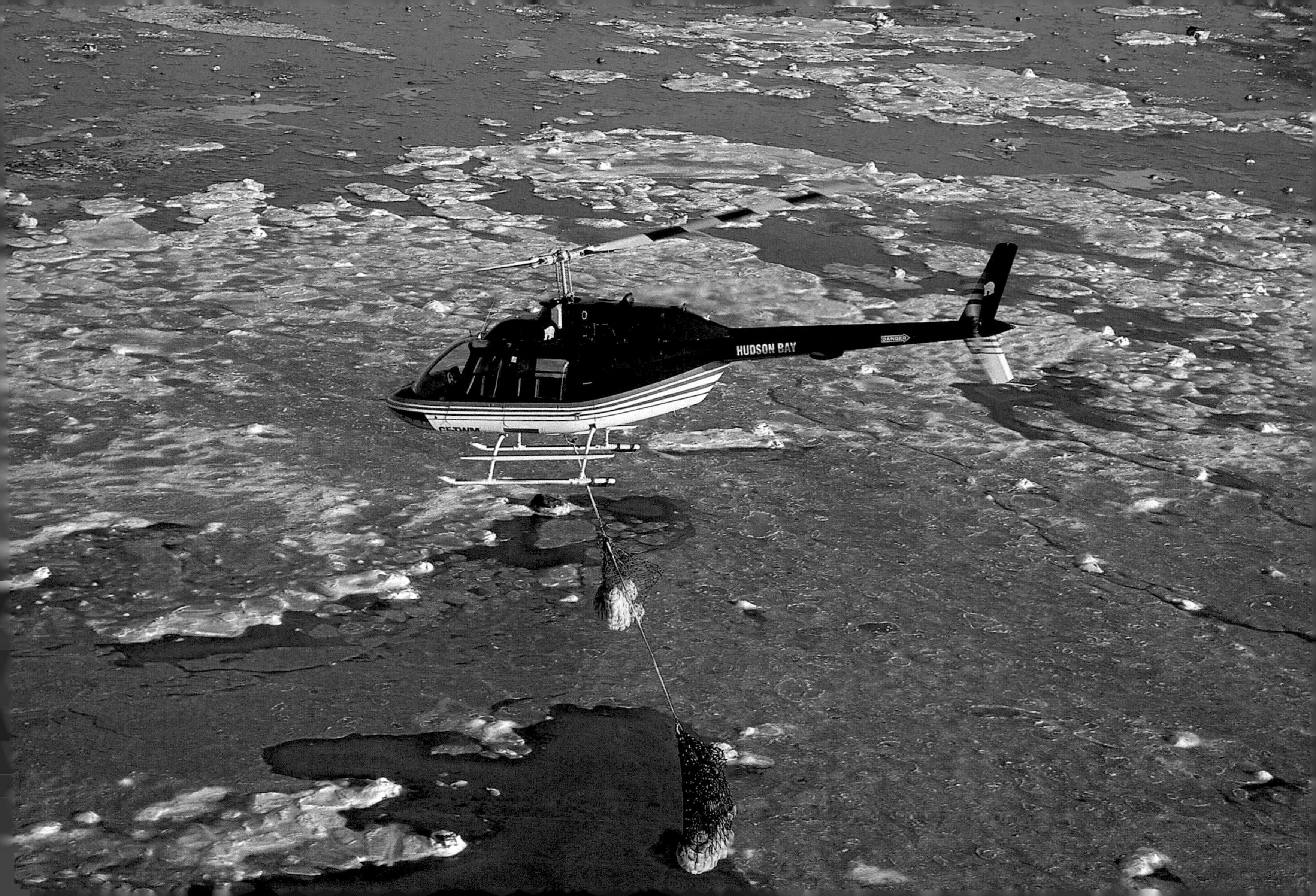
HUDSON BAY
DANGER

JANUARY

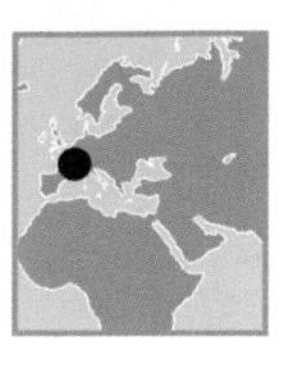

Boat graveyard at Kerhervy, Lanester, Morbihan, France (47°45' N, 3°20' W). Nestling in the hollow of the last meander of the Blavet river, where it slows down before flowing into the Scorff river at Lorient, dozens of wrecks lie in the marine graveyard of Kerhervy. The oldest ones, those of the tuna fishing boats from the Île de Groix (which rises out of the Atlantic a few miles offshore), have been lying in this bend in the river since 1920, and are sinking inexorably into the mud. The last boat was dumped in January 2001, when the *Ouragan (Tempest)*, a trawler from Port-Louis, went to its last resting place, joining the timelessly picturesque dandies that sank in the Blavet estuary. Countless vessels sail the planet's seas without going into retirement. Ships more than fifteen years old account for 40 percent of the world fleet, but for 80 percent of accidental shipwrecks. However, age is not a deciding factor: the oldest boats are mostly owned by operators who have tried to cut costs, both in equipment and in the crew's work conditions and training. And 80 percent of shipwrecks are due to human error.

22

JANUARY

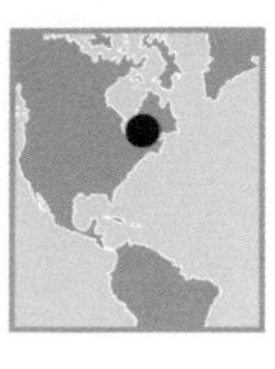

Forest of Saint-Hyacinthe, Montérégie, Québec, Canada (45°37' N, 75°57' W). One-third of the region of Montérégie—bordered to the north by the waters of the St. Lawrence river and to the south by the United States—is covered by mixed forest. Here the white and red pines, spruce, and balsam fir of northern forests mingle with the wild cherry, sugar maple, birch, and aspen of temperate regions. The Québec forest, which covers almost two-thirds of the province, contributes to the prosperity of Canada, which is the world's biggest producer of newsprint, second biggest of wood pulp, and third biggest of lumber. Canada's forests have diminished significantly as a result of long overexploitation, insect parasites, and damage by acid rain. Since 1992, the country has been trying to practice sustainable forestry in an effort to satisfy environmental, economic, social, and cultural demands. More than 123 million acres (50 million ha), or 12 percent, of Canada's forests are protected.

23

JANUARY

Gold prospectors' barges on the Caroni River, Bolivar state, Venezuela (5°20' N, 62°40' W).

On the Caroni River in southeastern Venezuela, gold prospectors use a tube known as a "sucker" to draw up gold-bearing deposits from the river bottom. The precious metal is sieved out while gravel and sediment are thrown away, leaving a long muddy trail behind the boats. The great majority of these prospectors use mercury to extract and purify the gold. It is estimated, for example, that 1 million prospectors lining the banks of the Tapajós River in northern Brazil use mercury, dumping 130 metric tons of contaminated waste in the surrounding land every year. In people, this heavy, highly toxic metal causes respiratory, gastrointestinal, and neurological problems and is usually ingested via contaminated fish or crustaceans. A quarter of the gold on the world market is obtained using these small-scale methods that rely on mercury and pollute hundreds of miles of watercourses.

24

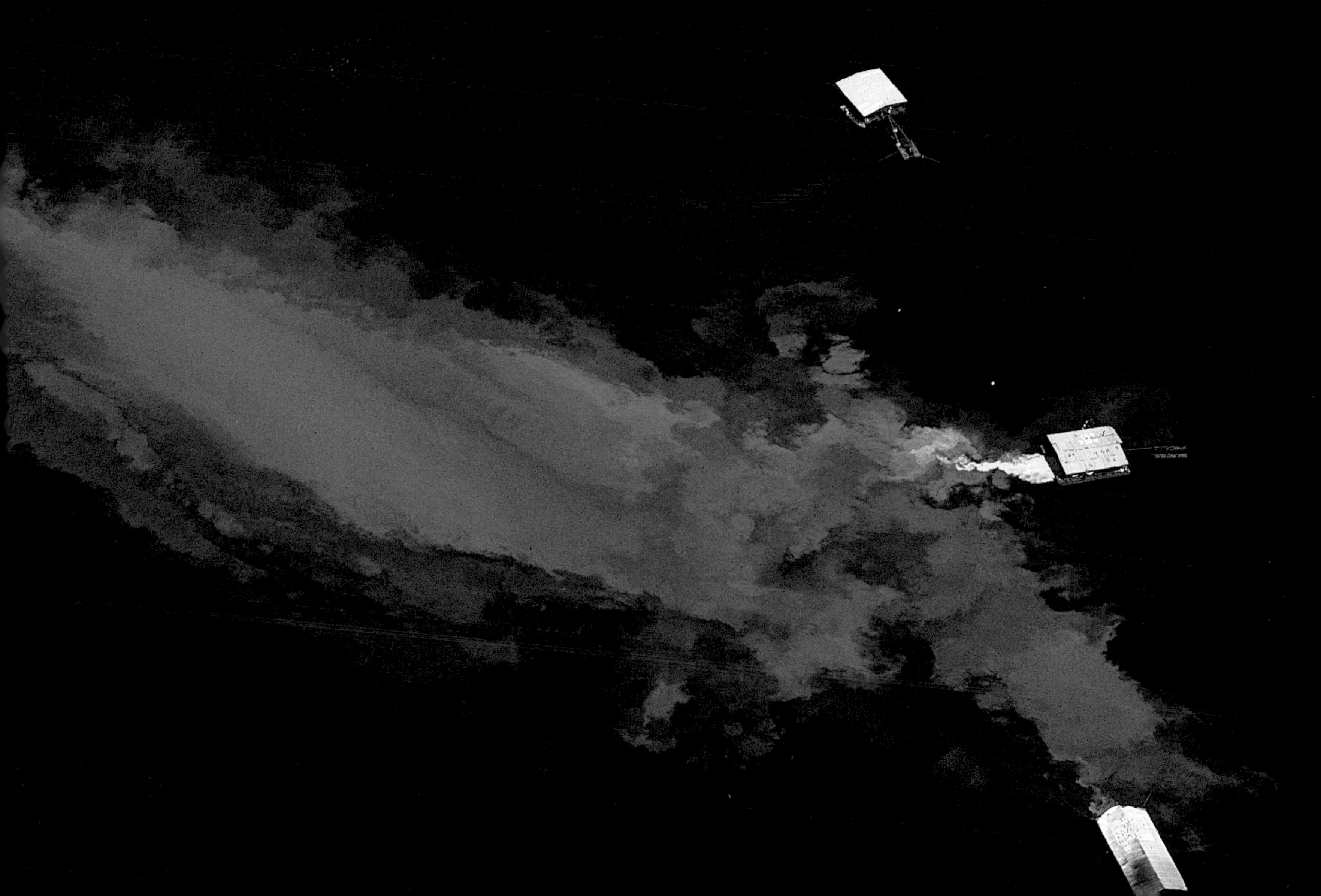

JANUARY

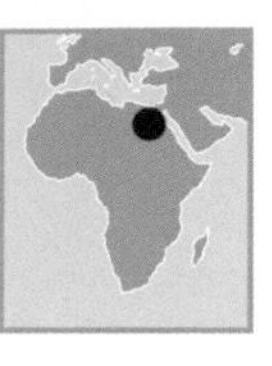

Outline of Birket Maraqi salt lake in the oasis of Siwa, Egypt (29°12' N, 25°31' E).

Under the burning sun of northwest Egypt, the evaporation of water from the shallower parts of this salt lake has cracked its bed of sand and mud, forming these extremely hard, rounded wrinkles. Here and there salt forms a white crust, tracing the outline of the bluish, stagnant pool. The salt concentration in the water is so high that no living organism can survive. However, the shores of these lakes are shaded by palm trees and olive trees, fed by the oasis's 230 freshwater springs. Thus, Siwa's 15,000 inhabitants grow 300,000 date palms and 70,000 olive trees. Fresh water is one of the scarcest resources on the planet, accounting for only 2.5 percent of the total volume of water on the Earth, and of that proportion, 77 percent is trapped as ice at the poles and in glaciers. Liquid fresh water is unequally distributed, being rare in the tropics but plentiful on the Equator and in temperate regions. Even where it is abundant, it is still precious. Its quality is constantly deteriorating as a result of contamination from excess organic matter, fertilizers, and other chemicals released by agriculture, industry, and the general population.

25

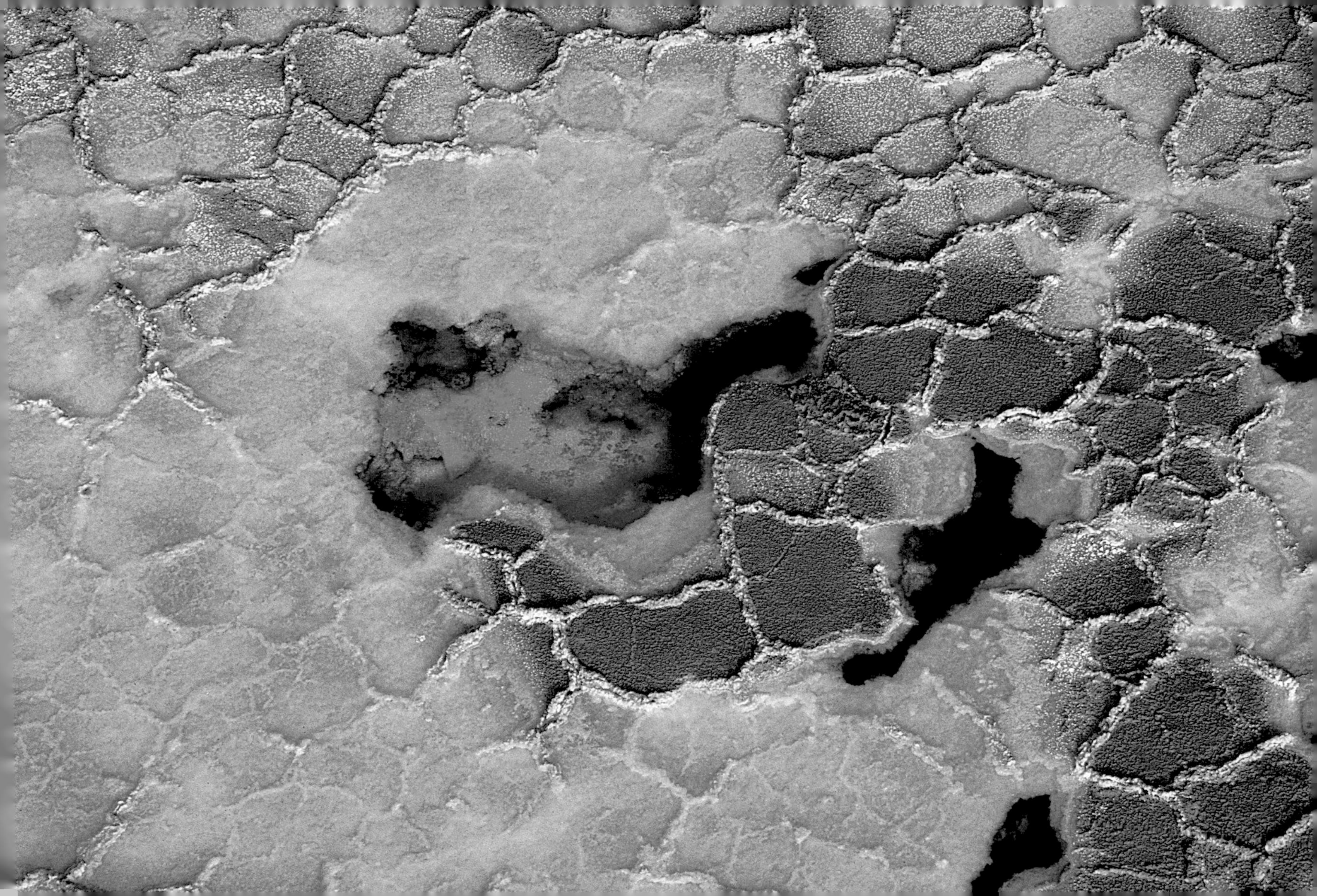

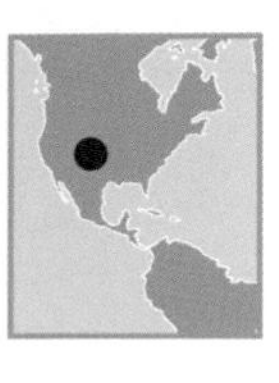

Highlands Ranch, Outskirts of Denver, Colorado, United States (39°44' N, 104°59' W).

These winding streets of identical houses do nothing to break up the monotony of the asphalt. The outskirts of Denver are a good example of the runaway sprawl of suburbs in North America. This phenomenon was triggered by postwar economic growth, which encouraged private home ownership and stimulated investment in roads. Since then, the number of people living in such areas has relentlessly grown—by 12 percent between 1990 and 1998—at the expense of the growth of city centers, at a rate of 4.7 percent over the same period. These networks of low-density suburbs make their residents totally dependent on their cars, one of the chief sources of greenhouse gases. This dependence is one reason that Americans generate the highest emissions of greenhouse gases on the planet. Although North Americans are only 5 percent of the world's population, in 1998, they produced almost a quarter of human-generated carbon dioxide.

26

JANUARY

Bullfight at the arenas in Seville, Andalucia, Spain (37°23' N, 5°59' W).
Hesitating between the picador and the matador, a bull paws the *albero*—hard, yellowish earth extracted from the nearby quarries of Alcalá de Guadaira—in Seville's Plaza de Toros de la Real Maestranza. Construction of these arenas began in 1761 and lasted 120 years. Bullfights are held in a slightly oval space 206.5 feet (63 meters) long, surrounded by tiered seating for 14,000 spectators. Two red lines, as the regulations demand, run 23 and 33 feet (7 and 10 meters) from the barrier, emphasizing its curve. Behind this life-saving fence, which can be vaulted over thanks to the white footboard at its base, runs the *callejón*, where matadors take refuge from the bull. Controversy continues to rage over this centuries-old tradition—which enthusiasts appreciate as an art and critics condemn as cruel butchery—heightened since 1997 by the Treaty of Amsterdam, which recognizes animals as "sentient beings" and demands that they be respected. Certain agricultural practices, such as force-feeding geese and the intensive rearing of animals, have also been condemned.

27

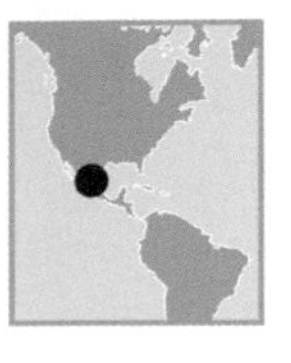

Tomato cultivation near Tepic, Nayarit state, Mexico (21°30' N, 104°54' W). The most widely cultivated vegetable in the world, the celebrated tomato originally came from Mexico. It has rich biodiversity, with 1,700 known varieties. In the sixteenth century, the great explorers brought the precious "golden apple" back to Europe. Italians integrated it into their national cuisine quite quickly, but the French, believing it to be poisonous, used it only as an ornamental plant for more than a century. Today the tomato is Mexico's fourth-biggest agricultural export, with a value of some $300 million a year. The country's neighbors in the North American Free Trade Agreement (NAFTA) can thus buy tomatoes cheaply all year round. But the widespread habit in Western countries of buying fruit and vegetables out of season produces heavy pollution, because the goods have to be transported or grown in heated greenhouses. The extra emissions of greenhouse gases thus produced are considerable. It is estimated that buying about two pounds (1 kg) of fresh green beans in the winter involves emissions of almost 28.6 pounds (13 kilograms) of carbon dioxide.

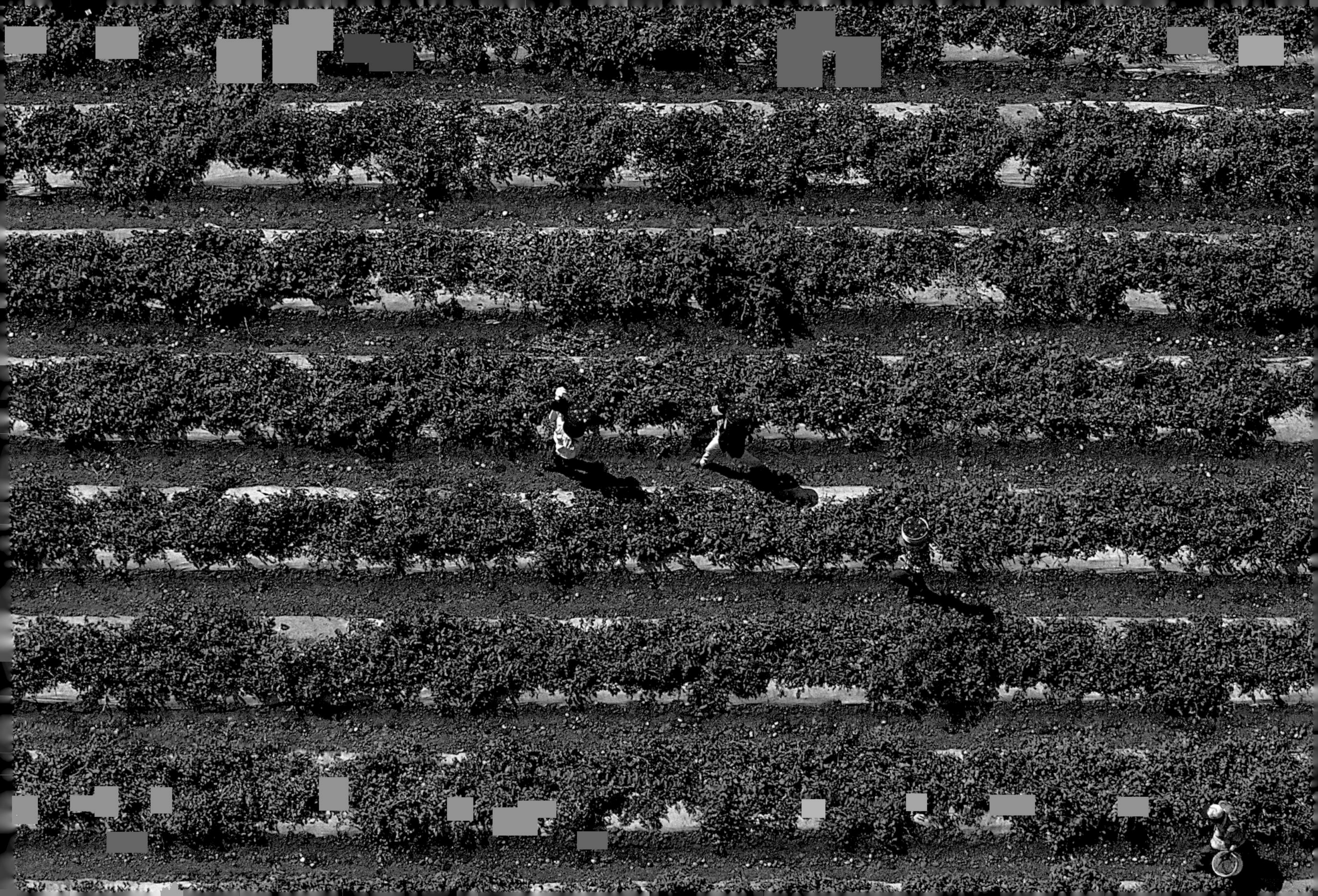

JANUARY

Tree plantation near Seix, Pyrenees, France (42°50' N, 01°30' E).

This plantation in the mountains of the Ariège *département* is part of the area's woodland management and helps to combat erosion of soil that has been stripped of vegetation. But these mountains have a secret. They are home to two female brown bears and two of their cubs, which were trapped in Slovenia and released here as part of the plan to reintroduce brown bears to the Pyrenees. Shooting, poisoning, and habitat fragmentation—especially due to road building—had got the better of this symbolic species. Preservation of biodiversity, and especially of threatened or vulnerable species, is one of the priorities that was set out at the United Nations summit held at Rio in 1992. In 2002, it was estimated that more than a fifth of mammal species, a quarter of reptile species, a third of fish, and almost 60 percent of insects were threatened or vulnerable—as well as more than half the flowering plants living on the planet.

29

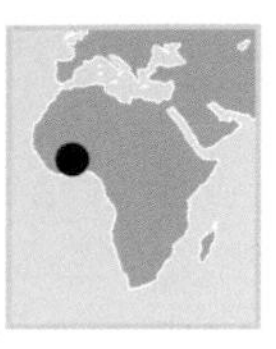

Dome of the Basilica of Our Lady of Peace, Yamoussoukro, Côte d'Ivoire (6°49' N, 5°17' W).

In 1983, Yamoussoukro replaced Abidjan as the official capital of Côte d'Ivoire. President Félix Houphouët-Boigny, who died in 1993, made his native village into a modern city with a grid of wide avenues—which are almost deserted—and every modern facility: international airport, luxury hotels, golf course, prestigious universities, and so forth. In Cote d'Ivoire, Catholicism is the second most popular religion, with Islam coming first. However, Yamoussoukro boasts the world's biggest basilica, Notre-Dame-de-la-Paix (Our Lady of Peace), consecrated by Pope John Paul II in 1990. The former president, who donated this building to the Vatican, insisted that he had financed the basilica's exorbitant cost out of his own personal fortune. This building was seen as a colossal waste by many Ivorians. It was highly controversial in a country that lacks schools and hospitals and has only nine doctors for every 100,000 inhabitants (compared to 413 in Norway).

30

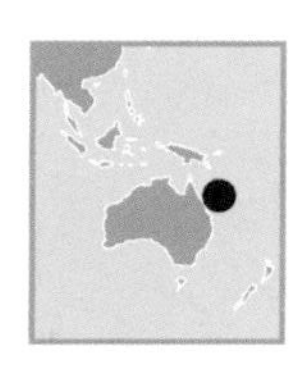

Great Barrier Reef, Queensland, Australia (16°55' S, 146°03' E). At a length of 1,550 miles (2,500 km) along the northeastern coast of Australia, with more than 400 types of coral, the Great Barrier Reef is the largest coral formation in the world. This rich, silent sanctuary of submarine life was declared a marine park in 1979 (comprising 15 percent of the world's protected sea surface) and a UNESCO World Heritage site in 1981. The Great Barrier Reef harbors more than 1,500 species of fish and 4,000 mollusks, as well as such animals as the endangered dugong (sea cow) and six of the seven species of sea turtle. Coral formations, the world's only relief that is biological in nature, are polyps that live symbiotically with photosensitive algae, zooxanthellae, which contribute to the development of the calcareous skeletons of their hosts. The coral reefs are essential to the protection of the coasts and of ocean fauna. They are sensitive to the smallest increase in water temperature, which can cause them to whiten. This phenomenon, which was particularly noticeable in 1998 (during El Niño), caused the loss of thousand-year-old corals. Many of the affected coral colonies are starting to regenerate, but the growing frequency of the whitening phenomenon, which could result from global warming, is disturbing.

CLIMATE CHANGE OR CLIMATE SHOCK?

Our ancestors, 10,000 or 20,000 years ago, may well have wondered about the climate of the future. Clearly, however, their own behavior could play no part in shaping it. Thus, they could calmly sacrifice a few mammoths in the belief that they were pacifying the forces of nature, and with little power to influence how the planet's climate would change a few centuries later.

Sadly, and to our detriment, we have acquired the power that they lacked to change the climate. For the first time in its history, humanity has become a climatic agent. How? By devising and using systems that emit vast amounts of greenhouse gases. These have boosted the greenhouse effect, which has been taking place naturally on the planet for several billion years. Without it, the average temperature would be 5 °F (–15 °C), and life as we know it would probably not exist.

FEBRUARY

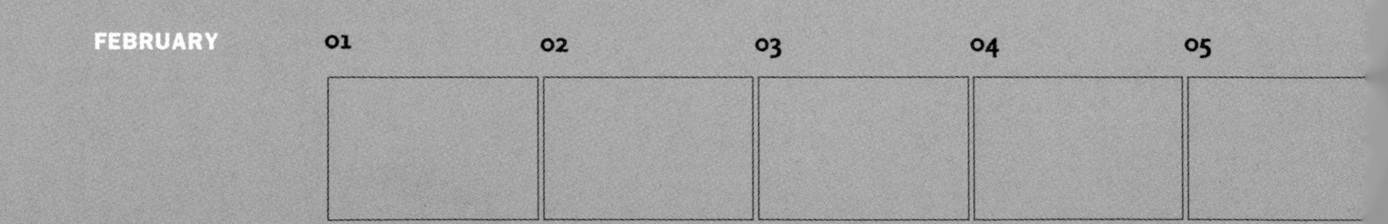

Greenhouse gases—chiefly composed of water vapor (0.3 percent of the atmosphere), carbon dioxide (0.04 percent), and methane (0.0002 percent)—let in the sun's rays so that they reach the Earth's surface. Our planet also emits rays, in the infrared range. Greenhouse gases are impervious to those infrared rays, and they prevent that energy from easily escaping back into space.

Our species has brutally broken the balance between incoming and escaping energies. Since 1850, we have increased the amount of carbon dioxide in the atmosphere by 30 percent and doubled the level of methane—two events unprecedented in at least the last 400,000 years. If we continue relentlessly increasing these emissions, the planet's average temperature—which had been stable for 10,000 years—could rise by several degrees over a century. This has been known for a long time. It was first explained in 1896 by a Swedish scientist, Svante Arrhenius, 1903 Nobel Laureate in Chemistry.

An increase of a few degrees in the planet's average temperature is no trifling matter. A few degrees means the difference between a "warm" period, such as the present, and an ice age. During ice ages of the past, northern Europe and Canada were covered by layers of ice several kilometers thick. The sea level was 120 meters lower, and France was a barren, frozen steppe incapable of supporting its present population of millions of people. Moreover, deglaciation takes some 10,000 years. We cannot know the exact result of an average temperature rise of a few degrees over one or two centuries. However, a "climatic shock" seems more likely than a matter of having to wear more or fewer sweaters in winter.

A rapid change in the world's average temperature could seriously disrupt the water cycle, bringing increasingly severe floods and droughts; more violent hurricanes; the melting, over five or six centuries, of the Greenland icecap and part of the Antarctic's, causing sea levels to rise by about twelve meters; and possibly the disappearance of the Gulf Stream over a few

06 07 08 09 10 11 12 13 14 15

decades, which could lower average temperatures in western Europe by five to six degrees centigrade, sounding the death-knell of all agriculture in France. We also risk the extinction of corals and the rapid migration of various tropical diseases, such as yellow fever, malaria, and dengue, along with other illnesses with the power to affect not only people but plants and animals besides. And, of course, we have no idea what other unpleasant surprises may await us. These are not fantasies dreamed up by science-fiction writers but possibilities that are soberly documented in the available scientific and technical literature.

It remains to be seen how our society will respond to sudden climatic upheaval. But it would do us good to remember that almost every aspect of the world around us is adapted to local climatic conditions. This is clearly true for massive industries like farming and forestry, but climate has just as much influence on the structure of our buildings, our communications capacities, and even the contents of our wardrobes.

If we want to act, what are we to do? Greenhouse gases have a lifespan of more than a century. Once emitted, they take only a few months to spread evenly throughout the atmosphere. Therefore, since they cause the same climatic disruption no matter where on the planet they were emitted, humanity must come together across national boundaries to agree on concerted reductions—otherwise, a single mischief-maker has the potential to wreck the efforts of all other countries. As important as it is for all nations to work together, every nation should aim to exemplify the virtue of leading by example, a virtue that, in my view, is too often forgotten.

How much should emissions be reduced? To halt the increase in amounts of carbon dioxide in the atmosphere, world emissions would have to be cut by at least half. If we assume each inhabitant of the earth has the same "emission rights"—which assumes, in turn, that the world is a completely equitable place—the French, to take one European example, would

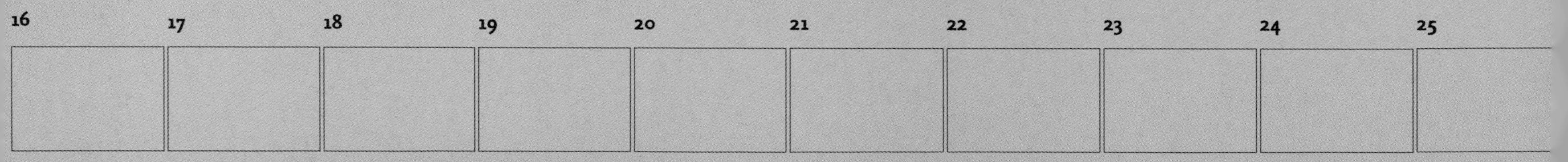

have to cut their emissions by a factor of four. This would mean that all French citizens would have to reduce their consumption of oil, gas, and coal to one-fourth of their current rate. (Nuclear and renewable energies remain available, with greatly different potential, depending on how they are developed.) In the United States, Americans would have to cut their emissions by a factor of twelve, using one-twelfth the resources they currently use. The Chinese, on the other hand, could increase their consumption, just barely, above present levels.

With so much technology at hand, it is easy for citizens of these countries to reach or exceed annual consumption ceilings. All it takes is for each individual to make a single transatlantic flight, use gas heating for a few months; drive about 3,000 to 6,000 miles (5,000 to 10,000 kilometers) in a car; buy almost any quantity of manufactured goods; or eat about 250 pounds (100 kilograms) of beef—agrochemicals and farm machinery mean almost everything we eat involves greenhouse gas emissions.

In short, stabilizing the disruption of our climate would involve a radical change in the way we run our society, not just some tinkering at the edges. The motto "Consume more and more!" would need to be replaced by "Emit less and less!" Our relentless growth, in terms of consumption of material goods, would probably not survive. But if we want to lead our children toward a world worth living in, do we really have a choice?

JEAN-MARC JANCOVICI
Expert on climate change, France

26 27 28 29

FEBRUARY

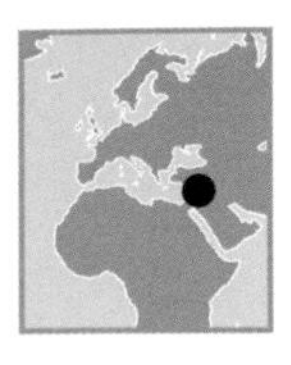

Masada, Judaean desert, Israel (31°18' N, 35°20' E).

Masada—"fortress" in Hebrew—was built by Herod between 37 and 31 BC on the western edge of the Judaean desert. Flavius Josephus's *History of the Jewish War* is the only written source reporting this place. In 66 AD, Josephus took part in the Jewish revolt against Rome. A few years later, after the Jewish surrender, Josephus's description of the revolt raised the fortress of Masada to mythical status. Jerusalem fell to the Romans in 70 AD, but three years later Masada was still holding out. It took more than 10,000 legionnaires to overcome the 967 besieged Zealots. Knowing they were defeated, the rebels faced a choice: honor or life. The citadel's last defenders drew lots to choose ten of their number who were to kill the rest before committing suicide themselves. Recent archeological investigation has confirmed the story in its essentials. Today, cadets joining the Israeli defense forces armored unit swear the oath "Masada shall not fall again" on this revered site, which has become a symbol of Jewish resistance.

01

FEBRUARY

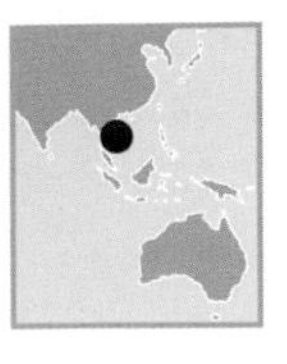

Floating village on Tonle Sap lake, Cambodia (13°00' N, 104°00' E).

In their floating houses, the inhabitants of Tonle Sap drift at the mercy of the lake's floods and follow the migratory movements of fish. Besides fishing, they practice a traditional form of fish farming that uses cages and open-top containers made from bamboo. The Tonle Sap lake is the world's fourth biggest freshwater fishery, yielding about 60 percent of the country's catch. Fish are generally eaten in the form of *prahok*, a fermented paste found all over Cambodia. Fish is an important food source because it provides 75 percent of the animal protein in the staple diet. Paradoxically, in this almost amphibian country where a third of the surface area is wetlands, it is estimated that 70 percent of Cambodians have no access to drinking water—the highest proportion in the whole of Asia. Worldwide, 4.5 million people die every year from diarrhea caused by drinking untreated water.

02

FEBRUARY

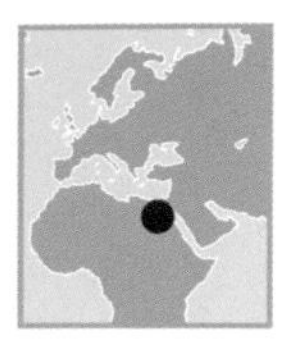

Ruins of the medieval citadel of Shali in the town of Siwa, Egypt (29°12' N, 25°31' E).

Founded in the thirteenth century, the citadel of Shali protected the inhabitants from pillagers for many years. But its walls built of *kershef*—salt bricks covered with clay and plaster—were no match for the violent rains of 1926. Three days of deluge left nothing but a mass of ruins, bearing witness to a vanishing style of architecture. Although some of the fortress' houses are still inhabited, modern Siwa is chiefly built of concrete, clustering around its old town. Since 1998, the United Nations Human Settlements Programme has encouraged young Siwa entrepreneurs to rediscover their cultural and architectural heritage. The region's governor, who also would like to see this heritage preserved, has required all new buildings around the citadel of Shali to be built using traditional methods.

03

FEBRUARY

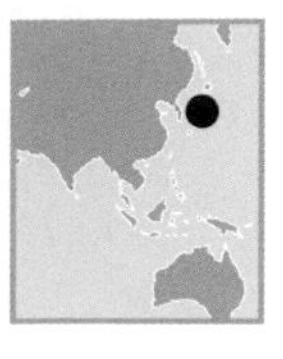

Osaka Palace, Honshu, Japan (34°40' N, 135°30' E).

The south coast of the island of Honshu forms a vast bay, within which nestles the island of Shikoku. Winding between these two elements of the Japanese archipelago, the Pacific Ocean becomes a calm inland sea, into which the Yodo river flows. At the river's mouth is Osaka, Japan's third biggest metroplex, with 11 million inhabitants. This economically thriving port city was at its zenith under Toyotomi Hideyoshi (1582–1598), during whose reign the palace was built. Now it is an industrial center, where traders are nimble with the *soroban*, the traditional abacus. Set in the heart of a great park east of the city, the present palace is merely a concrete replica of the original. It was built in 1931 and given a facelift in 1970 for the international exhibition, which drew 65 million visitors. After the first universal exhibition was held in London in 1851, several more took place during the twentieth century, acting as shop windows for progress and places to do business. Today such exhibitions have become prohibitively expensive, and their merits are being questioned. The city of Hanover spent 1.7 billion euros to host Expo 2000, which was expected to attract 40 million people; 18 million came.

04

FEBRUARY

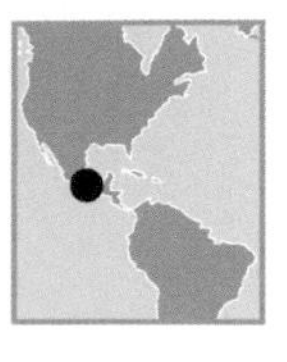

Birds flying over the lagoon near the mangroves of San Blas, Nayarit state, Mexico (21°60' N, 105°30' W).

Mexico contains a great diversity of climate and landscape, including deserts, mountains, hardwood forests, and, as here, lagoons alongside mangroves—a subtropical rainforest typical of alluvial coasts. Although Mexico boasts more than 900 species of cactus, 1,000 orchids, and the same number of mammals, it is the country's variety of birds that is most striking: no fewer than 10,000 species, including the celebrated *quetzal*, also known as the "bird of the Aztecs." Mangrove swamps are a favorite haunt of migrating birds. Here they find plentiful food: insects, mollusks, shellfish, shrimp, and small fish, which come to breed at the feet of the mangroves, a tree peculiar to this mixed ecosystem that is part sea and part land. Mangroves are essential to marine life. They also protect the shoreline, holding in place the sediments deposited by rivers and curbing the ocean's erosion by acting as a breakwater.

05

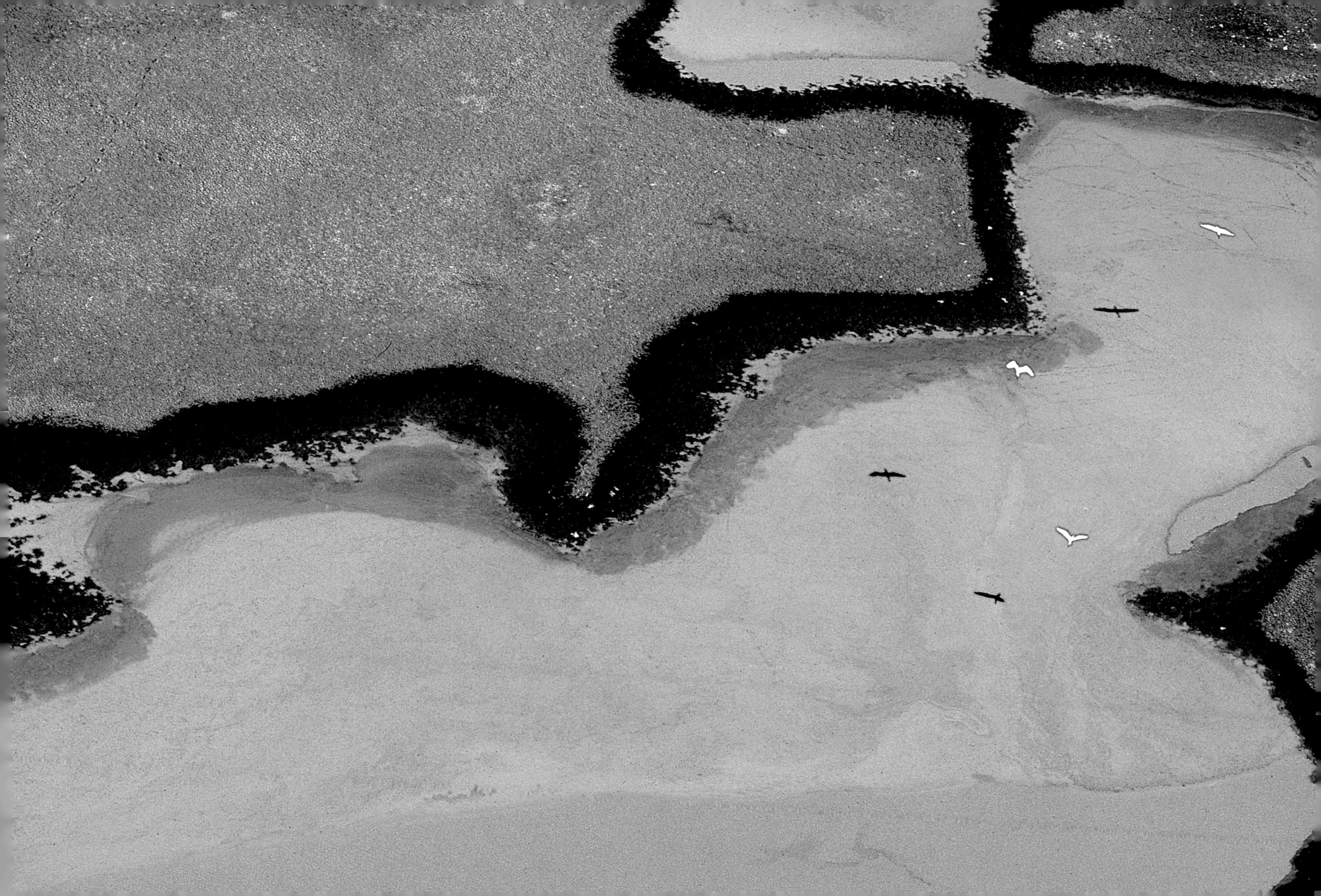

FEBRUARY

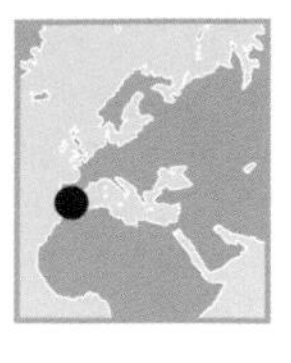

Details of the tarmac of the airport on Gibraltar, United Kingdom (36°08' N, 05°21' W).

On the tarmac of the Gibraltar airfield, at the southern tip of the Iberian peninsula, geometric traces of signs and the repairs in the asphalt create this strange piece of abstract art. The air-transport sector, with 28 million jobs and an estimated value of $1.4 trillion, figures prominently in the world economy. It is vulnerable to international insecurity, however, as was seen after the terrorist attacks in the United States on September 11, 2001: companies went bankrupt, and 120,000 jobs were lost. For the past ten years, air traffic has shown an average regular growth of 6 percent per year; 1.5 billion tickets were sold in 2000. If this trend continues, by 2010 some 20,000 planes will crowd the skies to transport 2.3 billion travelers. To avoid the congestion of air space, giant planes with as many as 600 seats are preparing for takeoff.

06

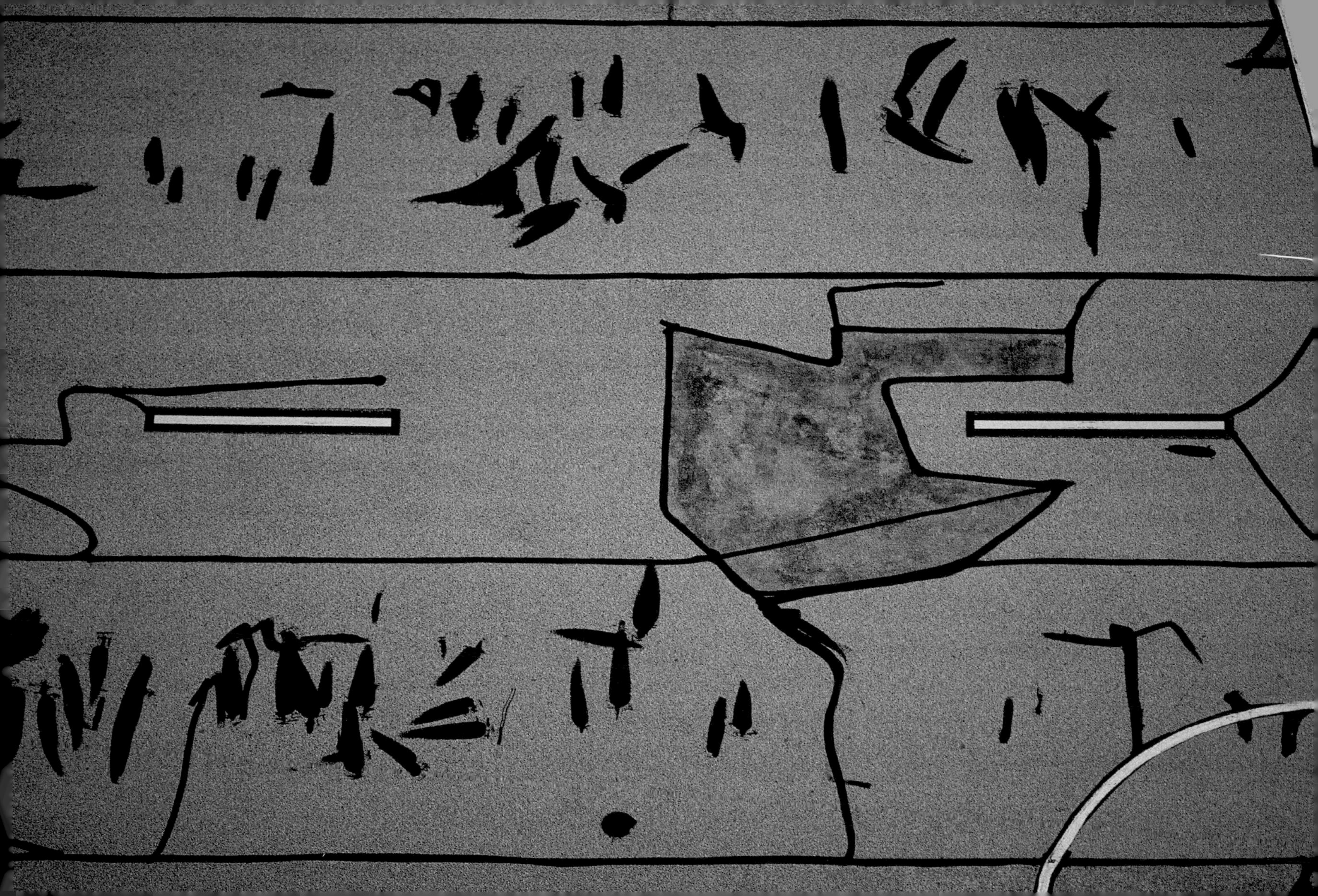

FEBRUARY

Fields on the Anta plateau, Cuzco region, Peru (13°29' S, 72°07' W).

The Cuzco region of Peru, which lies 11,800 feet (3,600 meters) up in the center of the Andes cordillera, contains the country's richest farmland. Wheat and barley are grown in the valleys, and alfalfa and potatoes grow on the hilltops. Five hundred years ago, this land fed the Inca Empire. To ensure good yields, its rulers divided up the land equitably, making regular censuses to record the makeup of households, ownership of domestic animals, and plots of land. Today, Peruvian agriculture is marked by inequalities between small and large landowners: 0.1 percent of farms (each larger than 1,000 hectares) own two-thirds of the arable land. To survive, tens of thousands of Andean peasants have turned to growing coca illegally. Although cocaine production fell by 68 percent between 1995 and 2000, Peru is still second only to Colombia, and drug traffickers have retaken control of some Amazonian valleys in the north of the country.

07

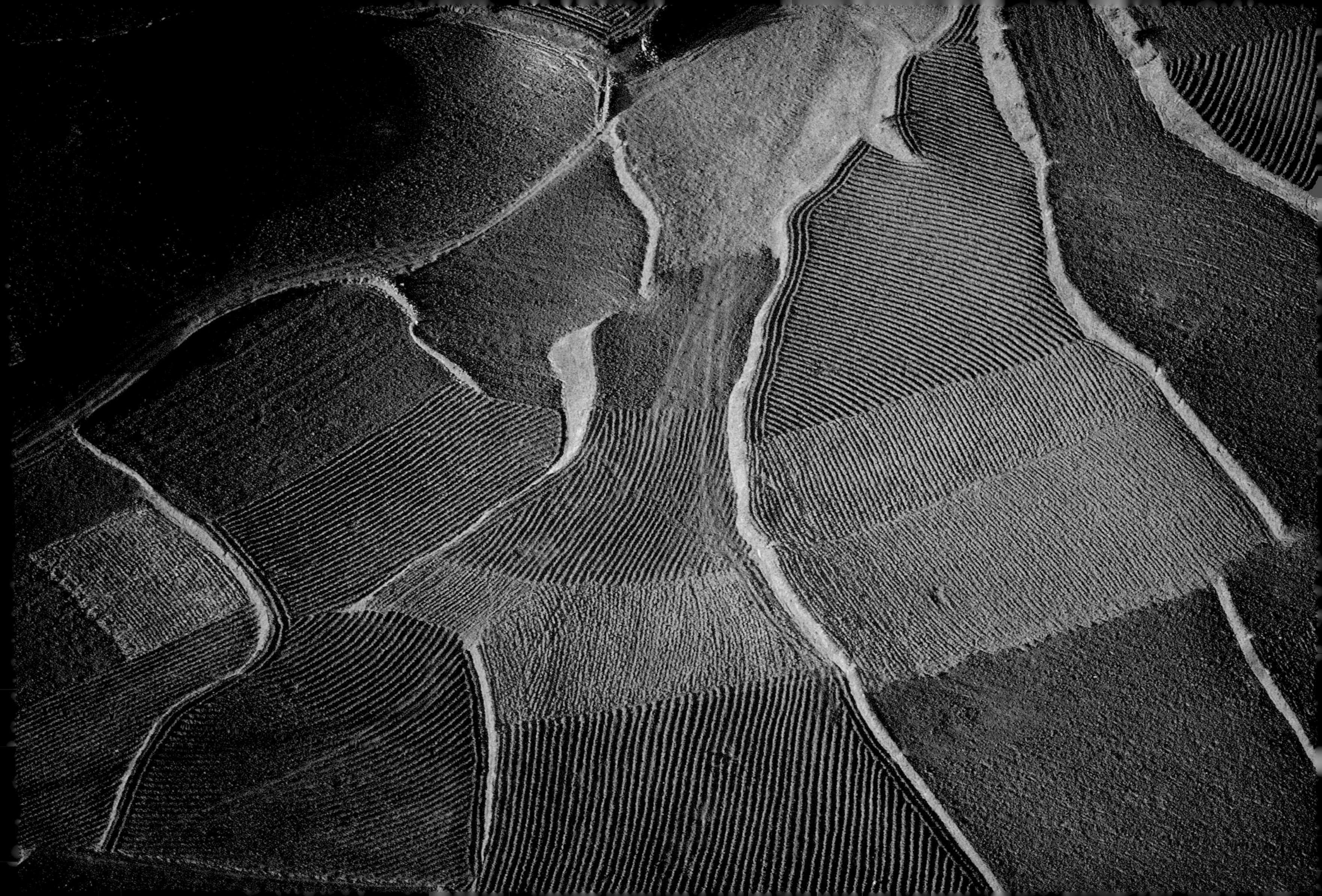

FEBRUARY

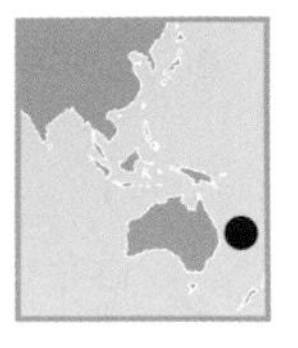

Neuika coral reef, New Caledonia, France (22°50' S, 167°25' E).

The string of coral reefs that girdles the deep blue lagoon of New Caledonia is under threat. It is turning pale and dying, attacked by pollution generated especially by nickel extraction on Grande-Terre, the territory's chief island. Every year, New Caledonia's opencast mines produce 118,000 metric tons of this metal—the tenth biggest production in the world. The industry is the archipelago's main economic resource, but it is a threat to the lagoon. Rainwater runs off from the mines into the sea, bringing waste that then settles on the surrounding coral. Many residents also suspect that the nickel treatment plants dump toxic waste and metals in suspension into the lagoon, and that these poison the coral reefs. These reefs are precious. Although they cover only 0.09 percent of the planet's seas and oceans, they are home to 2 million animal and plant species. But more than half are being damaged by human activity such as pollution, removal of coral, and dynamite fishing.

08

FEBRUARY

General view of a Himba village enclosure, Kaokoland region, Namibia (18°15' S, 13°00' E).
From the master hut, the village chief watches over the sacred fire adjacent to the *kraal*, or animal enclosure. The other villagers live in smaller huts, consisting of a frame of branches covered by a mixture of mud and cow dung. The 10,000 to 15,000 Himbas of Kaokoland live scattered in small clans to ensure the survival of their flocks in this desert region. But this dispersal has not always been an asset: in the nineteenth century, other ethnic groups took advantage of the Himbas' weakness to plunder the tribe, forcing them to live as hunter-gatherers as well as by begging, as their name indicates. ("Himba" means "those who ask for things.") Today, these shepherds have found their flocks once more. At first they suffered from tourism, but they have become aware of their exotic attraction and have refused to abandon their ancestral way of life while still selling jewelry or working as casual guides. Their chiefs have also learned to use the media to defend their rights. It is one way of combining tradition and the modern world.

09

FEBRUARY

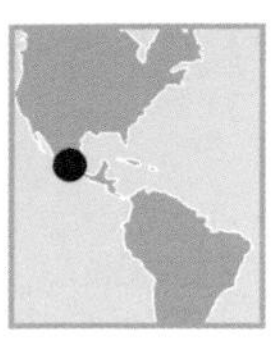

Salt drying on the coast of Quemaro, Jalisco, Mexico (19°63' N, 105°18' W).

Present in sea water in the ratio of 30 grams per liter, and in rock salt deposits of marine origin, salt is widespread throughout the planet and easily obtained either by mining or by evaporation in the sun. Only 20 percent of the world's production of 200 million metric tons is traded internationally. The United States produces 20 percent of the world's salt, and China 15 percent. Mexico, the seventh biggest producer, extracts 8 million metric tons of salt annually, exporting it to Japan and North America. Although Mexico's salt industry, on the country's Pacific coast, is highly mechanized, a few small-scale operations survive where the crystals are gathered by hand, as here. The chemical industry mops up 60 percent of the salt consumed in the world each year, and a further 10 percent is spread on icy roads. The remainder supplies the fishing industry (which uses it to preserve fish), the food industry, and is used as table salt. An individual eats 7 to 8 grams of salt per day. Too much salt is unhealthy and increases the risk of developing high blood pressure. Industrially produced food contains high levels of salt, with the result that the American population consumes too much: more than 4,000 milligrams per day, almost twice the recommended daily allowance.

10

FEBRUARY

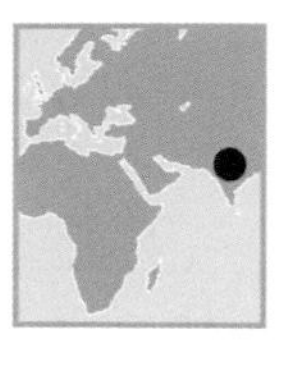

Cotton fabrics drying in the sun in Jaipur, Rajasthan, India (26°55' N, 75°49' W).

The state of Rajasthan in northeastern India is an important center of textile production, renowned for centuries for the crafts of dyeing and printing on cotton and silk fabrics. This activity is primarily practiced by the Chhipa, a community of dyers and painters who use ancestral techniques, decorating with wax and printing by stamping. Silkscreen printing, however, is increasingly used for larger-scale production, and chemical coloring is gradually replacing natural pigments. Craftsmen continue the multiple soakings to fix colors and dry their fabrics in the sun, as seen here in Jaipur, the state capital. The cottons and silks of Rajasthan have been exported since the Middle Ages to China, the Middle East, and Europe, and the international trade continues to flourish. Chhipa women perform this work; 32 percent of India's workforce is female, and this percentage is increasing. Recent decades have brought improved awareness of the rights of women throughout the world, but many countries are marked still by blatant inequalities between the sexes.

11

FEBRUARY

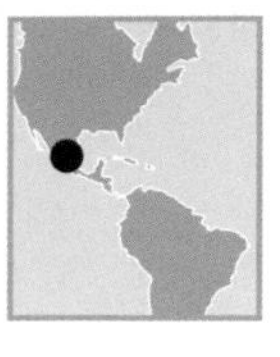

Market near Xochimilco district, Mexico, Mexico (19°20' N, 99°05' W).

This mosaic of brightly colored parasols hides a bustling, noisy market, set up for the day in a street of the capital. Shaded from the sun, stalls selling fruit and vegetables, herbal remedies, and spices sit side by side with others that sell cloth and craft artifacts. Mexico's flourishing markets are a national institution, held daily all over the country. Like their crafts, their traditional clothing, and the façades of buildings, the markets express Mexicans' love of vivid, bright colors, such as the brilliant pink known as "rosa Mexicana." Internationally, Mexico is a world champion of commercial vigor, with exports growing at the rate of 18 percent—with more than 85 percent of these going to the United States. But although economic liberalization led to a doubling in gross domestic product per capita between 1985 and 1999, it worsened inequality in the distribution of wealth. In rural areas, where agriculture cannot compete with imports, average earnings are as little as a quarter of the national average. The social unrest that has troubled the state of Chiapas since 1994 is partly due to this.

12

FEBRUARY

13

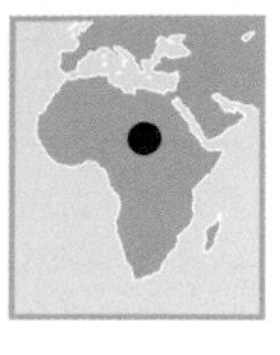

Nomad encampment, Lake Chad region, Chad (13°15' N, 15°12' E).

The nomadic herders of the Kanembu, Peul, and Fulbe peoples graze their livestock on Lake Chad's marshlands and fertile alluvial soils, as do the Buduma, who live on islands in the lake itself. At dusk, the herders light fires, as at this encampment on the lake's northeast shore. The livestock take cover of their own accord amid the thick smoke, avoiding the mosquitoes that infest the region and spread deadly diseases. But there is another threat to the survival of the Kuri breed of cattle, which now number 400,000 head. Endowed with impressive horns, which act as buoyancy aids, the breed is confined to the islands of Lake Chad, and its fate is closely dependent on that of the lake's waters. In 30 years, the breed's living space has shrunk considerably. The surface area of Africa's fourth biggest lake has decreased, from about 9,650 square miles (25,000 square kilometers) to 965 square miles (2,500 square kilometers), as a result of the droughts of 1972 and 1982, which the low rainfall could not make up for. Another reason is the drop in the flow of the Chari and Komadougou rivers, which feed into Lake Chad and have been partly diverted for irrigation. To save the lake from drying up completely, Chad, Niger, Nigeria, and Cameroon, which share its waters, plan to divert a central African river toward it.

FEBRUARY

14

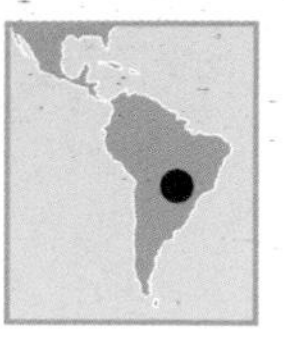

Herd of zebu on a road near Cáceres, Mato Grosso do Norte, Brazil (16°05' S, 57°40' W).

The Mato Grosso is one of Brazil's richest agricultural regions, where livestock and crops are raised on immense, extensive farms called *fazendas*. Almost two-thirds of the country's cultivable land is owned by less than 3 percent of the population; of those land holdings, half are not farmed at all. At the same time, more than 25 million landless peasants support themselves by itinerant farm labor. This situation has led to violent conflict, which has killed more than 1,000 people over the last 10 years. The struggle is driven by the Movimento dos Sem Terra (Landless Movement), which is fighting for fairer distribution of farmland. Since 1985, direct occupation of land has forced the state to grant ownership to more than 250,000 families. However, only agricultural reform can improve matters—but no government has yet dared to commit itself to it, for fear of going against the interests of rich landowners and the multinationals active in Brazil.

FEBRUARY

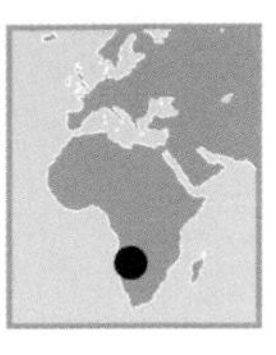

Boat run aground on the beach near Lüderitz, Namibia (26°42' S, 15°14' E).

The Benguela Current, moves north from the Antarctic and follows the coast of Namibia, where beaches alternate with reefs and shallows. The current causes a strong tide, violent turbulence, and a thick fog that conceals the contours of the coast. Thus, it is a passage feared by navigators sailing by on the way to the Cape of Good Hope, at the southern tip of Africa. Since 1846 Portuguese seafarers have called the shores "the sands of hell," and the northern part of the coast was given the evocative name Skeleton Coast in 1933. The rusted wreckage of boats as well as airplanes and all-terrain vehicles, along with skeletal remains of cetaceans (aquatic mammals such as whales) and even humans, are strewn along this melancholy shoreline. The wreckage is sometimes mired in the sand hundreds of yards from the water, as seen here near the city of Lüderitz, testifying to the violence of shipwreck. Although advancing rescue technology allows more lives to be saved than fifty years ago, the cost paid on the seas around the globe has been heavy: at least 65 fishing boats disappear daily around the world, and each week two large vessels are shipwrecked.

15

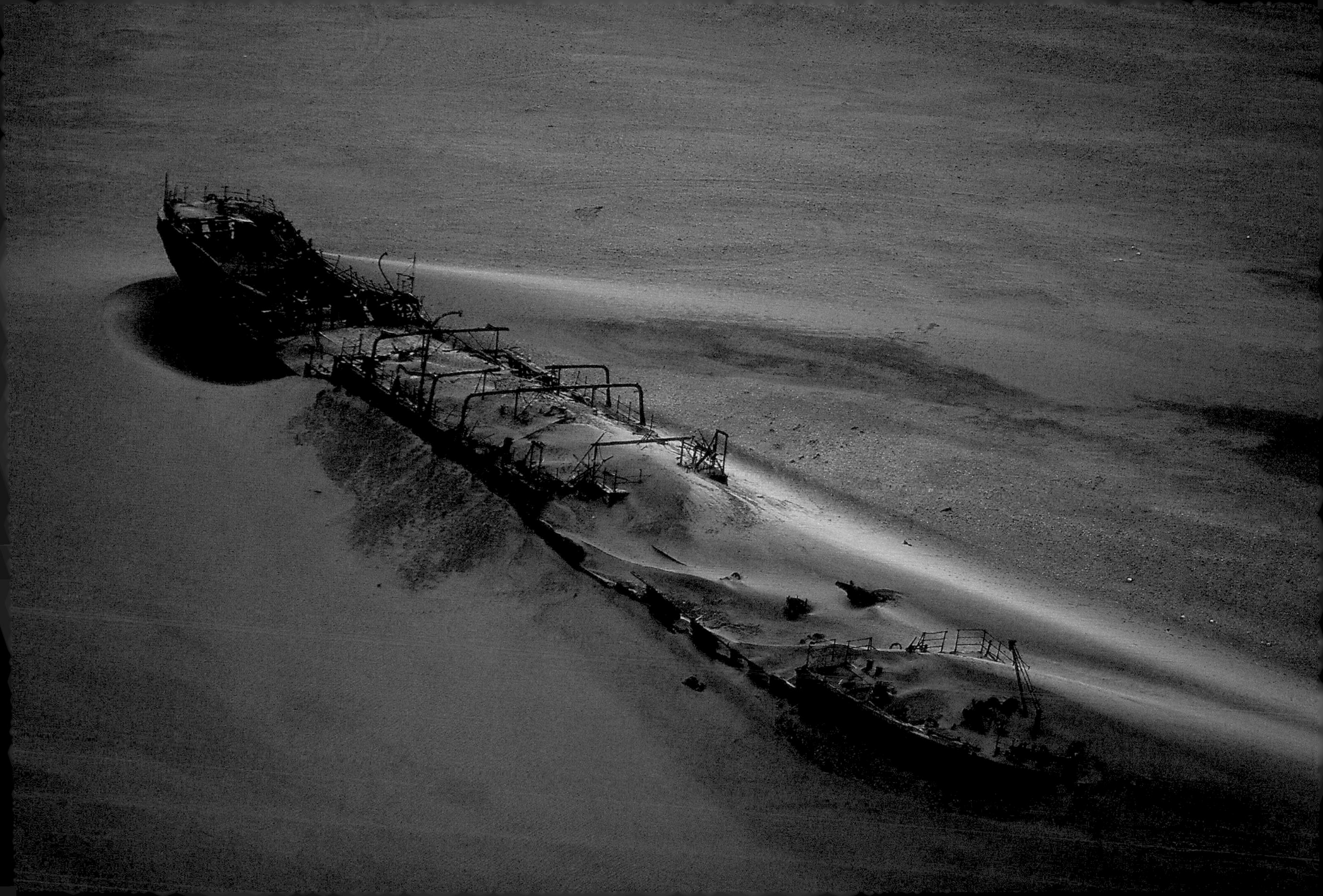

FEBRUARY

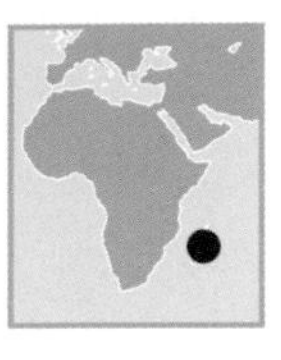

Convoy of carts southwest of Antananarivo, Madagascar (18°55' S, 47°31' E).

Every day, countless convoys of carts pulled by zebus take to the winding, muddy roads that lead to Madagascar's capital, Antananarivo. Because of the population's flight from the countryside to the city, Antananarivo has finally burst its borders and spread onto the surrounding marshy plain. Madagascar is one of the world's fifteen poorest countries, with gross domestic product per capita of about $250 (as compared to $21,300 for a citizen of the European Union). This poverty is essentially rural, and it affects small farmers: of the 70 percent of Malagasies who live beneath the poverty threshold, 85 percent are farmers. Several factors aggravate this situation: the unequal distribution of land, high rural birth rate, lack of skilled labor, and poor access to health services. Women are the most disadvantaged, since they can neither own land nor borrow money.

16

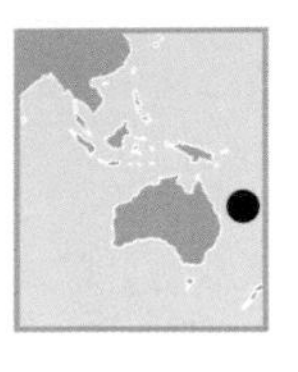

The islets of Nokanhui, south of Île des Pins, New Caledonia, France (22°43' S, 167°30' E).

In 1774, Captain Cook landed on a long island, which he named New Caledonia in a reference to his native country. The wild appearance and luxuriant vegetation of the surrounding islands and islets gave the impression of an oceanic Garden of Eden. This heavenly vision ended in 1863, when France turned several of the islands into prisons. No fewer than 22,000 people were "transported" there, including those deported for taking part in the Paris Commune of 1871. After they had served their terms, many remained on the islands, but the islands' first governor considered this population too small. From 1894 on, he invited more than 500 families, among whom he distributed 25,000 hectares of land. Native Melanesians were denied French citizenship until 1946, when the "indigenous law" was abolished, and only former convicts and colonizers were recognized by France. Today, the Kanaks have achieved independence, sharing sovereignty with the French government under the Matignon agreement of 1998.

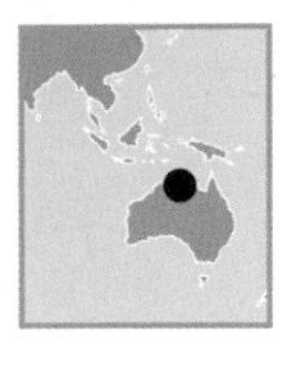

Marsh in Kakadu National Park, Northern Territory, Australia (13°00' S, 132°30' E).

Covering 7,720 square miles (20,000 square kilometers), Kakadu National Park is one of Australia's biggest. The park was added to UNESCO's World Heritage list in 1981 both for its natural and its cultural value. It has been inhabited for 40,000 years and constitutes a unique archaeological and ethnological record, containing traces of Neolithic hunters and fishermen as well as aboriginal rock paintings. The biodiversity of its marshy plains is rich and varied: the plains are home to almost 1,700 plant species, 10,000 types of fish, 117 reptiles, 280 birds, and 60 mammals. Because Australia has been separated from other continents for 150 million years, species have evolved here that exist nowhere else—for example, marsupials such as the koala and the kangaroo.

18

FEBRUARY

19

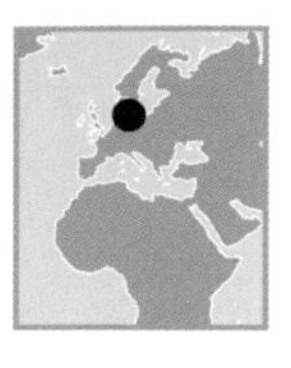

Middelgrunden offshore wind farm, near Copenhagen, Denmark (55°40' N, 12°38' E).

Since late 2000, the world's largest offshore wind farm to date has stood in the Øresund strait, which separates Denmark from Sweden. Its 20 turbines, each equipped with a rotor 250 feet (76 m) in diameter, standing 210 feet (64 m) above the water, form an arc with a length of 2.1 miles (3.4 km). With 40 megawatts of power, the farm produces 89,000 MW annually (about 3 percent of the electricity consumption of Copenhagen). In 2030 Denmark plans to satisfy 50 percent of its electricity needs by means of wind energy (as opposed to 10 percent today). Although renewable forms of energy still only make up 2 percent of the primary energy used worldwide, the ecological advantages are attracting great interest. Thanks to technical progress, which has reduced the noise created by wind farms (installed about one-third of a mile, or 500 m, from residential areas), resistance is fading. And with a 30 percent average annual growth rate in the past four years, the wind farm seems to be here to stay.

FEBRUARY

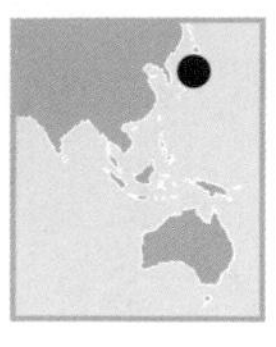

Kizuminami water tower, south of Kyoto, Honshu, Japan (34°41' N, 135°47' E).

Built in 1999 at Kizu, south of Kyoto, this 129-foot-high water tower (47 meters) holds enough water to supply 16,000 people. The cylindrical design draws its inspiration from bamboo, which is extremely common in the region. This type of giant grass, on which dinosaurs grazed 200 million years ago, can grow to a height of about 110 feet (40 meters) with a diameter of about 2 feet (60 centimeters), and numbers 1,250 species. It grows so rapidly (between 3 inches and 16 inches (75 mm and 400 mm) per day, with the record held by a native Japanese species which grows 1.2 meters in a day) that it can be harvested every two years. A single clump of bamboo can produce more than more than 9 miles (15 kilometers) of woody stems in thirty-five years of harvests. It has more than 1,500 different uses, ranging from construction (scaffolding, buildings, bridges), to the home (furniture, utensils), food (young shoots), and medicine. Its roots bind the soil together, protecting it from erosion, and research is being conducted into ways of replacing conventional wood sources with bamboo plantations. These could help curb the overexploitation of tropical forests in developing countries.

20

FEBRUARY

21

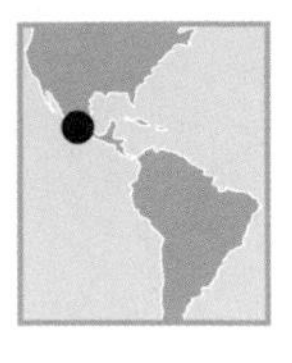

Vacationers swimming with dolphins, Puerto Vallarta, Jalisco State, Mexico (20°37' N, 105°15' W).

On the western Pacific coast, facing the Bay of Banderas, the town of Puerto Vallarta has been a mecca of Mexican tourism for decades. Here, in the city's amusement park, vacationers can swim with dolphins. Though this might seem a tempting idea, it is severely condemned by many groups, especially those concerned with protecting the cetaceans. The large number of visitors, and the close contact with the animals, increase the chances of diseases being transmitted between people and dolphins. In a more general sense, dolphins in parks where people can feed or touch them are stressed, often obese, and frequently injured. The visitors are not always safe, and they risk being bitten or hit inadvertently. Nature conservation organizations are constantly demanding the closure of such "petting pools."

FEBRUARY

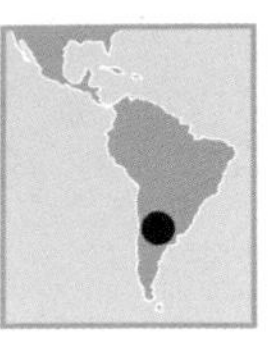

Autumn colors in Neuquén Province, Argentina (40°55' S, 71°37' W).

The Andes in the south of Neuquén Province are nicknamed "the Argentine Switzerland" because their landscape recalls that of the Alps. This temperate forest is unique in Latin America, and most of it lies in neighboring Chile. Wedged between the Atacama Desert to the north, the pampas to the east, and the ocean to the west, it is a botanical island, displaying a remarkable degree of endemism: almost 90 percent of its plant species grow nowhere else. As well as being highly varied, it is also beautiful in autumn, when the flaming red of the beeches contrasts with the dark green of conifers. But these two countries of southern Latin America have already lost almost half of this woodland. In Argentina, natural forest is often replaced by monoculture of pine or eucalyptus. These plantations are deeply impoverished biologically and, as a result, are vulnerable to illness and other problems. Nevertheless, in some countries they help keep deforestation in check and protect the soil from erosion.

22

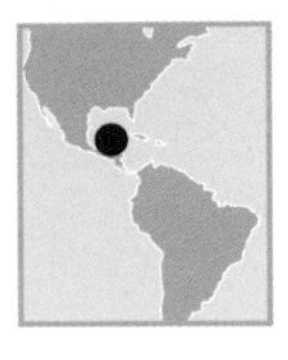

Punta Cancún, Cancún, Quintana Roo, Mexico (21°05' N, 86°46' W).

At the northeastern end of the Yucatán peninsula, a narrow strip of land separates a vast coastal lagoon from the Caribbean Sea. It was only in 1972 that the first buildings were erected on this extraordinary site; now it is Mexico's most important tourist destination and one of the world's biggest coastal resorts, linked to the United States and Europe by direct flights. Luxury hotel complexes are strung along the 9-mile (15-kilometer) seafront, and the development stretches for more than 40 miles (70 kilometers) south of Cancún, along the coastline known as the Maya Riviera. Tourism, which brought in $8.3 billion in income during 2000, is an important economic resource for the country. A mecca for mass tourism (3 million visitors per year, 74 percent of them foreigners), Cancún also hosted, in 1999, a North American convention on "sustainable tourism." This pioneering effort gives priority to conserving natural resources and local amenities, which are often altered by the environmental impact of development for tourism, and to ensuring that the profit from such development is fairly distributed.

23

FEBRUARY

Pattern of small walls on the island of Dugi Otok, Croatia (43°58' N, 15°04' E).

The bare ridges of Dugi Otok are fluted by kilometers of dry stones. These lines of walls tell the story of livestock grazing and farming on the islands of the Adriatic Sea. Sheep farming, which was widespread until the beginning of the twentieth century, led to serious deforestation. The walls were built not only to limit grazing areas (or plots where vines and olive trees still grow) but also to protect the hillsides from wind erosion and to reduce the amount of soil that rain washes away. Since the decline of sheep farming and agriculture in general, vegetation has been recolonizing this bare grassland. But this crowding of the environment is causing the disappearance of plant species adapted to this degraded "steppe" habitat typical of the Mediterranean region, where ecosystems have been shaped by human activity since the earliest times. Certain organisms have thus evolved in step with the changes made to the natural environment by agricultural ecosystems, sometimes called agrosystems.

24

FEBRUARY

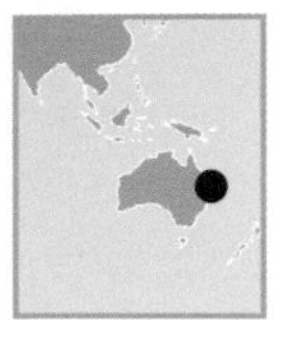

Sand Dune in the heart of vegetation on Fraser Island, Queensland, Australia (25°15' S, 153°10' E).

Fraser Island, off the coast of Queensland, Australia, is named after Eliza Fraser, who was shipwrecked on the island in 1836. At a length of 75 miles (120 km) and a width of 10 miles (15 km), it is the world's largest sand island. Yet on top of this rather infertile substratum, a humid tropical forest has developed in the midst of which wide dunes intrude, moving with the wind. Fraser Island has important water resources, including nearly 200 freshwater lakes, and has varied fauna such as marsupials, birds, and reptiles. Exploited for its wood since 1860, used for the construction of the Suez Canal, the island was later coveted by sand companies during the 1970s. Today it is a protected area and was declared a world heritage site by UNESCO in 1992.

25

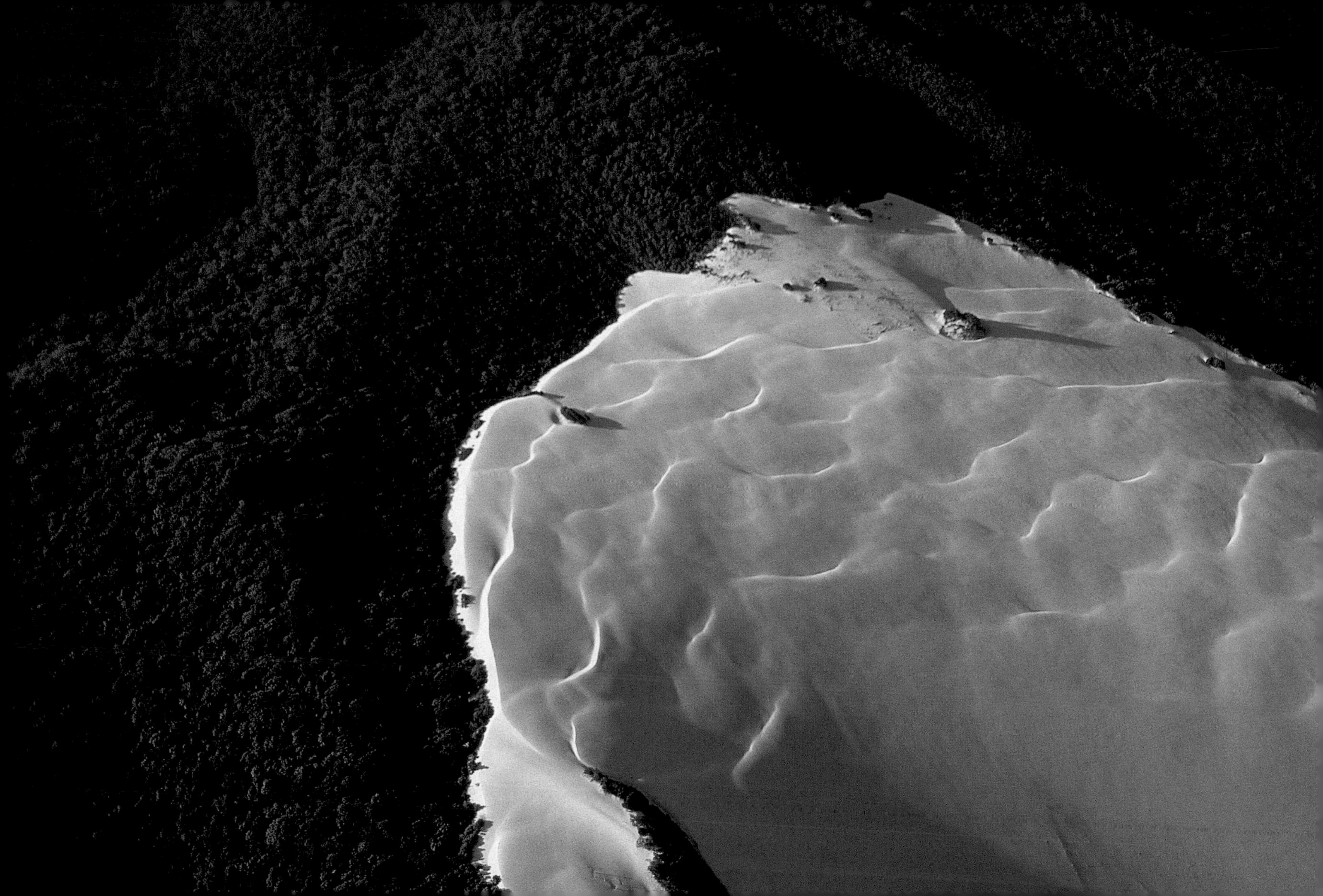

FEBRUARY

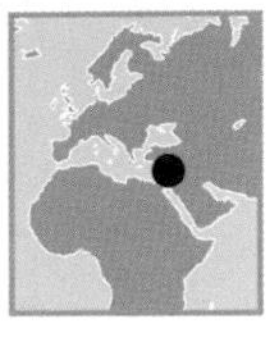

Fishermen on rocks, Ras Beirut, Beirut, Lebanon (33°53' N, 35°30' E).

The rocky headland of Ras Beirut, at the city's northwest corner, extends into a narrow pier where fishermen venture, undaunted by the rough waves. Apart from recreational fishing, the Lebanese practice small-scale commercial fishing from the many small ports dotted along the country's 130-mile (210-kilometer) coastline. Despite competition from Turkey, which catches 30 percent of the 1.3 million metric tons of fish taken from the Mediterranean every year, fishing retains an important social and cultural role in Lebanon. Unlike the north Atlantic, fishing around the shores of the Mediterranean is mostly small-scale, as the makeup of the fleet shows: 85 percent of fishing boats are coastal, 10 percent are trawlers, and 3 percent use seine nets. Nevertheless, several species—hake, sole, bass, and monkfish—are fished at a higher rate than stocks can replenish themselves. International agreements have recently been put in place to try to improve the coordinated management of the Mediterranean's fish resources.

26

FEBRUARY

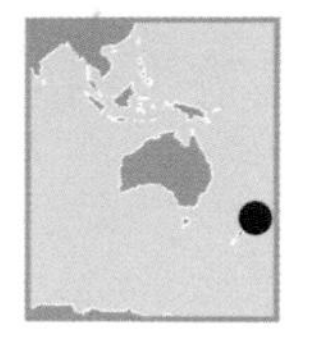

Wai-o-tapu geothermal area, North Island, New Zealand (38°20' S, 176°21' E).

Jets of hot steam rise constantly from the seething, bubbling surface of the "Champagne Pool" and the "Artist's Palette." In the Maori language, "Wai-o-tapu" means "sacred waters," which may be a mark of respect for these steaming waters. This tourist hot-spot is in the middle of an area of intense volcanic activity, positioned immediately above a 6.5-square-mile (17-square-kilometer) pocket of magma lying 2.5 miles (4 kilometers) below the earth's surface, permanently heating the underground waters. As they rise toward the surface, the water and steam absorb chemicals and minerals, which color the surface pools. Yellow indicates the presence of sulfur, red that of iron oxide, and delicate green points to a dangerously high concentration of arsenic. New Zealand, especially the North Island, is on the Pacific's famous "ring of fire," and makes use of this resource. In 1999, the country ranked seventh in the world for electricity production using geothermal energy.

27

FEBRUARY

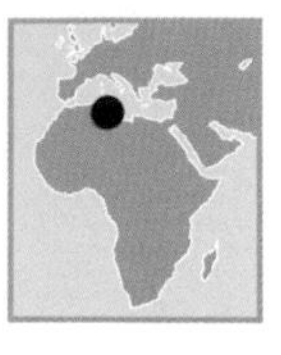

Valley of the Ksour, between Matmata and Tataouine, Tataouine governorate, Tunisia (33°00' N, 10°15' E).

Ksour, or *ksars*, are Berber fortresses typical of Tunisia's south. Arranged around a central courtyard and protected by walls measuring as high as 32 feet (10 meters), the *ksour* were built on hilltops for protection from attackers. They bear witness to the Berbers' long resistance to Arab invasion between the seventh and twelfth centuries. Most of these *ksour* were collective granaries for safe storage of cereals, oil, and animal fodder. For this reason they are divided into *ghorfas*, cells 13 to 16 feet (4 to 5 meters) deep and 6.5 feet (2 meters) high, stacked in several tiers. Others were used as dwellings before being abandoned in favor of the plains when peace returned to the valley. Now empty, these "hilltop granaries" still dominate the landscape, their size serving as a reminder that in former times a wetter climate than today's supported a sizeable population in a region that is now on the fringe of the Sahara Desert.

28

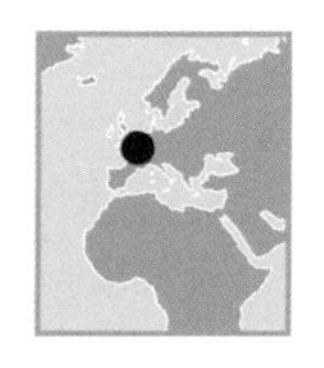

Horses in a meadow near Zelzate, Belgium (51°12' N, 3°49' E).
North of Ghent, near the Netherlands border, these horses dozing in the morning coolness form a picture that corresponds to the traditional image of the countryside. Ninety-seven percent of Belgians live in towns, and only 3 percent live in rural areas. In the European Union as a whole, 20 percent of the population lives in rural areas—a share that continues to shrink as the number of people employed in agriculture drops (21 percent of the European working population in 1960, 5 percent in 1996) and the exodus of young people to the cities. This trend is more pronounced in disadvantaged regions; in some areas, however, the trend is reversed by young couples and retired people who are attracted by the price of real estate. The countryside has lost the gloomy image it had in the 1960s and has become synonymous with authenticity, unspoiled nature, peace, and security. Throughout Europe, and especially in France, there are multitudes of new rural dwellers. Between 1990 and 2000, 500,000 city-dwellers left the Paris region, and 1 million people settled in Nice and Montpellier. These “new country dwellers” now account for 23 percent of France’s rural population.

SUSTAINABLE FOREST MANAGEMENT, CONSERVATION, AND RECOVERING

MARCH

01	02	03	04	05

Forests are complex ecosystems that need to be managed in a sustainable, balanced way. They take up a third of the planet's land surface, but this area is shrinking, and their ecosystems are being damaged. Yet they have a central role in answering the big questions that face us this century—such as finding the causes and effects of climate change, reducing world poverty, preserving fresh-water supplies, and protecting biodiversity. One of our biggest challenges is discovering how to meet the often conflicting demands of people who depend on forests for a whole range of products and services. Without considerable scientific and technological progress, this will soon be impossible.

A total of 232 million acres (94 million ha) of forest—or 0.2 percent of their surface area—were lost every year during the 1990s. This decline, especially alarming in developing countries, primarily affected tropical forests, which account for almost half the world's forested area. These were being cleared at the rate of almost 1 percent per year.

In developed countries, on the other hand, forests are generally on the increase. The exception is the northeastern Mediterranean, where countries typically suffer forest fires, and the southeastern Mediterranean, where forests are being cleared or damaged.

A report by the United Nations Food and Agriculture Organization entitled *The State of the World's Forests 2003* highlights the important roles forests play in climate change—as a source of carbon dioxide (CO_2) when they are destroyed or damaged; as a source of biofuels that can replace fossil fuels; and, when they are sustainably managed, as carbon "sinks." For trees, like all plants, build organic matter from the CO_2 in the air, absorbing most when they are young. Forests contain more than half the world's carbon. Forest management also raises other questions, and these will have to be dealt with as we prepare for talks on the next commitment period of the Kyoto Protocol on climate change, which starts in 2005.

The capacity of forests to contribute to the fight against poverty—especially rural poverty in developing countries—has aroused renewed interest in recent years. Often, the support forests offer to poor households, in the form of fuel, food, and economic resources, is not counted when national statistics are compiled. Effective programs will depend on our ability to obtain better knowledge of the relationship between forests and rural livelihoods. We must also be able to find solutions—through industry, governments, and local people—that increase the revenue forests produce. Local people depend on natural forests and stand to suffer most from their destruction or degradation; consequently, if they can secure a certain level of political influence, they can play a big role in conservation measures.

The warnings sounded at the end of the twentieth century concerning the depletion of freshwater resources are proving correct, to the point that lack of water is threatening food security as well as human health and livelihood. Although they are no panacea, forests offer both economic and environmental advantages, particularly in those basins that are drained by river systems. However, forests must be managed together with water resources. This is especially true in basins containing wooded mountain areas, which are among the most important regions of the planet for the production of fresh water but which also originate many floods and landslides.

By looking at water as an economic resource—rather than as something that comes to us for free—policymakers and institutions can provide incentives and means to preserve fresh water. Land use, for instance, could be modified in a given region. This kind of cooperation among sectors is vital, along with a broader-based economic analysis, for creating a new perspective that considers water as part of the economy. This could help reduce inequality between payers and beneficiaries by adequately rewarding local populations that improve forest areas or reduce leaks downstream.

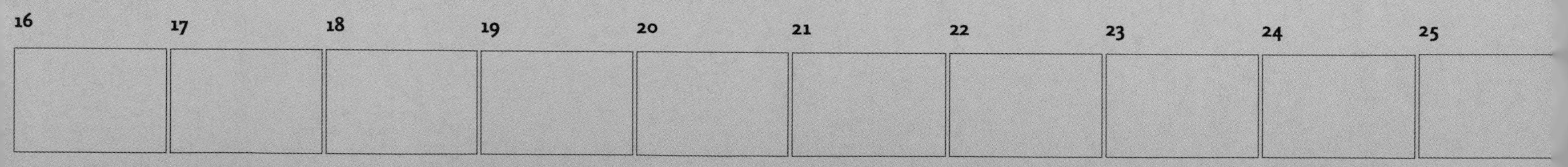

An estimated half of the world's species—a huge preponderance of its biodiversity—are to be found in forests, especially those in the tropics. The range of goods and services a forest is deemed to provide must take into account balanced use of these resources on a national scale. Use must be determined based on a dialogue among government, the private sector, universities, nongovernmental organizations, and local communities. One study, conducted throughout thirty-nine sites in Asia and the Pacific, suggested that a strategy based on community enterprise can safeguard forests if it is combined with other factors, such as access to markets, and if the company in question adapts to changing conditions.

However, to improve current practice in forest management, it is essential to develop indicators that measure the effects of human actions on biodiversity. This is extremely difficult. There is no single yardstick against which to measure diversity, and ecosystems vary enormously. Moreover, forest management practices do not necessarily affect all elements of diversity equally: they may benefit some, and harm others.

Therefore, sustainable forest management, and the ability of any forest to meet the human demand for goods and services, will depend on scientific and technological advances. However, there are not enough resources to maintain and increase research efforts, and the disparities are large between developed and developing countries; the public and private sectors; and the different economic activities connected with forests. In many tropical countries, most forest-based economic activities involving large numbers of people are unregulated, and there has been very little research into these sectors.

If the forest sector continues to grow weaker in the areas of science and technology, the gulf is likely to widen between developed and developing countries.

HOSNY EL-LAKANY

Assistant Director-General of the FAO Forestry Department, Italy

MARCH

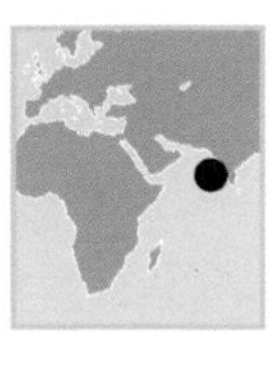

The Eye of the Maldives, atoll of North Mali, Maldives (4°16' N, 73°28' E).

The Eye of the Maldives is a *faro*, a coral formation on a rocky base that has sunk, concealing all but a ring-shape reef that encircles a shallow lagoon. Coral can only form in water of a relatively high temperature, and thus atolls develop principally in intertropical regions. The lowest country in the world, with a high point not exceeding 8.25 feet (2.5 m), the Maldive archipelago has suffered the devastating effects of several tidal waves. Locally, dike projects have also begun. Its 26 large atolls include 1,190 islands, nearly 300 of which are inhabited either permanently or seasonally by tourists. After the construction of the first resort on the island of Kurumba in 1972, tourism in the Maldives expanded rapidly: 80 resorts exist today and 300,000 tourists visit each year. Tourism is the world's leading industry. In 2000 the global total was almost 700 million tourists, and tourism yielded $476 billion in revenues. As tourism grows, it is essential to ensure that countries realize an economic advantage from tourism without destroying their natural and cultural patrimony.

01

MARCH

Working-class area of Belfast, County Antrim, Northern Ireland (Ulster), United Kingdom (54°35' N, 5°55' W).

Invaded in the twelfth century by England, Catholic Ireland rose against English rule in 1916 and was granted independence by the Crown in 1921, with the exception of the province of Ulster in the north of the island, where most of Ireland's Protestant minority lives. Since 1972, violent clashes between the province's Protestants and Catholics, who feel they suffer discrimination, have caused thousands of deaths. Ulster, with a surface area of 5,404 square miles (14,000 square kilometers), has 1.6 million inhabitants, of whom 500,000 live in Belfast. The city has a Protestant majority, but many Catholics live in its industrial outskirts, where the rows of identical houses recall the *corons* (nineteenth-century miners' houses) of northern France. About 2 million houses and apartments are built in Western Europe every year. Half of these are houses. The Republic of Ireland (Eire) has the highest proportion of households living in houses—92 percent, compared to a European Union average of 52 percent—and 75 percent of Irish people are owner-occupiers (in Spain, the figure is 82 percent). At the bottom comes Germany, the only country in the Union where fewer than half of households own the premises where they live.

02

MARCH

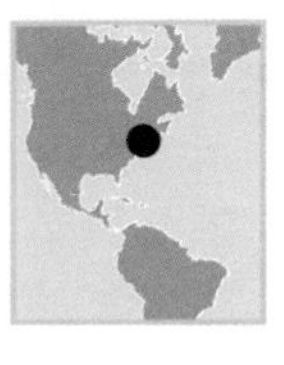

Wollman Rink in Central Park, New York, United States (40°45' N, 74°00' W).

Among New Yorkers' greatest pleasures is their freedom to escape the city simply by stepping into Central Park. This green space, which covers more than 842 acres (341 ha) between Fifty-ninth and One-hundred-and-tenth Streets, has been at New York's disposal ever since 1859, when the city spent more than $5 million on what was then just a stretch of wild, muddy marshland. In summer it is a relaxing haven for roller skaters and cyclists; in winter, ice-skaters can use the rink in the park's heart. Central Park is so much a part of Manhattan that few people realize it is entirely man-made. The park's architects, Frederick Law Olmsted and Calvert Vaux, could hardly have imagined when they were planning it how important the park would become to New York's identity—no more than they could have envisioned the flood of more than 250,000 people who would wander through its paths on spring weekends. At the time, Olmsted and Vaux were launching a plan to make leisure more democratic and to bring it within reach of all regardless of social barriers.

03

MARCH

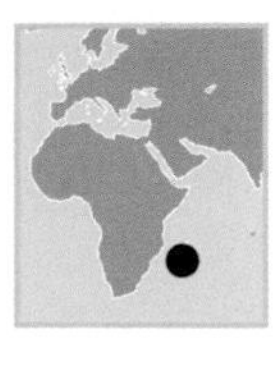

Locust infestation outside Ranohira, near Fianarantsoa, Madagascar (22°27' S, 45°21' E).

Madagascar's cereal crops and pastures have been chronically destroyed for centuries by invasions of migratory locust (*Locusta migratoria*) or red locust (*Nomadacris septemfasciata*). Several miles long and numbering as many as 50 billion insects, the hordes move at a rate of 25 miles (40 km) per day, laying waste all vegetation in their path. To eradicate this scourge, authorities have resorted to massive spreading of insecticides by airplane or helicopter. However, toxicity to humans and the environment, as well as the development of resistance in harmful insects, have shown the limits of this procedure. A recently discovered natural pesticide made from mushrooms might provide an organic method of eliminating these swarms of locust.

04

MARCH

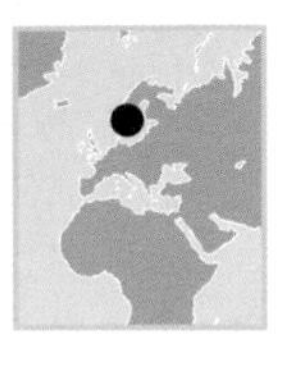

Gullholmen village, north of Göteborg, Sweden (58°10' N, 11°24' E).

The string of islands off the western coast of Sweden is famous for its sunny climate and its fishing villages. Gullholmen, which was home to a community of fishermen as early as the thirteenth century, retains its original appearance, with small red houses huddled together. At the end of the nineteenth century, this large village experienced its golden age, when herrings were exported to the United States in large numbers. Rich businessmen took advantage of this trade to establish themselves in this village, underlining their rank by building vast white villas, which can still be seen on the small island. This class divide has disappeared from this country that, since the economic crisis of 1930, has based its social system on equal opportunities. A social-democratic state par excellence, Sweden provides one of the best levels of social provision in the world and levies high taxes. In 2002, unemployment stood at 3.9 percent. After its neighbor, Norway, Sweden is considered to be the world's most advanced country from the social and humanitarian point of view.

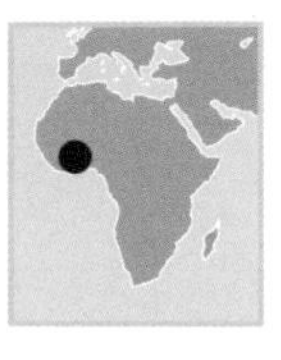

Washing laundry in a creek, Adjamé district in Abidjan, Côte-d'Ivoire. (5°19' N, 4°02' W).

In the neighborhood of Adjamé in northern Abidjan, hundreds of professional launderers, *fanicos*, do their wash every day in the creek located at the entrance of the tropical forest of Le Banco (designated a national park in 1953). They use rocks and tires filled with sand to rub and wring the laundry, washing by hand thousands of articles of clothing. Formerly a fishing village, Adjamé has been absorbed gradually by the metropolis of Abidjan, and it is now a working-class district. Abidjan is the economic and cultural center of the country, yet some parts of it are without running water or electricity. It has undergone staggering urban growth: its population has increased fiftyfold since 1950 and today it has more than 3 million residents, one-fifth of the national population. The city has seen a proliferation of dozens of small trades, such as these *fanicos*, which offer the only means of subsistence for the poorest groups.

NATATION
COUPE
SOCAP

MARCH

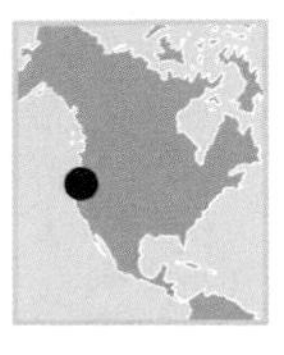

Farming near Pullman, Washington, United States (46°42' N, 117°12' W).

Known as the Evergreen State, Washington has been raising wheat for decades, striving to adapt the grain in order to protect a soil made fragile by the erosive agricultural practices of earlier times. The development of "agrobusiness," an alliance of agriculture, industry, science, and financial investment, encourages technological innovations aimed at improving productivity and helps keep the United States the leading exporter of cereals (about 35 percent of the total) as well as corn (40 percent) and soya (nearly half of world production). The use of biotechnologies, especially in the production of corn and soya, has led to the creation of varieties that are resistant to parasites, and herbicides that are believed to increase yield. Although these genetically modified organisms (GMO) are still the subject of prohibitions and sharp controversy all over the world, notably because of the limited knowledge of their effects on health and the environment, their cultivation is widespread in Argentina, Canada, and especially the United States, where half of all soya is genetically modified.

07

MARCH

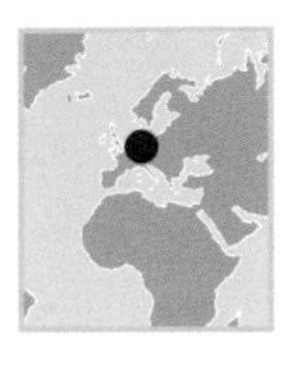

Site of Dachau KZ concentration camp, Bavaria, Germany (48°15' N, 11°27' E).

The small town of Dachau, about 12 miles (20 kilometers) from Munich, nestles at the foot of its castle. Until 1933, German painters and writers liked to come to stay in its vast Bavarian-style houses. Here, on March 22, 1933, a few days after winning the election, the Nazi party opened its first concentration camp, Dachau KZ. The next day, the first inmates began to arrive—opponents of the National Socialist regime, communists, monarchists, dissident Nazis, and personal enemies of the new governing class—followed, in increasing numbers, by Jews, Gypsies, and homosexuals. In all, more than 200,000 inmates were condemned to forced labor within Dachau's walls: of these, 76,000 died there. Of the camp's thirty-four dreary blocks, only traces now remain among the cypress trees. In the last decade, civil strife has claimed 3,600,000 victims worldwide—a far greater number than casualties claimed by wars between countries—either because they were dissidents or simply because of their ethnic origin.

08

MARCH

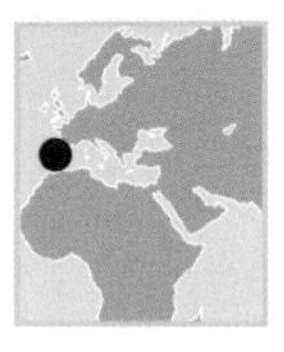

Skyscraper in Benidorm, Costa Blanca, Valencia, Spain (38°32' N, 00°08' W).

Europe is the world's biggest tourist destination, receiving 58 percent of the 700 million tourists who go on vacation every year. The Mediterranean Sea is a big factor in the Continent's popularity, and some 250 million vacationers a year enjoy the many charms of its shores. The high-rise buildings of Benidorm, a seaside Manhattan offering countless entertainments, show one side of the Mediterranean. The skyscrapers lining the seafront behind two long golden beaches have swallowed up what was, until the 1950s, still a fishing village. The exceptional amount of sunshine (more than 300 days per year) enjoyed by southwest Spain has profited the Costa Blanca, whose coastline, on the Gulf of Alicante, is densely populated. Benidorm alone accommodates up to a million people in the high season. Humankind is choosing in increasing numbers to live by the sea: thirteen of the nineteen cities on the planet with more than 10 million inhabitants are on the coast. This urban pressure on coastlines is growing faster than the facilities for treating waste-water. Only one city in ten has, like Benidorm, a waste-water treatment plant.

09

BANCAJA

MARCH

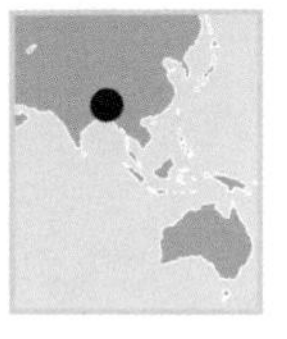

Flooded village south of Dhaka, Bangladesh (23°43' N, 90°25' E).

Bangladesh lies in the immense delta of the Ganges and Brahmaputra rivers, and half of the country is flooded by the monsoon each year. Yet for drinking water, Bangladesh's 135 million people need to look underground. During the 1970s, international organizations such as UNICEF funded the drilling of thousands of wells that were supposed to provide "healthy" water for the population. But these operations led instead to what the World Health Organization has described as the "largest mass poisoning of a population in history." More than a quarter of the wells have been contaminated by high levels of arsenic, up to seventy times the national standard for drinking water. Planners never tested for arsenic, which has occurred naturally in the delta's alluvial deposits for millennia, before drilling the wells. This "natural" pollution is thought to have exposed 75 million people to poisoning that can lead to breast cancer, kidney and liver diseases, respiratory problems, and death. United Nations researchers have forecast 20,000 deaths per year.

10

Crop circle, Wiltshire, England, United Kingdom (51°15' N, 1°50' W).

Is this "land art" or a sign from extraterrestrials? The question seems absurd, yet every spring this old debate is eagerly rekindled by the British press. On the arable plains of Wiltshire, these graphic designs, known as crop circles, occur in greater numbers every year. In 2002, the English countryside was adorned by 108 mysterious spirals, ellipses, rosettes, fractals, and even portraits of stars. Everyone has their own theory about these designs, which are very large and appear overnight. Self-styled scientific experts point to telluric, electrical, or magnetic forces, while others attribute them to the action of secret military satellites, and many see them as messages from outer space. But in fact we are indebted to young, talented artists, who are imitated all over the world, for these works. As for the farmers who are victims of their nights of crazy activity, some of them come off with something a little better than just trampled corn. In 1996, one crop circle attracted more tourists than the nearby site of Stonehenge, netting $57,500 (50,000 euros) for the lucky landowner.

MARCH

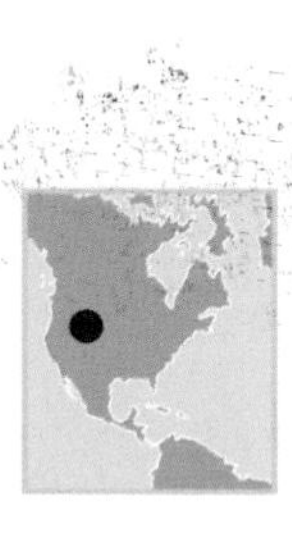

Grand Prismatic Spring, Yellowstone National Park, Wyoming, United States (44°26' N, 110°39' W).

Situated on a volcanic plateau that straddles the states of Montana, Idaho, and Wyoming, Yellowstone is the oldest national park in the world. Created in 1872, it covers 3,500 square miles (9,000 km^2) and contains the world's largest concentration of geothermic sites, with more than 10,000 geysers, smoking cavities, and hot springs. Grand Prismatic Spring, 370 feet (112 m) in diameter, is the park's largest hot pool in area and third-highest in the world. The color spectrum for which it is named is caused by the presence of cyanobacteria, which grow faster in the hot water at the center of the basin than at the periphery where the temperature is lower. Declared a Biosphere Reserve in 1976 and a UNESCO world heritage site in 1978, Yellowstone National Park receives an average of 3 million visitors per year. The continent of North America, which contains the five most visited natural sites in the world, is visited by more than 70 million tourists per year—one-tenth of world tourism in numbers, but one-fifth of world tourism in revenues.

12

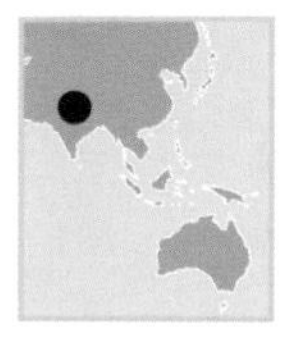

The stupa of Bodnath, Buddhist temple, Katmandu, Nepal (27°43' N, 85°22' E).

The city of Bodnath is home to one of the holiest Buddhist temples in Nepal, especially venerated by the thousand Tibetan exiles who live in this neighboring country. The stupa, which is a reliquary in the form of a tumulus topped by a tower, holds a bone fragment of Buddha. At a height and width of 132 feet (40 m), the temple is one of the largest in Nepal. Everything in the architecture of this sanctuary is allegorical, representing the universe and the elements (earth, air, fire, and water). The Buddha's eyes are fixed on the four cardinal points; the various stages in the acquisition of supreme knowledge, Nirvana, are represented by the thirteen steps of the tower. On religious holidays the monument is decorated with yellow clay and hung with votive flags. In the number of followers (350 million, 99 percent of them in Asia), Buddhism ranks behind Christianity, Islam, and Hinduism. In Europe today Buddhism has 2.5 million adherents; in France the number rose from 200,000 in 1976 to 700,000 twenty years later.

13

MARCH

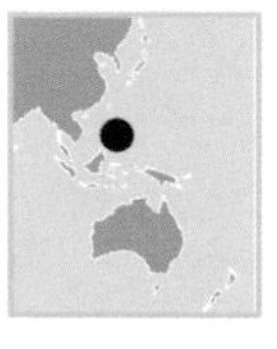

Gold mining near Davao, island of Mindanao, Philippines (7°04' N, 125°36' E).

Gold prospectors in Mindanao occupy precarious shelters of branches and cisterns clinging to the sides of mountains. The slopes, endlessly mined over the years, are made fragile by the network of mining tunnels, which often collapse under the torrential downpours of monsoons, killing many miners. This precious metal is often extracted by means of rudimentary tools, such as hammers or scissors, at a rate of 88 pounds (40 kg) per day. Since prehistoric times 150,000 tons of gold have been mined across the entire globe, one-third of which was used for the fabrication of objects, one-third was hoarded by governments, and the remainder was lost. Today almost 2,500 tons are extracted each year throughout the world, primarily in South Africa (20%), the United States (15%), and Australia (13%).

14

MARCH

15

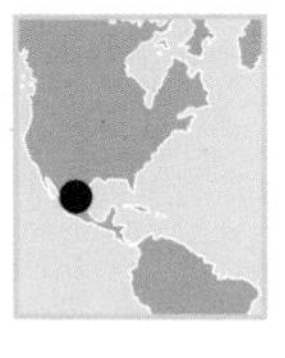

INFONAVIT housing, Toluca, Mexico state, Mexico (19°17' N, 99°40' W).

With 103 million inhabitants—compared to 70 million just twenty years ago—Mexico faces a demographic explosion and an acute housing shortage. Mexico state alone gains 1,000 new inhabitants every day. By offering subsidized loans to buyers of these standardized houses, built on the outskirts of towns, the INFONAVIT system allows families to acquire accommodation at a reasonable price. For many Mexicans, however, INFONAVIT remains a distant dream. In Toluca, a large industrial center, 79 percent of people of working age do not have a steady job; in the country as a whole, 40 percent of the population lives below the poverty threshold. Nevertheless, despite wide inequalities, the general situation is improving. Over the last twenty years, average life expectancy has risen from sixty-six to seventy-three, and illiteracy has been halved.

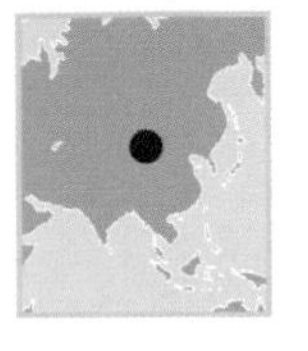

Flock of Kashmir goats in the Khustaïn Nuruu reserve, Mongolia (45°50' N, 106°50' E).

Mongolia is home to 2.5 million people and as many horses. In the saddle from infancy, nomadic herders lead their livestock to pasture. More than a third of the animals are Kashmir goats, from whose backs the precious wool is combed for export. Herding animals is one of the most sustainable ways of managing the resources of the arid steppes, where the extent of available range makes up for the thinness of the close-cropped grass and thickets. However, in 2003, after three consecutive years of *dzud*—a combination of summer and autumn droughts with bitter winters—almost 64 percent of the country's 24,000 wells became unusable, and most rivers dried up. Plant cover is becoming increasingly thin and scarce, while the effects of overgrazing threaten the plains, despite their vast size. A tenth of Mongolia's livestock—between 2 and 3 million animals—have already died from lack of food and water.

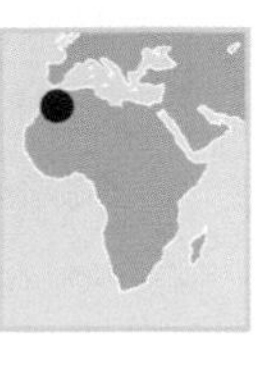

Boat under construction, Larache, Morocco (35°12' N, 6°10' W).

Only 50 percent of the world's timber is cut for industrial use, and almost 80 percent of this is in developed countries. For more than half the planet's population, wood is the sole energy source for heating and cooking, especially among poor people, such as the inhabitants of certain African countries. In Tanzania, Uganda, and Rwanda, 80 percent of total energy consumption comes from wood. Tree felling for fuel is one of the chief causes of deforestation in Asia, Africa, and Latin America. The situation is critical in developing countries, where alternatives to firewood are hard to come by for both rural people and the poor in cities. Worldwide—and taking reforestation into account—22 acres (9 million ha) of forest are lost every year.

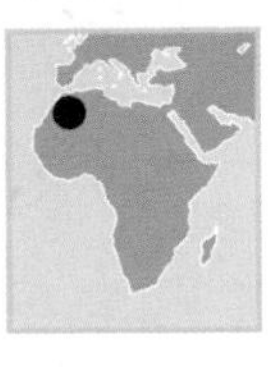

Village and fields in the Rheris valley, Morocco (31°35' N, 4°40' W).

The Rheris River flows south of the High Atlas, near the Algerian border. Many fortified villages nestle in the valley, sheltered from the heat and dust. This extremely arid region of southern Morocco has developed an ancient irrigation system using *rhettaras*—underground canals that draw water from springs and ground water sources and carry it to crops or gardens. This system is labor intensive, and it has gradually been abandoned in favor of new methods; these, however, can clog the soil and contaminate it with salt, as well as using large amounts of water. As a result of the spread of such techniques, agriculture is responsible for 70 percent of the world's consumption of fresh water, and every year 2 percent of irrigated land is lost to agriculture because it has become too saline. For this reason, methods such as the rhettaras or drip irrigation are now being studied anew in an effort to limit wastage. Irrigation remains essential because although only 20 percent of cultivated land is irrigated, it provides 40 percent of our food.

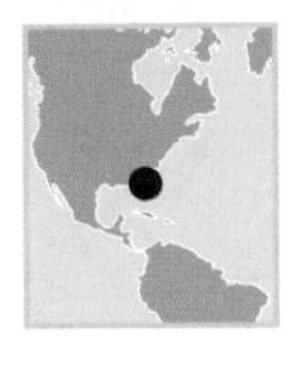

Herd of cows crossing Lake Kissimmee, Florida, United States (27°55' N, 81°16' W).

Sometimes cows roam along the shallow margins of Lake Kissimmee, in Florida. Their pastures adjoin the areas of land often flooded by this sheet of water, and their enclosures, often poorly maintained, are not always strong enough to keep them from straying. By their very numbers, these animals produce pollution. Their droppings contain phosphorus, a natural fertilizer that encourages the proliferation of microscopic algae, which are harmful to both plants and fish because they consume oxygen in the water. To try to address this problem, known as eutrophication, measures were taken in 1980 to keep livestock away from watercourses in Florida. They also limit the use of phosphorus-enriched fertilizers, which are carried from fields into freshwater basins by rainwater. The restoration of wetlands, which has been planned since 1993, could also greatly improve matters, since wetlands are more efficient than water-treatment plants at absorbing phosphorus from rainwater runoff.

MARCH

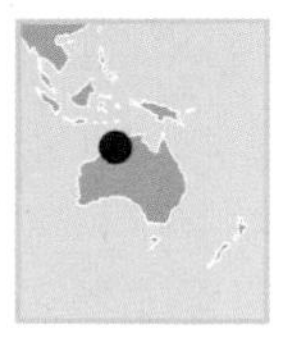

Estuary crocodile, Buccaneer Archipelago, Kimberley, Western Australia (16°17' S, 123°20' E).

The thousands of islands and islets of the Buccaneer Archipelago in northwestern Australia block the entrance to King Sound, an inlet of the sea. They are largely uninhabited, apart from a scattering of communities of aborigines. Covered with tropical rainforest and mangrove swamps, these islands comprise a habitat that is half land and half water, where salt and fresh waters mingle with the ebb and flow of tides with a range of up to 40 feet (12 m), the highest on the Australian continent. The estuary crocodile, also known as the saltwater crocodile, is the archipelago's most distinguished resident. These fearsome carnivores are born in fresh water. The adult males then drive the young animals toward more salt waters where, unusually for reptiles, they can live thanks to glands that excrete salt. Though highly sought-after for its skin, the crocodile is doing well in Australia because its natural habitat has been preserved and because it is also farmed.

20

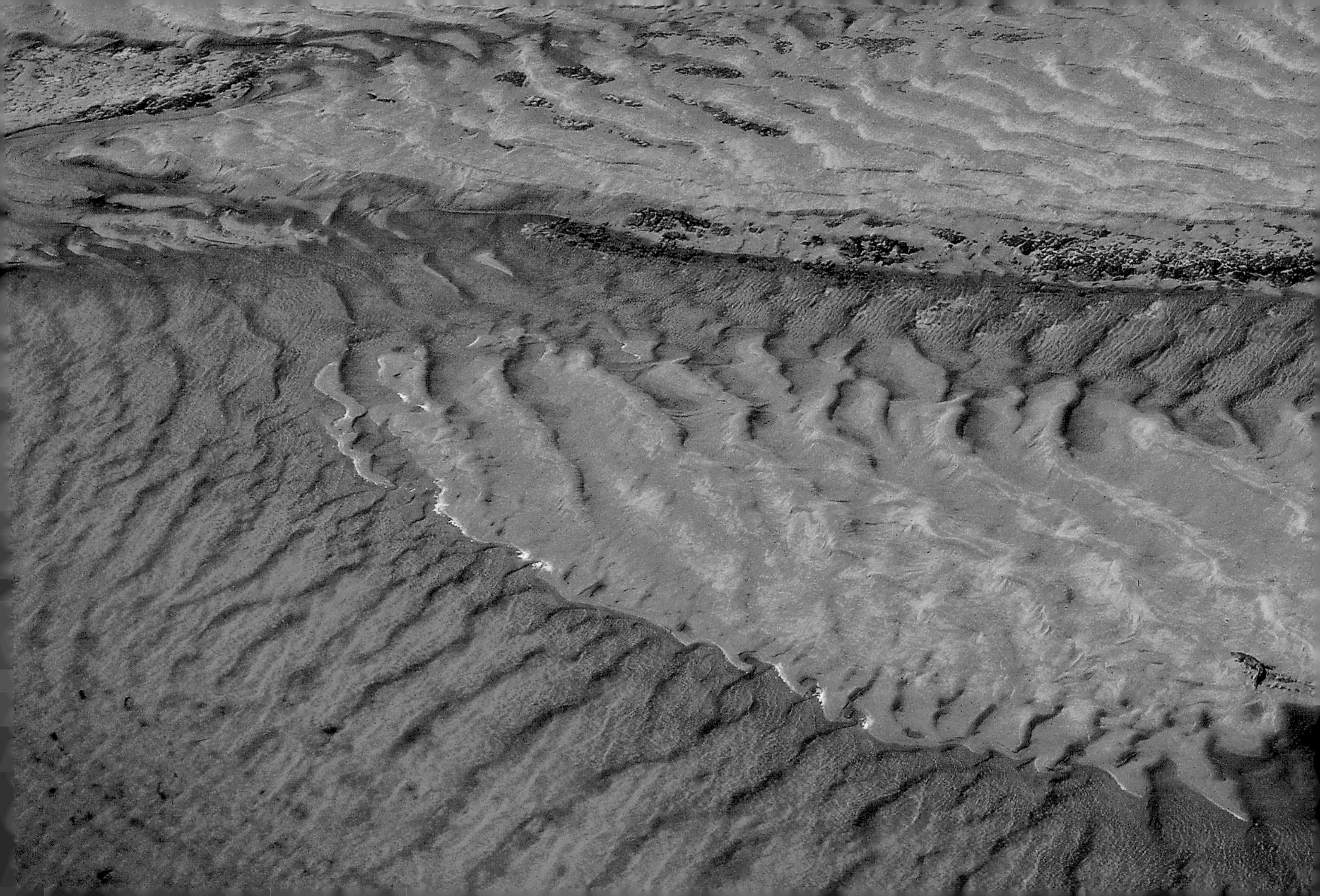

MARCH

Haram-al-Sharif, Jerusalem, Israel (31°45' N, 35°15' E).

Haram-al-Sharif, or the Esplanade of the Mosques, is the Muslim sanctuary that makes Jerusalem Islam's third most important holy city, after Mecca and Medina. However, it is also built on the ruins of the Temple Mount, where the Tablets of the Law were kept. Only traces remain of the temple's western wall, of which the Wailing Wall is the sole visible section. This place, which is equally holy to Jews and Muslims, forcefully illustrates the difficulty of resolving the Israeli-Palestinian conflict. In July 2000, the Camp David II negotiations, which aimed at a peace accord brokered by the United States, foundered over the status of Jerusalem in general and Haram-al-Sharif in particular. U.S. President Bill Clinton proposed that the Palestinians should have sovereignty over the Christian and Muslim areas, including the esplanade, and that Israel retain sovereignty over the ground beneath them. The Palestinian delegation considered this symbolic sovereignty totally illegitimate.

21

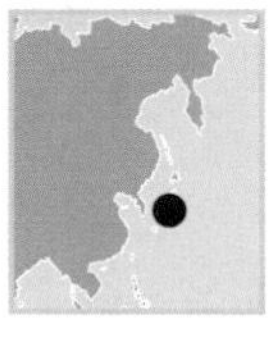

The Genbaku Dome, epicenter of the 1945 atom bomb explosion, Hiroshima, Honshu, Japan (34°24' N, 132°27' E).

The shell of the former Industrial Promotion Hall, the Genbaku Dome is the only building in the city center that partly survived the atom bomb dropped by the U.S. Air Force on Hiroshima on August 6, 1945. It has been preserved untouched to bear witness to the brutality of this act of war, and since 1996 has been on UNESCO's World Heritage list. When the bomb was dropped, 200,000 people died, and 40 percent of the city was completely destroyed. This part of Hiroshima, dedicated to the memory of the tragedy, includes the World Peace Memorial Cathedral, the Peace Park, and the Peace Memorial that contains, among other relics of the atrocity, tiles vitrified by the heat and watches that stopped at 8:16 AM, the moment the explosion happened. Three days after the Hiroshima bomb, on August 9, the Americans dropped a second atom bomb, which completely destroyed the city of Nagasaki and led to Japan's surrender. Nuclear weapons have transformed international relations. Although their use in 1945 was a barbarous act, the terror they subsequently inspired held in check the ideological confrontation between the United States and the Soviet Union during the Cold War, which lasted from 1945 until *perestroïka* in 1989.

22

Surfer at Copacabana Beach, Rio de Janeiro, Brazil (22°58' S, 43°11' W).

The beach stretches for miles on either side of the mouth of the Baia de Guanabara, the bay at the foot of Rio de Janeiro. The Atlantic waves that wash these sands can reach anywhere from 3 to 10 feet (1 to 3 m) in height. Sparkling under a perennial sun, they are an irresistible draw to the thousands of surfers who come to revel in the rollers of the cariocan coast. Captain Cook was the first European to see anyone surfing; he made his observations in 1770 at the Sandwich Islands (now Hawaii), where it was an initiation rite for young warriors. Surfing became established as a water sport in California at the beginning of the twentieth century, and in the 1960s came to symbolize an almost mythically pressure-free lifestyle. Increased leisure time in developed countries was one of the most important socioeconomic developments of the last century: in France, paid leave increased from two weeks a year in 1936 to five in 1982. However, the number of hours worked per year varies widely from country to country: in Japan, the average worker spends 2,100 hours every year on the job; in the United States, the figure is 1,800; in Sweden, 1,500.

MARCH

The oil tanker *Le Prestige* off Galicia, Spain (43°00' N, 11°50' W).

On November 19, 2002, *Le Prestige,* a twenty-six-year-old vessel, ran aground off the west coast of Spain, spilling 77,000 metric tons of fuel oil. Thousands of volunteers cleaned up the pollution that washed up on beaches. This disaster was a reminder that more than a third of the 40,000 large vessels now in service do not meet regulations, that our oceans suffer a major shipwreck on average every three days, and that at least half of all large oil tankers are more than twenty years old, even though ships of that age are lost at a rate ten times higher than newer vessels. Spectacular though they are, oil spills are unfortunately only the tip of the iceberg. Illegal dumping releases ten times the amounts of hydrocarbons, and two-thirds of marine pollution comes from the land. In the Mediterranean, 85 percent of the waste-water from cities is released untreated into the sea.

24

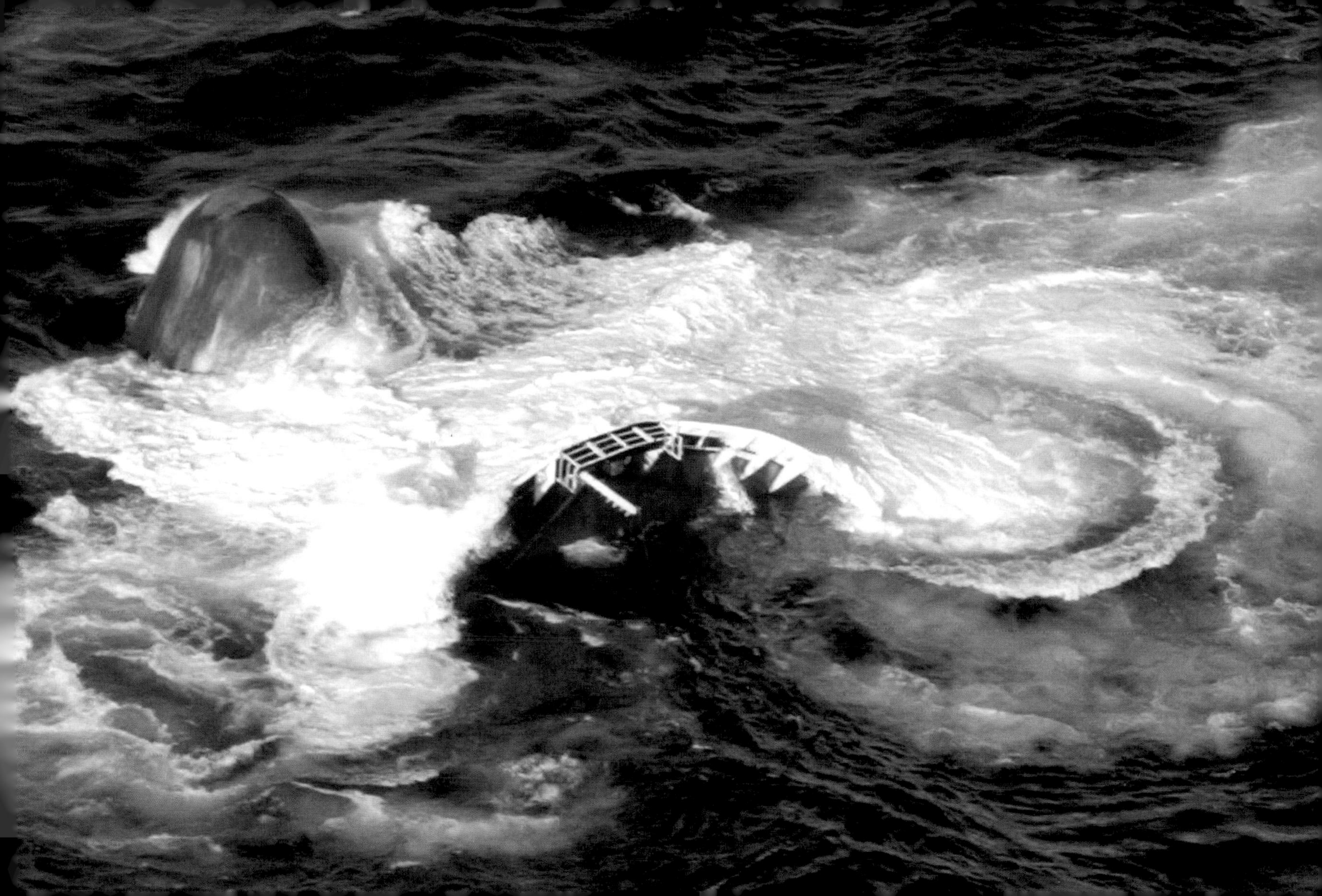

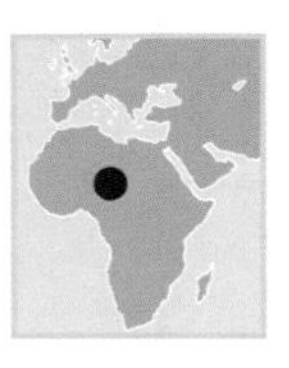

The crab claw of Arakaou, Ténéré Desert, Niger (18°96' N, 9°57' E).

The dunes of the Ténéré Desert are more than 655 feet (200 m) high and appear to be charging toward the Aïr Mountains. Driven by the wind, they are swallowed up by the open claw of Arakaou, a collapsed extinct volcano 6.2 miles (10 km) in diameter. The contrast between the bright sand and the dark mountain is striking. This landscape consists exclusively of minerals, yet 20,000 years ago, the region was a lush prairie, populated by antelope and rhinoceros. Since then the climate of the Sahel has become dry, and the desert has taken over. This is due to the natural fluctuation of the climate but also to human activity, such as overgrazing and forest clearance, which has contributed, and still contributes, to the speeding-up of this process. In the last 50 years, the Sahara Desert, of which the Ténéré *erg* (sand desert) is a part, has increased by 250,900 square miles (650,000 km^2)—an area equivalent to that of France—at the expense of Africa's fertile lands.

25

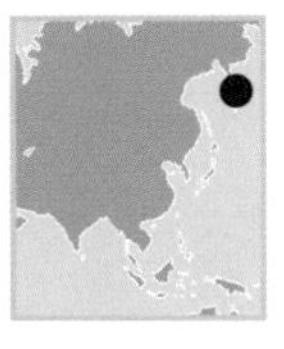

Snow-covered flanks of Kronotskaya Volcano, Kamchatka Peninsula, Russia (56°00' N, 160°00' E).

At the eastern tip of Siberia, Kamchatka Peninsula spreads over nearly 145,000 square miles (370,000 km^2). This region of Russia is ruled by nature, and humankind is barely present (the population density is below 1 person per km^2). The peninsula is geologically very young (less than 1 million years) and has 160 volcanoes, including thirty that remain active; they were declared a UNESCO World Heritage site in 1996. Kronotskaya Volcano is one of the highest, at 11,570 feet (3,528 m). The 3,500 square miles (9,000 km^2) of the Kronotski Reserve are home to several protected species: the Kamchatka brown bear, lynx, sable, and fox. Facing Kamchatka across the Bering Strait, Alaska offers a similar landscape. Twenty-six thousand years ago small groups of people crossed the strait, at that time dry land, and gradually populated all of the Americas. The Sioux, the Inca, and the Guarani are all descendants of the people from Kamchatka.

MARCH

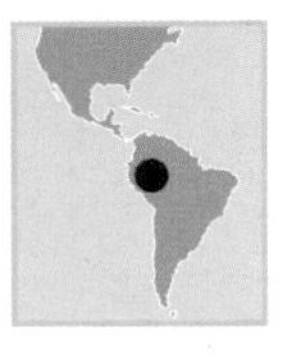

Hiram Bingham Road, leading to Machu Picchu, Cuzco region, Peru (13°05' S, 72°35' W).

In 1911, the American historian Hiram Bingham discovered Machu Picchu, an Inca city that had escaped the Spanish invasion of the sixteenth century. In 1948, a modern road was built bearing his name: the Hiram Bingham Road. Punishingly steep and less than 20 feet (5 m) across, it climbs in tight hairpins for 5 miles (8 km), rising 2,623 feet (800 m) up the precipitous side of the Salcantay mountain, toward the supreme "skyscraper" of the Incas. This access road to Inca culture allows us to rediscover something of what the Spanish conquest plunged into oblivion, even though the extent of what has been lost can never be measured. Some Indians pine for the lost social structure, which was based on the Inca principles of mutual collaboration (*ayni*) and of community living and work (*ayllu*). In 2001, the Peruvian government, under President Alejandro Toledo, took a further step toward this culture. A Quechua educated in the United States, the president made the reintroduction of the Inca languages one of his key goals.

27

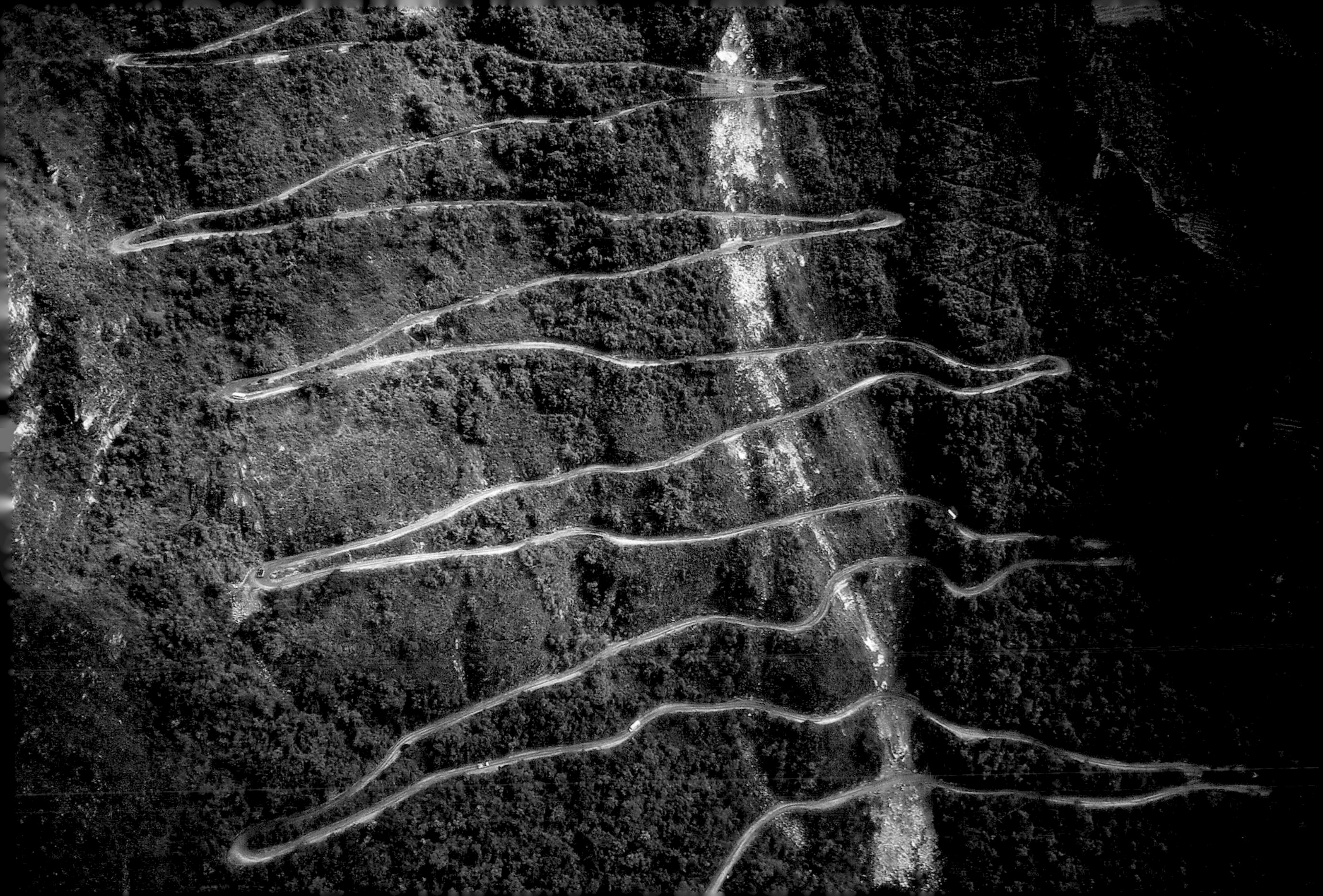

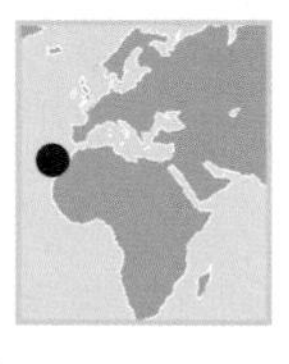

Vineyards, region of Geria, Lanzarote, Canary Islands, Spain (28°48' N, 13°41' W).

Lanzarote, one of seven islands in the Spanish archipelago known as the Canaries, lies closest to the African continent. Agriculture is difficult because of the island's desert climate and the total absence of streams and rivers on its territory of 313 square miles (813 km^2). Its volcanic origins, however, have provided the island with a fertile black soil made up of ash and lapilli over a substratum of fairly impermeable clay. Residents have developed a distinctive viticultural technique to adapt to these original natural conditions. Vine stocks are planted individually in the center of holes dug in the lapilli in order to draw on the accumulated moisture, shielded from the dry winds from the northeast and Saharan regions by low, semicircular stone walls. Spain's total wine production represents nearly 12 percent of the approximately 80 million gallons (300 million hl) of wine produced worldwide each year, ranking third among producing countries after France and Italy.

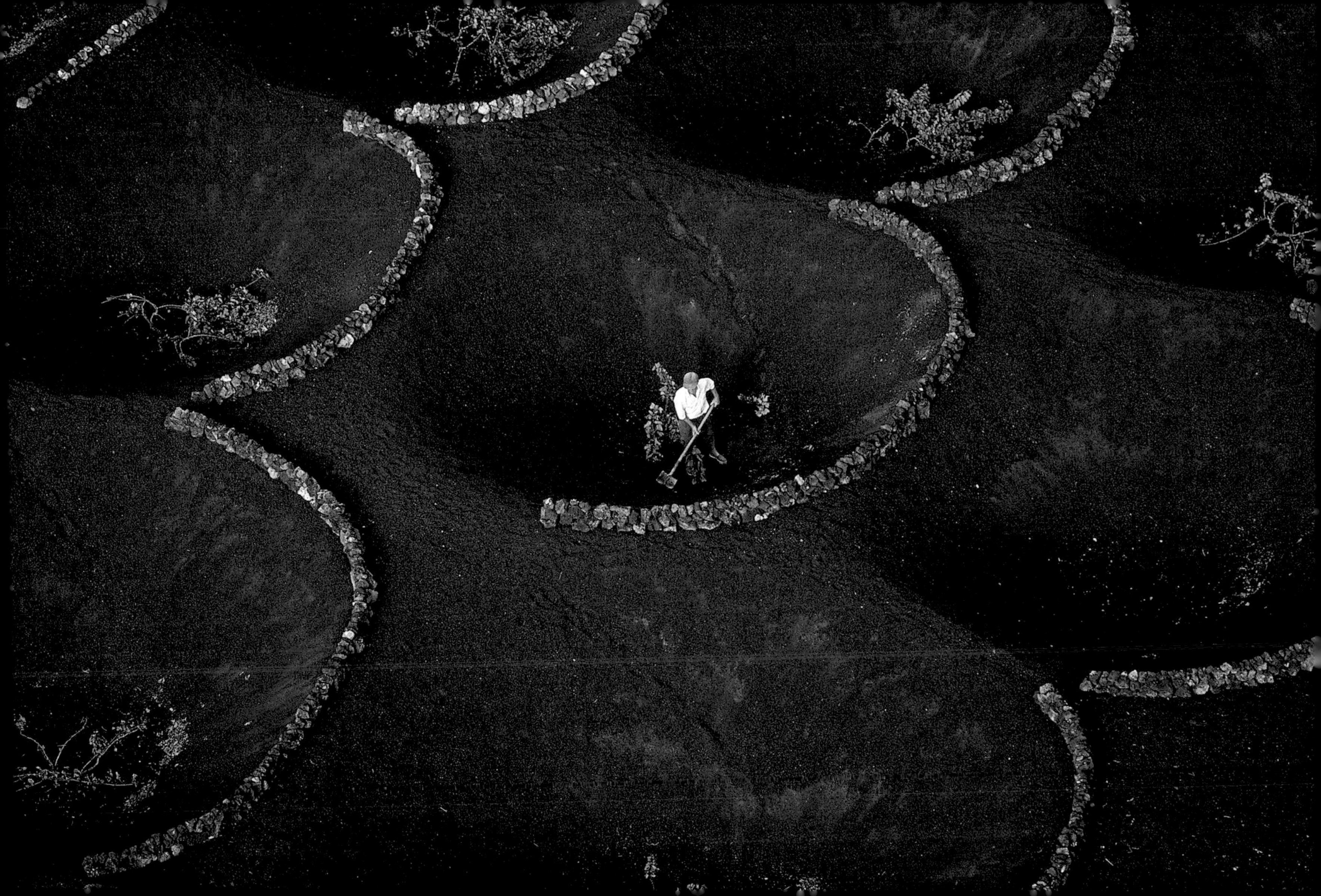

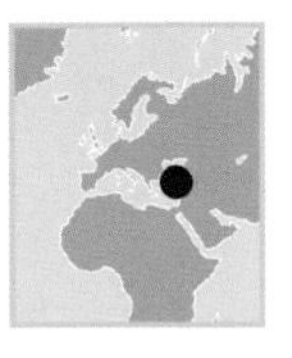

Agricultural landscape between Ankara and Hattousa, Anatolia, Turkey (37°34' N, 38°13' E).

The great plateaus and arid steppes of Anatolia are the cradle of Turkish civilization and were inhabited by nomads for centuries. Since 1977, this region, through which the Tigris and Euphrates rivers flow, has been the site of the Southeastern Anatolia Project (GAP), which envisages the construction of twenty-two dams and nineteen hydroelectric power stations by 2010. This project would increase the area of irrigated land in Turkey by 50 percent, and the country's output of electricity would double. The diversion of the waters of these two rivers is causing serious tension between Turkey and its neighbors in the Mesopotamian basin: Syria, Iran, and Iraq. In 2001, the United Nations Environment Programme raised the alarm for this internationally important ecosystem, in which hundreds of miles (or kilometers) of mountain valleys are to be flooded, while the lower regions are parched by huge drainage projects, reduced river flows, and damage caused by war. We are thus seeing the disappearance of one of the world's great wetlands systems—the Mesopotamian marshes, which straddle Iraq and Iran.

MARCH

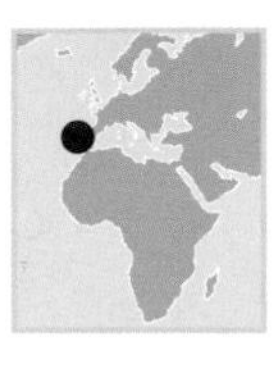

Monument to the Discoveries on the bank of the River Tagus, Lisbon, Portugal (38°43' N, 9°08' W).

Inaugurated in 1960 under the dictator Antonio Salazar for the 500th anniversary of the death of Henry the Navigator (1394–1460), the Monument to the Discoveries proudly towers 164 feet (50 m) above the Tagus estuary. The celebrated navigator stands at the bow, a caravel in his hand, and crowding in his wake come the great maritime explorers of the fifteenth and sixteenth centuries: Vasco da Gama, who rounded the Cape of Good Hope in 1498; Ferdinand Magellan, who crossed the Pacific Ocean in 1521; and Pedro Álvars Cabral, the first European to sail to Brazil. The first to be freed from Moorish domination, Portugal had a fifty-year start over its neighbor, Spain, in its attempt to conquer the world, and played an important role in voyages of discovery. Portuguese sailors were pioneers in all the world's seas and allowed the Portuguese empire to reach as far as the Indies by the beginning of the sixteenth century. On their expeditions to distant lands, the Portuguese conquerors—followed by the Spanish, Dutch, British, and French—took with them the Christian faith and scientific ideas, which they often disseminated by means of war and slavery.

30

MARCH

31

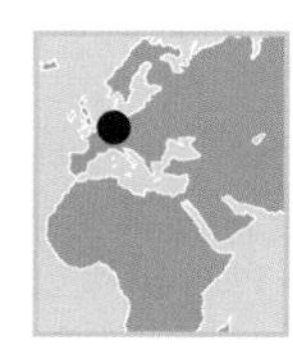

Bottle racks near Braunschweig, Lower Saxony, Germany (52°20' N, 10°20' E).

Not far from Braunschweig, in northern Germany, an avalanche of mineral water, beer, fruit juice, and carbonated drinks of all kinds is spread out in a grocery store's warehouse lot. Bottled water, estimated as a $22 billion market per year, leads all competitors in the world beverage industry. This most basic of bottled drinks is increasingly successful; world consumption is growing by 7 percent a year (15 percent in the Asian and Pacific region). One and a half million tons of plastic are used to contain the 89 billion liters of mineral water distributed in the world every year. Alcoholic drinks are still abused all over the world, a symptom of despair and social malaise caused by poverty and unemployment. In Russia, alcoholism keeps male life expectancy at 59 years, as opposed to 72 for women.

BIODIVERSITY AND THE BIOSPHERE

The biosphere is the envelope around the Earth's surface. Just a few kilometers deep, it is relatively thin compared with the planet's radius—more than 3,900 miles (6,300 km)—in which all living organisms occur. Although some microbes exist in deeply buried rocks, and microbial spores circulate high in the atmosphere, most organisms are restricted to the sunlit surface regions between the upper waters of the ocean and the alpine regions of the world's mountains. Here, green plants and other photosynthetic life forms capture the energy of sunlight and use it to make the organic compounds that are the building blocks of all organisms. Biological diversity, or "biodiversity," refers to the variety of these organisms, including their genes and the ecosystems in which they interact.

Biodiversity is valued at many different levels. Materially, people derive food, medicine, and other products from wild or domesticated environments that are biologically diverse. Diversity at the genetic level is particularly important in this context, and it is exploited to produce the distinct varieties and breeds of plant and animal life used in agriculture. Biodiversity underlies the life-supporting services provided by ecosystems, such as pollination, water purification, and the recycling of dead material and other waste. Many people prize diversity for its own sake, and their lives are enhanced by the knowledge that healthy

APRIL

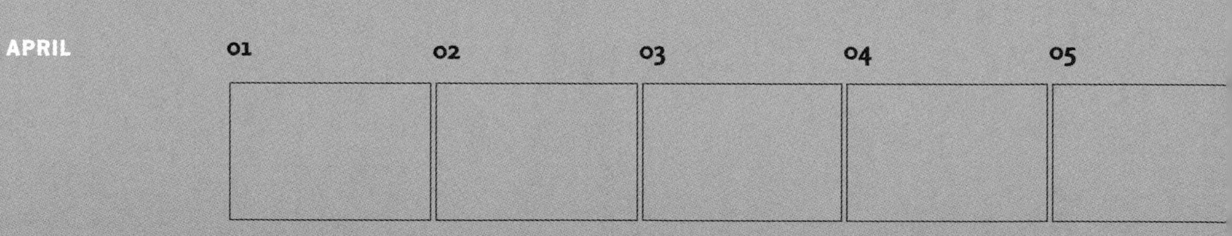

ecosystems, with a variety of species, remain on Earth. Ultimately, the human species cannot be extricated from a biologically diverse environment. We are bound by evolutionary history to the apes and those preceding lineages that stretch back to the beginning of life; we are bound by physiology to green plants and other organisms that manufacture the foodstuffs humans cannot make for themselves.

The planet Earth is more than 4,500 million years old, with an undisputed fossil record providing evidence of living organisms dating from about 3,500 million years ago. For the first 3,000 million years or so, life apparently consisted solely of simple microscopic organisms. A range of invertebrate animal types and early plants made their relatively sudden appearance some 500 to 550 million years ago.

Extinction appears to be the fate of every species. As the fossil record shows, the average lifespan of any given species is about 4 million years. The various species that are now living may comprise fewer than 5 percent of all the species that have ever lived. The rate of extinction has varied greatly through time, sometimes peaking during phases of mass extinction during which most lineages disappeared. However, new species have always evolved to fill the ecological gaps, more than compensating for those losses. Despite regular extinctions, the global variety of organisms appears to have increased the planet's biodiversity, as species found new ways to make a living from available ecological opportunities, and, incidentally, created further opportunities for other species to consume or live on or inside them. Global biodiversity may have reached its peak before the onset of the Pleistocene glaciations, some 1.8 million years ago.

Early humans appeared in Africa during the later Pliocene, a little more than 2 million years ago. Anatomically modern humans are known in Africa from about 100,000 years before the present, and these had reached Eurasia by at least 40,000 years ago. By the end of the Pleistocene, some 10,000 years ago, modern humans had spread to all continents except Antarctica. These migrations were subsequently to have an unforeseen and ever-increasing impact on the world's species and ecosystems.

There is a strong association between the arrival of humans in new lands and the subsequent extinction of larger species of wildlife ("megafauna"). About forty species of larger mammals, birds, and reptiles were lost from Australia about 46,000 ago, a few thousand years after humans arrived there. Saber-tooth cats, giant ground sloths, mammoths, and other large species disappeared from the Americas as humans dispersed over the continent. Giant birds and mammals were similarly lost soon after humans arrived in New Zealand, Madagascar, and the main Caribbean islands. Excess hunting is suspected to have been a major cause of these losses—a pressure that is still an important threat to many species today—and the low reproductive rate of many larger species makes them susceptible even to relatively low-intensity hunting by indigenous peoples.

Until the end of the Pleistocene, humans derived their food from hunting, fishing, and gathering of wild resources. About this time, some 10,000 years ago, agriculture was developed in Eurasia, Africa, and the Americas. Agriculture is a systematic technique of diverting more of the biosphere's production into human bodies. As farming methods have been developed, an increasing proportion of the Earth's surface has been transformed from natural habitats, often high in biodiversity, into agricultural landscapes, typically low in biodiversity. Increasing human need means an ongoing high rate of land conversion, with cropland and grazing land replacing forests and extending upslope into areas of fragile mountain habitat. The habitat loss and fragmentation that comes from agricultural expansion, combined with land that is converted for roads and settlements, is now a leading factor in the widespread decline of the planet's biodiversity.

Many scientists accept that the current rate of extinction is between ten and a hundred times greater than the general background rate suggested by the fossil record. Information is incomplete, but it does suggest an irregular rise in species loss

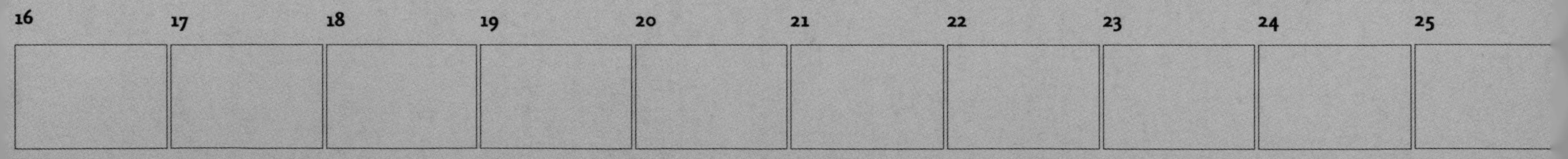

through the eighteenth and nineteenth centuries, as people spread in greater numbers to increasingly remote parts of the world. At least sixteen mammals and twenty-five birds became extinct during the last third of the nineteenth century, and numbers were similar in the early twentieth. Most were species with small ranges, such as those restricted to islands. While the number of known extinctions appears to have fallen, perhaps in part because of conservation action, the number of species at risk of extinction has risen: at present, some 24 percent of mammals and 12 percent of birds are threatened. In many ways, the widespread loss of populations and gradual erosion of genetic diversity is more significant than the extinction of localized species. This broad reduction in biodiversity mostly contributes to ecosystem degradation, and it is liable to impact most directly on the livelihood of the world's rural poor and disadvantaged.

A major international agreement, the Convention on Biological Diversity, has been in force for the past decade. By providing a set of objectives, recommended actions, and a funding mechanism, the convention has helped many countries to reduce the impact on their biodiversity as they work toward sustainable development. Nations present at the 2002 World Summit on Sustainable Development committed themselves to the goal of significantly reducing the current rate of biodiversity loss by 2010, and this renewed stimulus for action is cause for hope. But many observers believe that maintaining Earth's present biological diversity, and the integrity of the biosphere as a source of sustenance and wonder, will demand unprecedented change in the way that humans use the world's resources.

BRIAN GROOMBRIDGE
Senior Programme Officer at the UNEP (United Nations Environment Programme)–WCMC (World Conservation Monitoring Centre), Cambridge, England

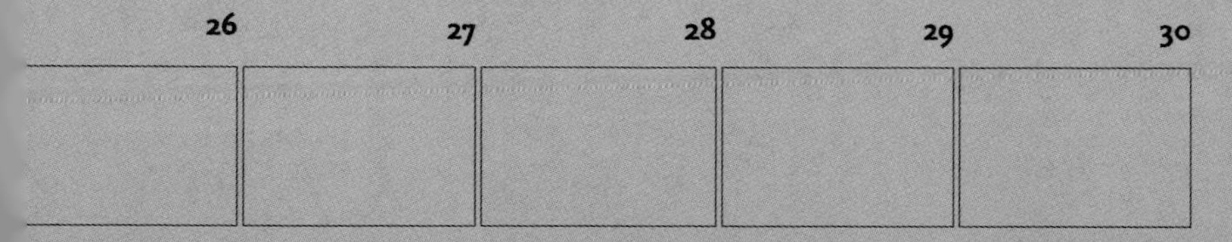

APRIL

Purmamarca, village in Jujuy province, Argentina (23°43' S, 65°29' W).

Northwest Argentina is the realm of the *puna*, a vast, cold desert of high valleys and plateaus that foreshadows the Bolivian *altiplano.* In this land of multicolored rocks, the Indian village of Purmamarca huddles at the foot of the *cierro de los siete colores*—"mountain with seven colors." Its population, descended from marriages between Indians and Spanish colonizers, lives chiefly from rearing livestock and from small market gardening plots on the valley floors. The inhabitants of Jujuy province, where half the land belongs to four large landowning families, face particularly harsh conditions. In the early 1990s, they rebelled against the federal government. The economic depression afflicting the country since 2001 made matters worse: in 2002, 63 percent of Jujuy's population lived in poverty, and almost 30 percent lacked the most basic necessities. During that same year, the region's public hospital admitted more than 700 children per day who showed the symptoms of malnutrition.

01

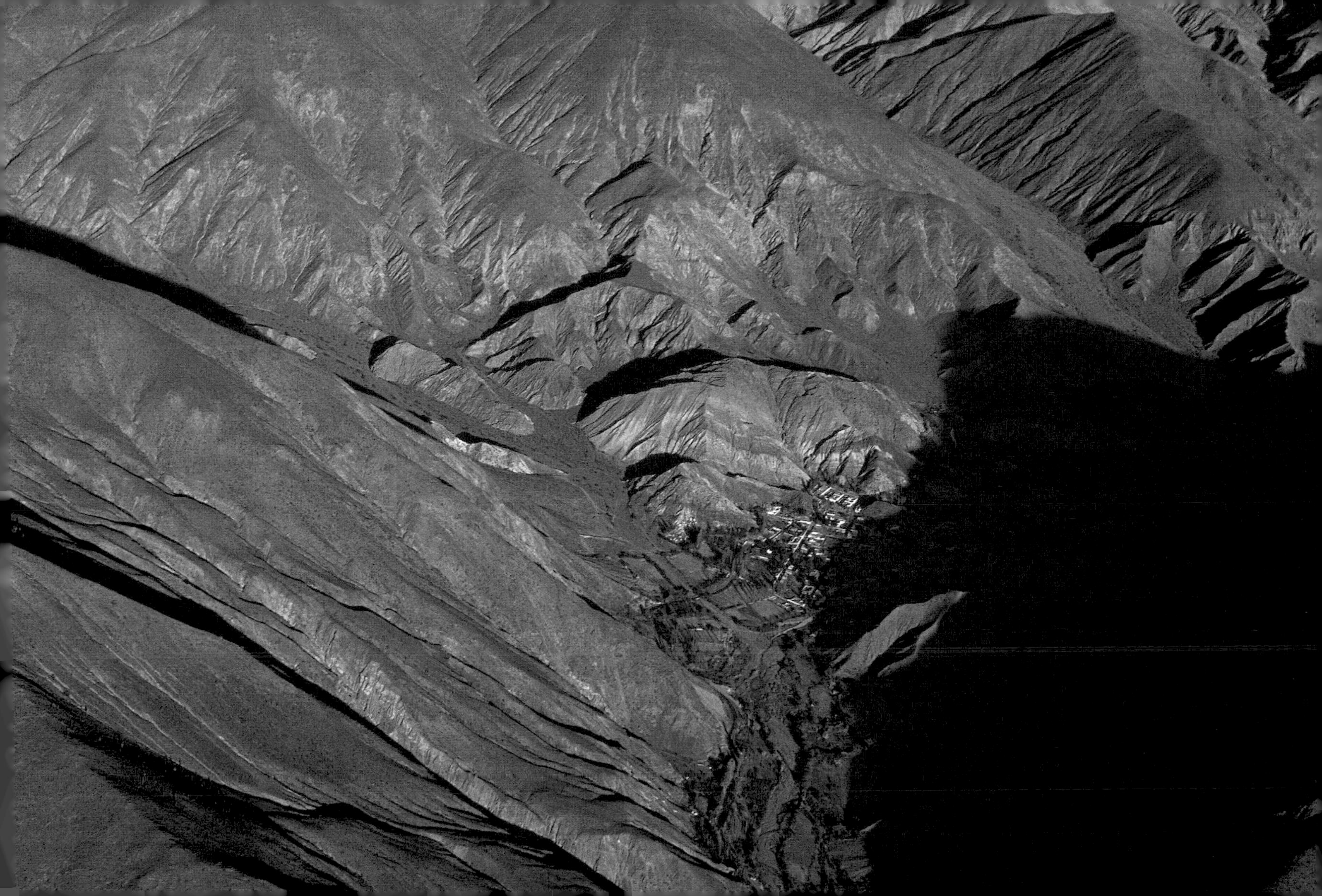

APRIL

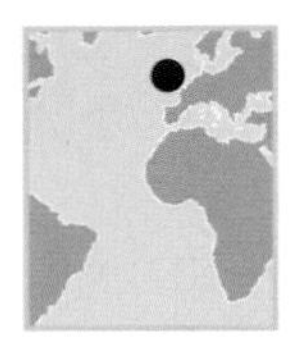

Dun Conchuir fort on Inishmaan Island, County Galway, Ireland (53°04' N, 09°34' W).

The fort of Dun Conchuir, rising above the close-cropped grass, was built centuries ago on this island some 30 miles off the Irish coast. The Aran Islands contain some of Europe's most magnificent prehistoric remains. This trio of islands—Inishmore, Inishmaan, and Inisheer—with their high rocky cliffs protect the Galway coast from the violent winds and currents of the Atlantic. For centuries, their inhabitants have helped keep the soil fertile by regularly spreading a mixture of sand and seaweed on the rock to produce the thin layer of humus needed for farming. To protect their plots from wind erosion, the islanders have built a vast network of almost 7,500 miles (12,000 km) of low walls, which give the land the appearance of a vast mosaic. Growing numbers of tourists are visiting the Aran Islands—whose inhabitants live chiefly by fishing, farming, and livestock—drawn especially by their wealth of archaeological remains.

02

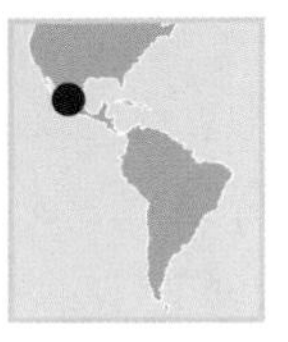

Church at the foot of the Paricutín volcano, San Juan Parangaricutiro, Michoacán, Mexico (19°27' N, 102°14' W).

The Transverse Volcanic Axis, which runs along the country's southwestern coast, contains more than 300 volcanoes, the most recently formed being Paricutín, which rises to more than 9,100 feet (2,800 m). In February 1943, a Michoacán farmer noticed plumes of smoke rising from a field of maize, heralding the arrival of the volcano. Within a few months, a 1,500-foot (450-m) cone of volcanic ash had risen on the spot, and lava flows had engulfed the surrounding houses. The young volcano remained active for nine years without causing any casualties. However, of the hamlet of Paricutín, only the name survives—now given to the volcano—and all that remains of the village of San Juan Parangaricutiro is the church tower and nave, protruding from a bed of black, solidified lava. Visitors, tourists, and occasionally pilgrims—as here, on Easter eve—brighten this lunar landscape with their presence. Of Mexico's population, 90 percent is Catholic, and religious festivals and rites play a dominant part in the country's culture. The Virgin of Guadalupe is Mexico's patron saint, and on her feast day, December 12, some 1,500 processions take place across the country, the biggest attracting up to 100,000 faithful.

APRIL

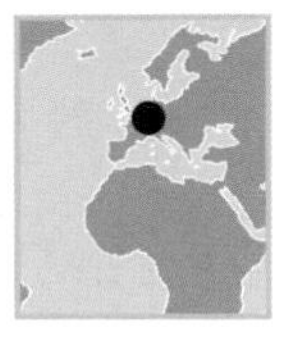

Telecommunications station at Raisting, Bavaria, Germany (47°80' N, 11°07' E).

On the shore of the peaceful lake of Ammersee, southeast of the Bavarian capital Munich, the station of Raisting aims its powerful parabolic antennae at the telecommunications satellites that dot the invisible "information superhighways." Access to telecommunications networks is unequally shared across the planet. The 20 percent of the world's people who live in rich countries have 74 percent of all telephone lines (compared to the 1.5 percent of telephone lines accessible to the 20 percent who live in the poorest countries), and 95 percent of computers with Internet access are in rich countries. Finland now has more mobile phones (572 for every 1,000 people) than landlines. The number of personal computers is growing at the rate of 10 percent per year in the European Union, but it still lags behind the United States, which has 460 computers for every 1,000 people. In poor countries, there are now 162 television sets per 1,000 inhabitants (there were 95 in 1990), and Internet use is continuing to grow. This new means of communication offers new ways of gathering and exchanging information between the peoples of the planet.

04

APRIL

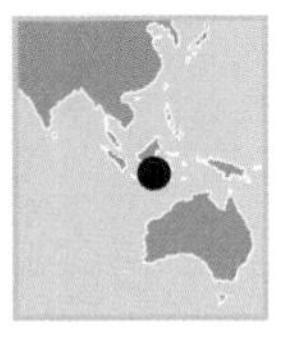

Gathering seaweed, Bali, Indonesia (8°17' S, 115°06' E).

Water has just as much potential as dry land for producing commercial goods. We need only replace *ager* (field) with *aqua* (water) to go from agriculture to aquaculture—or from chicken farming to fish farming. With its rows and furrows, seaweed under cultivation looks very much like fields of grain. This is a shrewd way of making a profit in spaces that are hard to exploit, which are numerous on an island dominated by volcanoes and bordered by a narrow coastal plain. Growing seaweed also provides extra nutrients for Indonesians, 27 percent of whom are still below the poverty threshold. The average annual revenue per inhabitant is less than 10 percent of that of Switzerland.

05

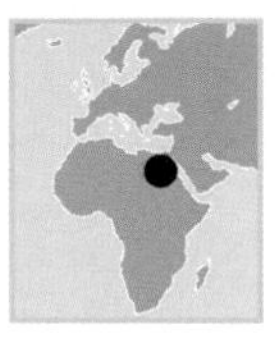

The Unfinished Obelisk, Aswan, Egypt (24°01' N, 32°58' E).

Fixed forever to its rocky bed, this obelisk is condemned to remain horizontal, meaning it will never fulfill its social role as a great symbol. The obelisk broke while it was being hewn out and was abandoned in its granite matrix. The strange destiny of the biggest obelisk of all—it weighs 1,200 metric tons and is 138 feet (42 meters) long—is thus to remain in its quarry at Aswan, where it nevertheless contributes to tourism, Egypt's biggest currency earner. Despite sharp declines due to the Gulf War in 1991, and to a terrorist attack at Luxor in 1997—which killed sixty-two people, fifty-eight of whom were tourists—the industry earned $4.3 billion from 5 million visitors in 2000. For countries whose economies depend heavily on tourism, many of which are developing countries where the industry's role is growing, such events can have serious consequences even if they take place at a great distance. Tourist numbers fell again after terrorist attacks in the United States in September 2001. In 1998, Egypt lost trade worth $2 billion.

APRIL

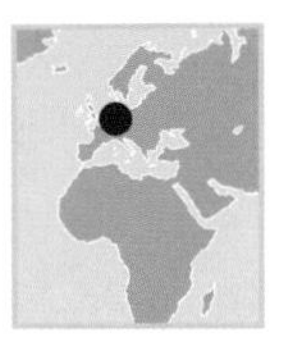

Windmill amid fields, North Holland province, Netherlands (52°57' N, 4°45' E).

The Netherlands is aptly named, with a third of its surface area—the *polders*—lying between 3.25 and 13.12 feet (1 m and 4.5 m) below sea level. Beginning in the fourteenth century, these areas were reclaimed from the sea, from lakes, and from marshes. They are drained using a complex system of dikes, locks, and canals, along which stand the windmills used to pump the water. From the nineteenth century onward, the windmills were replaced by steam pumps and later by electric pumps. Polders provide excellent farmland and are densely populated. However, as the climate becomes warmer, some of this land could disappear, as sea levels are likely to rise by an average of about 1.5 feet (some 50 cm). Thus, 6 percent of the Netherlands' surface could vanish under water. Other countries in the world are also in serious danger—Bangladesh could lose 17 percent of its land area—and their governments are looking at ways to shore up their defenses.

07

APRIL

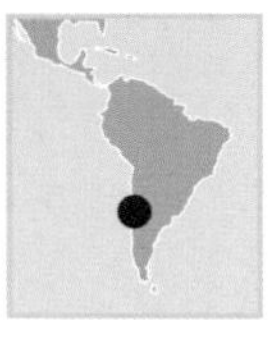

Sand dune at the entrance to the Valley of the Moon, Chile (22°52' S, 68°19' W).

Covered with a glittering crust of white salt, the Valley of the Moon reflects that celestial body's light so strongly that it seems to want to vie with it. Tourists willingly make a detour into the Atacama Desert—the world's most arid region, after Antarctica—to see its celebrated sunsets from the top of the vast gray sand dune that crosses the valley. Ecotourism is in fashion. Unfortunately, in some parts of the world it damages the environment. Water and energy resources are overused to supply tourist accommodation facilities. Natural sites are damaged, and trash is left at the places visited. The World Tourism Organization advocates sustainable tourism, which takes into account ecological and human factors as well as economics. Such a commitment would especially benefit southern countries, which are host to many holiday clubs but receive less than 30 percent of the revenue these generate, because equipment and staff often come from northern countries.

08

APRIL

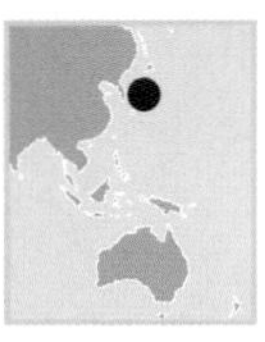

Detail of Himeji Castle, west of Osaka, Honshu, Japan (34°49' N, 134°42' E).

The 500,000 people who live in the city at the fortress' feet call it the Castle of the White Heron, because of its immaculate white walls. Its eighty-three buildings and labyrinthine lanes, alleyways, and secret passages have been on the UNESCO World Heritage list since 1993. Although it is Japan's largest castle—dating from the feudal age—it came to be in a time when fortresses were already being superseded. Built during the 1570s by Toyotomi Hideyoshi, the Japanese empire's all-powerful general, it was completed at the beginning of the seventeenth century by Ikeda Terumasa, son-in-law of the shogun Tokugawa Ieyasu. It combines military efficiency with an aesthetic refinement that prefigures the castles of the court. For example, the top floor of the keep bears *shachinoko*, mythical dolphins that protect the roof from fire. In the West, medieval fortresses had given way to handsome Renaissance castles almost a century earlier, for wars by then were being fought with cannons, against which fortress walls were useless. Thus, the castle was transformed from a defensive structure to a work of art and place of leisure.

09

APRIL

10

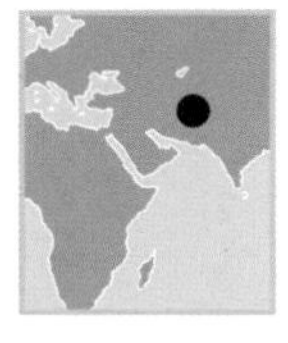

Grounded boat, Aral Sea, Aralsk region, Kazakhstan (46°39' N, 61°11' E).

In the early years of the twentieth century, when the Aral Sea in Kazakhstan covered an area of 26,000 square miles (66,500 km^2), it was the world's fourth-largest inland body of water. After the construction in the 1960s of a vast irrigation network for the cultivation of cotton in the region, the flow of the Amu Darya and Syr Darya rivers, which feed the Aral Sea, diminished to a disturbing degree. The sea lost 50 percent of its area and 75 percent of its water volume, and its shores shranks 40 to 50 miles (60 to 80 km)—leaving behind the hulls of small boats that had once fished its waters. As a direct consequence of water diminution, the salinity of the Aral Sea has continually increased in the course of the past thirty years, today reaching about 10 ounces per quart (30 grams per liter), three times its original salt concentration, causing the disappearance of more than 20 species of fish. Salts carried by the winds burn all vegetation within a radius of hundreds of miles, contributing to the desertification of the environment. The example of the Aral Sea, although among the best-known examples, is not unique: 230,000 square miles (600,000 km^2) of irrigated land all over the world, 75 percent of it in Asia, have excessive salt, which reduces their agricultural productivity.

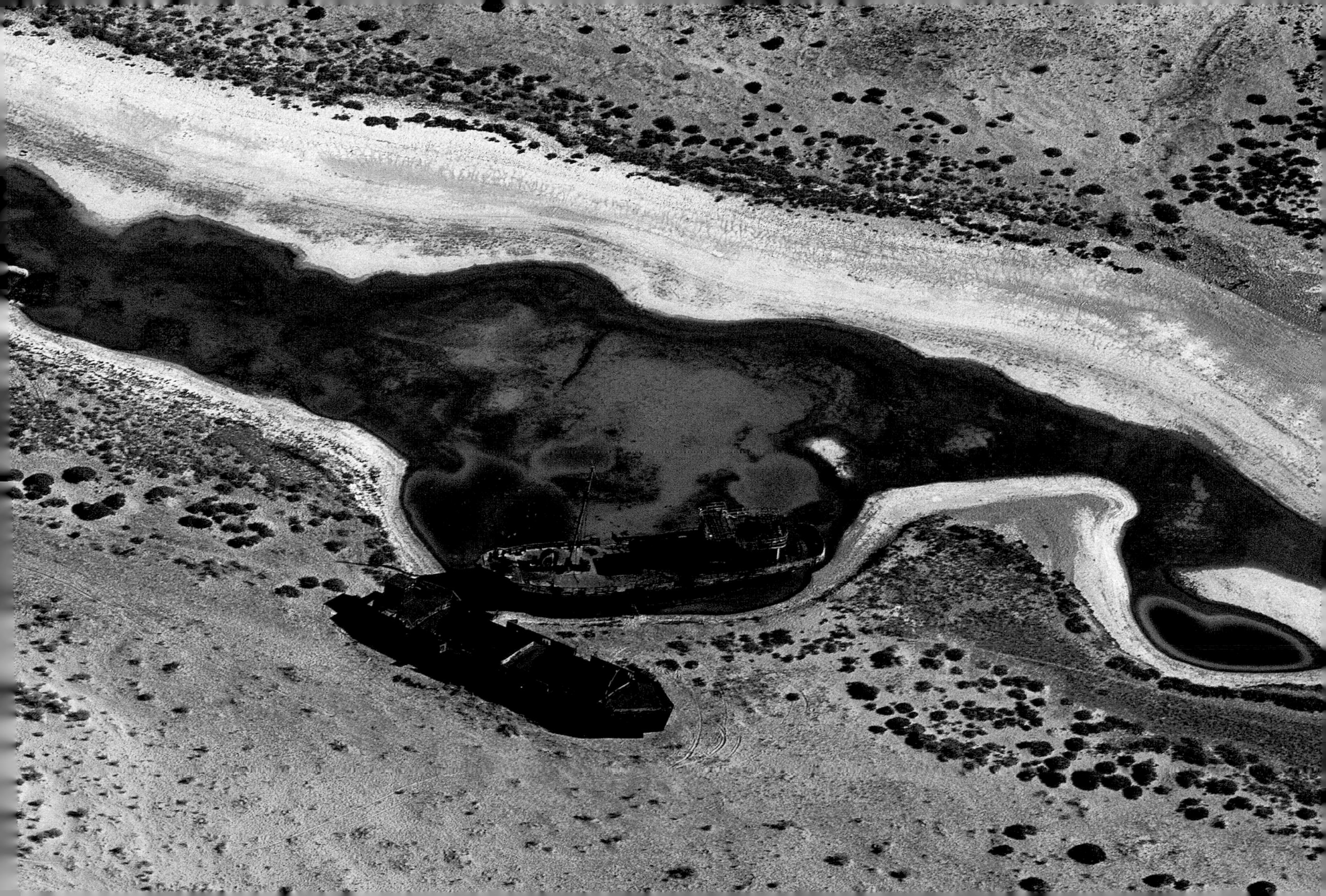

APRIL

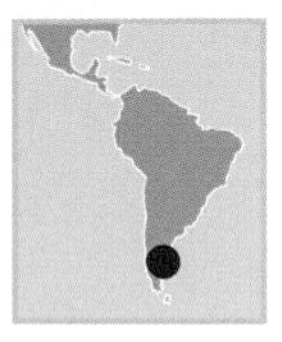

Lake Argentino, Santa Cruz province, Argentina (50°13' S, 72°25' W).

In the heart of the Patagonian Andes, Lake Argentino is the biggest in the country, with a water surface of 574 square miles (1,560 km^2). Some of the forty-seven glaciers of the Los Glaciares National Park come to expire on its shores. The park was set up in 1937 and added to UNESCO's World Heritage list in 1981. As they break up, these giants release icebergs that have a slightly turquoise hue because the ice in them is old and extremely dense. As they melt, the icy blocks give the water a distinctive milky blue color that the Argentines call *dulce de glaciar*—"glacier cream." Patagonia's glacial sheet is the world's third biggest after those of Antarctica and Greenland. Its area has shrunk by 193 square miles (500 km^2) over the last fifty years, during which global temperatures have risen by 33.08°F (0.6°C). Upsala, the national park's biggest glacier, is thought to have receded at an average of 197 feet (60 m) per year for the last sixty years—and the rate is accelerating. The retreat of the glaciers could be especially dangerous to water supplies in arid regions.

11

Roman city, Baalbek, Lebanon (34°00' N, 36°13' E).

In the heart of the Bekaa Valley, a fertile but stiflingly hot plain, lies the town of Baalbek with its remarkable Roman remains, on UNESCO's World Heritage list since 1984. Under the Roman empire, this ancient Phoenician town—dedicated to Baal, the god of storms—became a sanctuary sacred to Jupiter Heliopolitanus, the sun god. Today, six isolated columns keep alive the memory of the temple of Jupiter that, at 288.5 feet (88 m) long and 157.3 feet (48 m) wide, was the biggest religious building in the Roman world. It was begun during the reign of the emperor Augustus (27 BC–AD 14) but was only completed in AD 60. This grand place of worship bore witness to Rome's desire to display the supremacy of its religion in the cradle of Christianity. Since then, the town has become the headquarters of Hezbollah, an organization of Shia Muslim fundamentalists that has inflamed Lebanon's already tense relations with Israel.

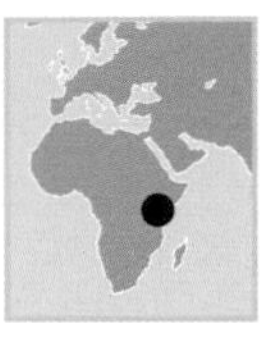

Crystalline formation on Lake Magadi, Kenya (1°50' S, 36°15' E).

The great fracture of the East African Rift, almost 4,500 miles (7,000 km) long, was formed by a tearing of the earth's crust 40 million years ago. Bordered by high volcanic plateaus, this vast sunken trench has created a string of great lakes (including Turkana, Victoria, and Tanganyika) as well as smaller bodies of water such as Lake Magadi, the southernmost in Kenya. This lake is fed by the rainwater that washes off the surrounding volcanic slopes. It carries mineral salts and therefore has a high sodium carbonate content. The deposits of crystalline salts, or licks, thus formed allowed the installation of Kenya's oldest mine. In places, the lake's dark surface contrasts with the shimmering crystals, which produce colored shapes when they mix with the briny water. This is an inhospitable environment, but it is not devoid of life: millions of small flamingoes come to feed on microalgae, shrimps, and other crustaceans that teem in the lake.

APRIL

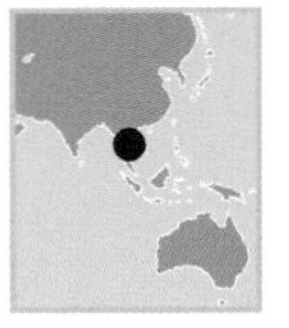

Women in a rice paddy, Siem Reap province, Cambodia (13°22' N, 103°51' E). The young girl on the left, who is gathering rice stems, wears a *krama*, or traditional Khmer headgear. In 2002, rice paddies covered almost 7,720 square miles (20,000 km^2), or 11 percent of the country's surface area. After a drastic reduction under the Khmer Rouge regime, which killed at least 1 million people, this sector of the economy has recovered, and rice production has increased from 538,000 metric tons in 1979 to more than 4 million today. This recovery was vital for a population closely dependent on agriculture: 85 percent of Cambodians grow rice for food. However, the floods of 2001 wiped out 15 percent of the land planted with this precious crop. Also, the 8 million land mines thought to be scattered around the country are a serious obstacle to agricultural development. They maim sixty people every month.

14

APRIL

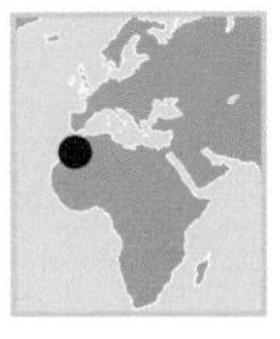

Animal hides spread out in an abandoned cemetery,

Fès region, Morocco (33°55' N, 4°57' W).

As long ago as the Middle Ages, the town of Fès supplied Europe with morocco, which is either sheep- or goatskin leather. Animal skins are still tanned in the traditional way, using natural substances obtained from plants, but increasingly this technique is being replaced by a more industrial process that uses chromium. This toxic metal pollutes the waste-water from the process, which is frequently released directly into sewers. To tackle the problem, the Moroccan government and the Fès local authority are building a new district on the edge of town, equipped with plants for treating water and recycling chromium. This will accommodate the polluting tanneries, preserving the environment and increasing their productivity. The building of this district is part of a much wider program for regenerating the city. For Morocco's cultural capital, which has been on UNESCO's World Heritage list since 1981, is in a sorry state. Half its buildings are in need of repair, its historical and religious monuments are full of cracks, and its cemeteries are abandoned.

15

APRIL

16

Storage area at the Daimler-Benz factory, Wörth am Rhein, near Karlsruhe, Rhineland-Palatinate, Germany (49°03' N, 8°16' E).

In the parking lot of a car factory in Wörth am Rhein, a town close to Karlsruhe, gleaming trucks wait to take to the road. In Europe, 44 percent of goods travel by truck, compared with 31 percent in 1970. Over the same period, rail's market share fell from 21 percent to 8.4 percent. The road transport sector, which has grown by 35 percent over the last decade, is increasing inexorably despite the fact that it exacerbates global warming. Moreover, the number of motor vehicles in Western Europe has doubled over the last 30 years. However, low-pollution cars, public transport and bicycles in town, and transport of freight by rail, river, and sea offer sustainable alternatives.

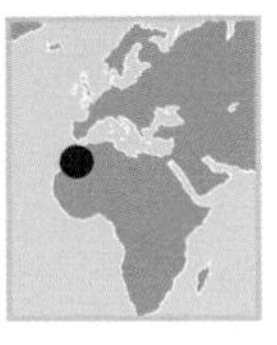

Sugar cane fields, Gharb plain, Morocco (34°45' N, 6°00' W).

Nearly 125 million metric tons of cane or beet sugar are consumed in the world every year. In 2000, Morocco produced about 4 million metric tons, which makes it a small producer compared with giants such as Brazil, India, and China, but also the United States, France, and Germany. In some industrialized countries, the importance of sugar production to the economy leads them to protect their domestic markets by heavily taxing imports or subsidizing domestic producers. This protectionism tends to lower sugar prices, penalizing developing countries whose economies depend heavily on growing crops for export. A fall in sugar prices generally leads to lower earnings and worsened living and working conditions for workers. In certain countries, the average life expectancy of a sugarcane cutter is no more than thirty years.

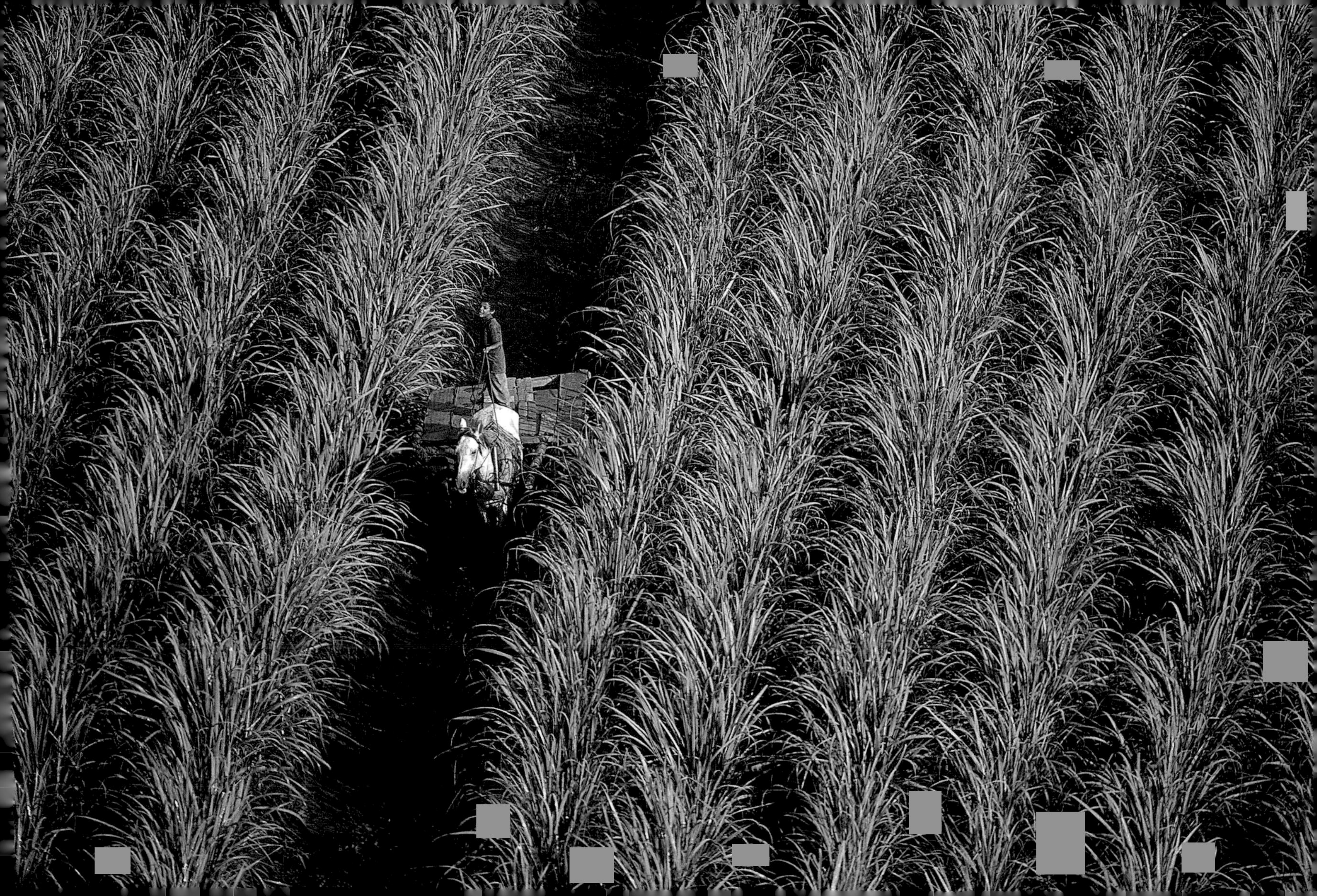

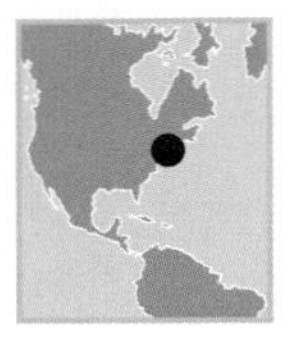

The Golf Club at Chelsea Piers, New York, United States (40°43' N, 74°01' W). Opened a few years ago, Chelsea Piers is New York's biggest sports complex. It is in the heart of Chelsea, a district that bears the marks of New York's industrial past in the form of elevated railway lines, cast-iron facades, and a population descended from nineteenth-century immigrants. All sports can be played at Chelsea Piers, from basketball and volleyball to boxing. Golf, far from being the privilege of the wealthy, has long been beloved of the middle and working classes, with the result that the United States has more golfers than any other country—hence the proliferation of small-scale golf courses used for practice. However, this passion has an ecological and human cost. In developing countries, building a golf course often involves displacing villages and confiscating land. And modern methods of maintaining grass turf use water in large quantities, along with fertilizers and especially pesticides. Golf greens hold the dismal record of being the biggest consumers of water per square foot (or meter).

LEASING

APRIL

19

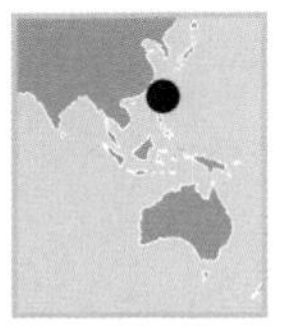

Traditional cemetery near Pingtung, Taiwan (22°40' N, 120°29' E).

According to Chinese tradition, the world of the dead has an influence on that of the living. It is good, therefore, to look after an ancestor's well-being if you want to ensure happiness for your family. How is this done? By building the ancestor's last resting place according to the rules of feng shui. Taiwanese tombs, set against a protecting hillside, should ideally look out over an open landscape, such as the sea or a lake. It is also recommended that they be built in an "armchair" shape and covered with grass that surrounds the tombstone. This arrangement, synonymous with comfort and dignity but also with power and wealth, creates a platform that can be used for reflection. Every year, on the festival of Qingming, families gather at the cemeteries to maintain the tomb and share a picnic with their ancestors. Now, however, space is at a premium in Taiwan, and traditional cemeteries are expensive. Many choose to cremate their dead, which still allows families to gather on the great lawns that are planted in front of the columbaria, or vaults where the cinerary urns are kept.

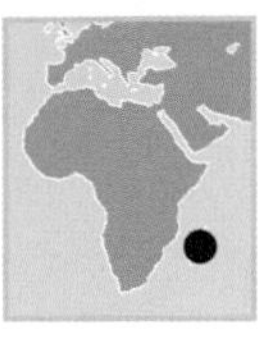

Tsingy of Bemaraha, Morondava region, Madagascar (18°47' S, 45°03' E).

The strange mineral forest of Tsingy of Bemaraha stands on the western coast of Madagascar. This geological formation, called a *karst*, is the result of erosion, as acid rains have gradually dissolved the stone of the chalky plateau and carved out sharp ridges that can rise to heights of 95 feet (30 m). This nearly impenetrable labyrinth (*tsingy* is the Malagasy term for "walking on tip-toe") shelters its own unique flora and fauna, which have not been completely recorded. The site was declared a nature reserve in 1927 and a UNESCO World Heritage site in 1990. Madagascar is a 230,000-square-mile (587,000 km^2) fragment of earth produced by continental drift, isolated for 100 million years in the Indian Ocean off the coast of southern Africa, and has thus developed distinctive and diverse animal and plant species, sometimes with archaic characteristics. It has an exceptional rate of endemism: more than 80 percent of the approximately 12,000 plant species and nearly 1,200 animal species recorded are indigenous to the island only; but close to 200 Madagascan species are in danger of extinction.

20

APRIL

Atomium, Brussels, Belgium (50°50' N, 4°20' E).

Towering 334 feet (102 m) above the Heysel plateau in Brussels, the Atomium was built for the Universal Exhibition of 1958. A symbol of man's mastery over the atom and the race toward scientific progress of the 1950s and 1960s, it represents a steel crystal molecule enlarged 165 billion times. Its steel structure, covered with aluminum, consists of nine spheres 60 feet (18 m) in diameter that contain, among other things, a restaurant and an exhibition hall. These are connected by 95-foot (29-m) tubes through which visitors can walk. Since the 1970s, universal exhibitions have increasingly invited their visitors to reflect on what is and has been happening to humanity. In June 2000, 190 countries took part in the Twenty-First Universal Exhibition in Hanover, Germany, whose central theme was "Man, Nature, Technology: A New World Is Dawning." The event, which was held under the banner of "sustainable development"—a principle dealt with at the Earth Summit held in Rio in 1992—attracted almost 18 million visitors.

21

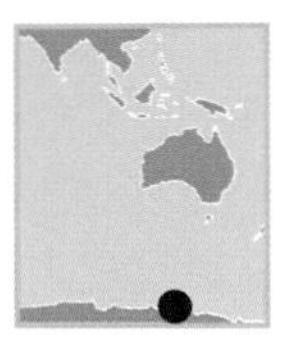

Icebergs and Adélie penguins, Adélie Land, Antarctica (South Pole) (66°00' S, 141°00' E).

The Antarctic covers an area of 6.37 million square miles (16.5 million km^2)—or thirty times the size of France—plus an additional 580,000 square miles (1.5 million km^2) of glaciers that stretch out into the sea. The sixth continent is a unique observation point for atmospheric and climatic phenomena; its ancient ice, which trapped air when it was formed, contains evidence of the Earth's climate as it has changed and developed over the past millions of years. Sea ice at the poles tends to melt—a natural phenomenon that is probably accentuated by global warming. The average thickness of Arctic sea ice has decreased, from 10.23 feet (3.12 m) in the 1960s to 5.9 feet (1.8 m) in the 1990s. The threat also affects mountain ecosystems, which fall victim to rises in temperature. The melting of glaciers is a threat to many inhabited regions; Kilimanjaro has lost 55 percent of its glaciers in forty years. In Peru, in 1970, falling ice pinnacles and rocks from the Huascarane glacier killed at least 15,000 people. If present trends continue, a large number of mountain glaciers, including those in Montana's Glacier National Park, will vanish within the next one hundred years.

APRIL

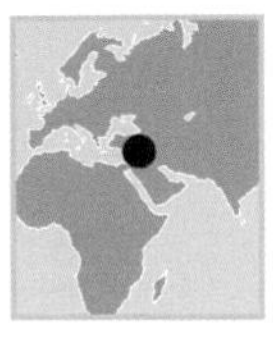

Al-Dayr, Petra, Maan region, Jordan (30°20' N, 35°26' E).

Jordan occupies a strategic position between the Mediterranean and the Red Sea. In the seventh century BC the Nabataeans, a people of merchant nomads, settled here. They carved a city out of the pink and yellow sandstone of the cliffs in the southern part of the country and made it their capital. They called it Petra, the Greek word for "rock." Through the trade of rare products (incense from Arabia, spices from India, gold from Egypt, silk from China, and ivory from Nubia) and taxation of caravan routes, Nabataean civilization extended its influence far beyond the Transjordan region before it fell to the Romans in 106 AD Al-Dayr, standing at the top of the city, was built between the third and first centuries BC Because of its imposing stature (138 feet high and 148 feet wide, or 42 x 45 m), it dominates the approximately 800 monuments of Petra. Petra was declared a World Heritage site by UNESCO in 1985, but a new threat has begun to menace the cliffs in the past few years: mineral salts dissolved in groundwater that reaches the base of the monuments become encrusted on the stone and make it fragile. Wind adds to the progressive degradation of the monuments.

23

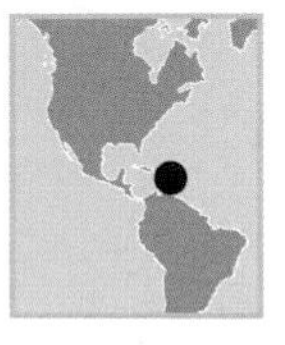

Islet and sea bed, Exuma Cays, Bahamas (24°00' N, 76°10' W).

Two James Bond films were set in Exuma Cays, *Thunderball* and *Never Say Never Again*. The white sand, emerald sea, and deep underwater caves of the Bahama islands chain provide an ideal movie backdrop. Consisting of 700 islands and rocky coral islets, or cays, the region attracts travelers in search of sun and sand, as well as nature lovers. More and more people are flocking to see the brightly colored fish, exotic birds, and distinctive Bahamian iguanas in the Exuma National Park. Although ecotourism in the Bahamas is in its infancy, the government would like to encourage it to grow. This should bring more income to the local population, who will thus be able to make the most of their knowledge of the natural world, rather than taking poorly paid jobs in holiday complexes. Worldwide, such facilities are usually owned by foreign investors who pass on less than 30 percent of the revenue generated to the host country.

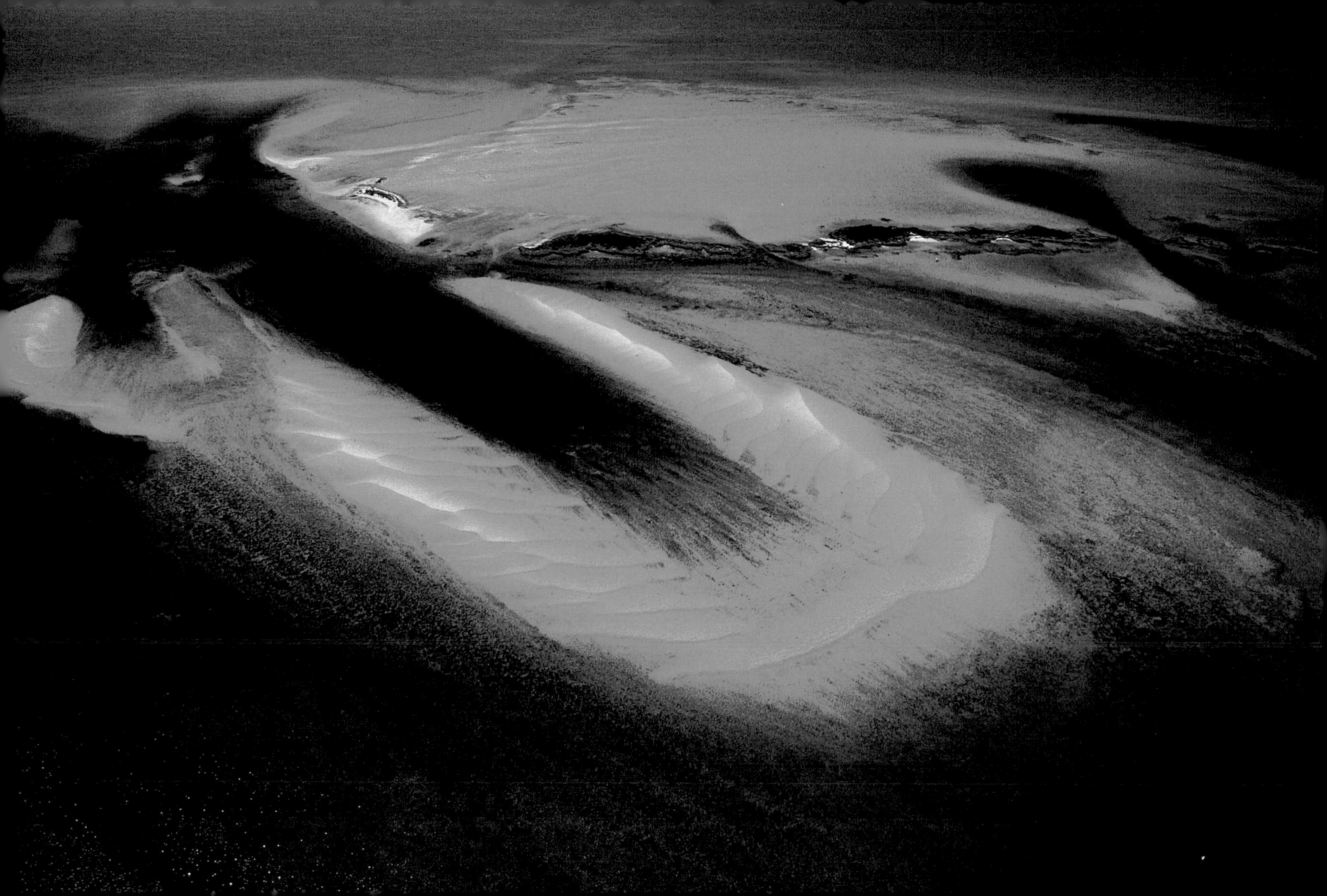

APRIL

Waste from the copper mine at Chuquicamata, Chile (22°19' S, 68°56' W).

This giant scallop shell is made of earth. A crane deposits the earth in successive, slightly curved lines giving the appearance of sheets of sand lined up side by side. This earth is extracted with the copper, but it is separated from the ore by sieving. The metal is refined in the Chuquicamata foundry that, thanks to newly installed equipment, can now filter out 95 percent of the sulfur dioxide (SO_2) and 97 percent of the arsenic that the process releases. A 1992 law aimed at reducing air pollution has required Codelco-Chile, the state-owned company that runs the mine and its facilities, to invest tens of millions of dollars to modernize them. This has not, however, prevented the company from increasing production capacity; indeed, starting in 2004, the Chuquicamata mine will have ramped up to produce 750,000 metric tons of copper per year, compared with 630,000 metric tons in 2001.

25

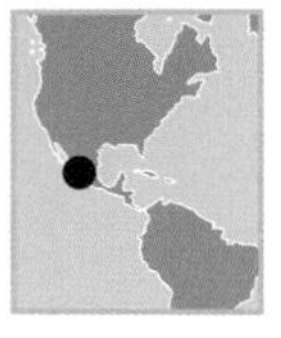

"Western Sun" villa, Costa Careyes, Jalisco state, Mexico (19°22' N, 105°01' W).

The style of the Italian architect Gianfranco Brignone is on display at Costa Careyes, a stretch of the Jalisco coast on the western side of Mexico, which is dotted with ten or so of his luxurious creations. Their modern conception gives pride of place to traditional, brightly colored facades and rustic materials. In these environmentally advanced and sophisticated buildings, palm roofs and adobe walls are an effective replacement for air conditioning. The pink "Western Sun," surrounded by its swimming pool, and its sister villa "Eastern Sun," decorated in yellow, cling to the cliff, standing guard on either side of a bay. This precipitous coastline, suspended between the Pacific Ocean and the tropical forest, is the new Promised Land for a handful of extremely wealthy people. The wealth of the world's three richest people is greater than the total gross domestic product of the forty-eight poorest countries. Just 4 percent of the combined wealth of the 225 biggest fortunes (which are worth a total of $1 trillion) would be enough to pay for education, food, and basic health care for the planet's entire population.

26

APRIL

Church in the town of Samara, Russia (53°13' N, 50°10' E).

At the confluence of the Samara and Volga rivers, the industrial town of Samara stretches some distance with its machine industry, petrochemical, aeronautics, and food industry plants. Samara—formerly called Kuibyshev—was where the Soviet government took refuge during the Nazi invasion. Later, the town reverted to its Russian identity—witness this recent church, which rises in front of the apartment blocks of the industrial city. In 1914, more than 50,000 churches were still active; by 1941, after closures and demolitions, fewer than 1,000 were still open to worshipers. State and religion were reconciled in 1988, during the millennium celebrations of the baptism of Russia. Since then, churches have been opening constantly in the country. Russian Orthodox, Catholics, Baptists, Jews, and even Hare Krishnas now have legal status. Today's freedom of worship might almost have effaced yesterday's religious persecution. But the repairs, restoration, and new reconstruction are a visible reminder that yesterday was not long ago.

27

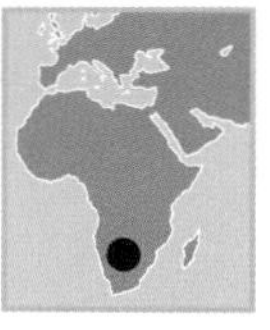

Marsh in the southern Okavango Delta, Botswana (18°45' S, 22°45' E).

Antelope take advantage of the dense growths of papyrus and thick reeds that thrive in the Okavango Delta to escape the hunter's rifle. Every year, 45,000 tourists treat themselves to a luxury safari holiday in the region. They are prepared to pay the heavy taxes imposed by the state of Botswana but only on the absolute condition that the delta is preserved as an earthly paradise. Forbidden in this lush area, agriculture is banished to the drier, less fertile lands in the south of the country, whose meager yields barely cover 10 percent of Botswana's cereal demand. Establishing some farming around the Okavango could rectify this problem. Although the area has been designated, under the Ramsar Convention, a wetland of international importance since 1996, there is no reason why farming should be proscribed. The convention demands only that the natural balance should be taken into account, to allow the development of sustainable social and economic activities.

APRIL

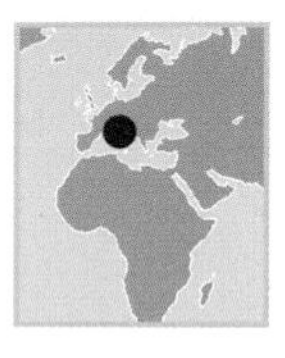

Island of San Giulio, Lake Orta, Piedmont, Italy (45°47' N, 8°24' E).

On the little island of San Giulio, in the center of Lake Orta, the westernmost of northern Italy's lakes, are a Romanesque basilica and a Benedictine convent. Here, as in the rest of Italy and all over Europe, ecclesiastical architecture shapes village and urban landscapes. Indeed, until the nineteenth century, such buildings were the principal vectors of expression of architectural styles—from Romanesque to Gothic, Baroque, and Neo-Classical, and even Cubist in the twentieth century. New religious buildings are increasingly rare today. The close of the twentieth century saw the inauguration, in 1989, of the prestigious and controversial Notre Dame de la Paix (Our Lady of Peace) in Yamoussoukro, capital of the Republic of Côte d'Ivoire. Nevertheless, with 2,000 years of history behind them, churches still own vast amounts of property, especially in Europe. In Paris, for example, the Church is the second-biggest landowner, and it is also one of the biggest landowners in Italy.

29

30

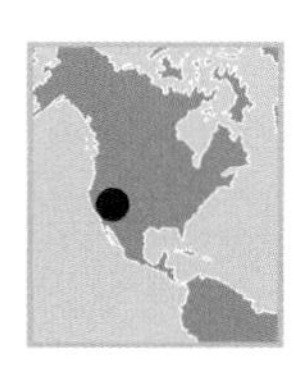

Aircraft graveyard at Davis-Monthan Air Force Base, near Tucson, Arizona, United States (32°11' N, 110°53' W).

At the Davis-Monthan base, more than 5,000 American decommissioned military aircraft await dismantling, conversion into drones (unmanned reconnaissance aircraft), or sale to other countries. Among them is the Grumman A-6E, an aircraft with the capability to hit its target in zero visibility that was used, notably, in the 1991 Gulf War. Today, it has been replaced by more efficient models that help increase the military might of the United States. The latest demonstration of its power—in March 2003 against Iraq—took place without the endorsement of the United Nations' Security Council. During the crisis, the United Nations was the center of a heated public debate, in which the whole world became involved. Thus, on February 15 and 16, 2003, millions of people—the largest numbers ever seen—demonstrated all over the globe, opening the door to a new form of globalization: the globalization of debate.

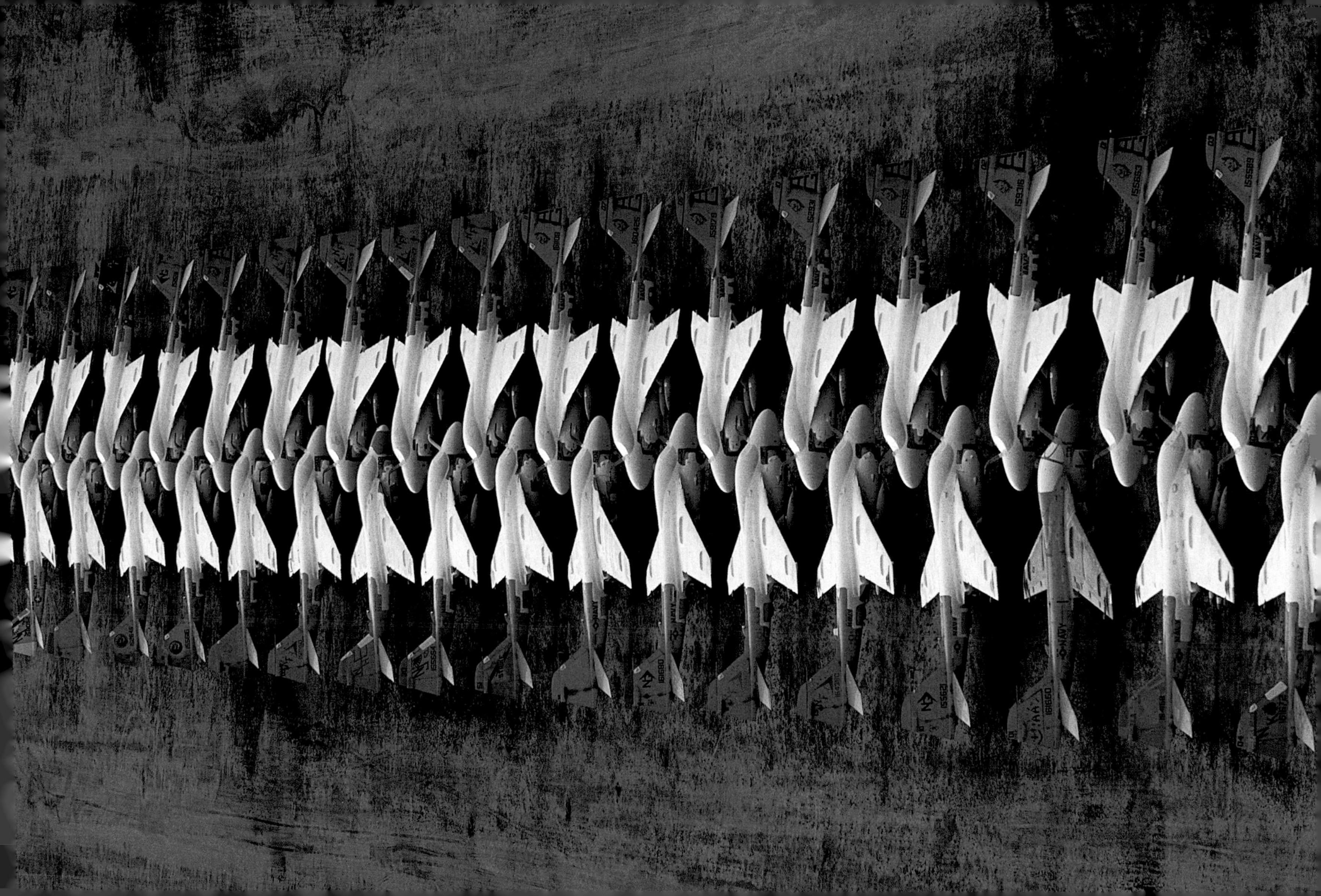

OCEANS AND SEAS

Our planet is one of oceans and seas. They are the major life-supporting part of the Earth. They drive our climate. Wherever we are on the Earth, we are affected by the sea.

The satellite photographs of the Earth to which we have become so accustomed show that it is, indeed, “the Blue Planet.” Oceans and seas cover seven-tenths of Earth’s surface. The biosphere—the part of the planet that supports life—extends throughout the oceans and the seas. Even 4,000 meters below the surface of the sea, on the abyssal plain in the Atlantic, there is life. There are also life forms in the sea that are completely different in their basis from those on land. Life on land, and the vast majority of life in the sea, depends on the radiant energy of the sun, converted in the first place by plants. But in the depths of the sea, life forms have recently been found that derive their energy from thermal vents that bring energy up from the molten core of the planet. In all its forms, life in the oceans is very diverse: there is as much species variation in a sea loch as in a tropical rain forest. The oceans are, therefore, a vast reservoir of life.

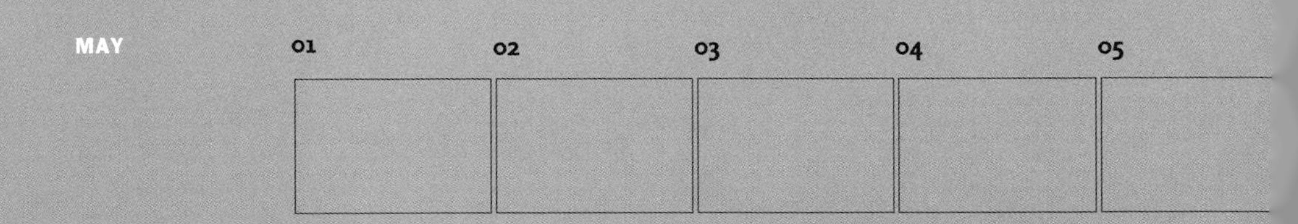

The oceans and seas form a huge circulatory system that, coupled with the circulatory system of the atmosphere, girdles the planet. Each influences the other, and together they drive the climate that we must live with. New York, in the United States, has very cold winters and scarcely tolerable summer humidity. Lisbon, in Portugal, is at the same latitude and is just as much a seaport, but it does not suffer such climatic extremes. The difference between them lies in the ocean currents.

Such circulation is not stable. Oscillations in the great circulatory systems of the Pacific Ocean affect the whole planet. From time to time, these oscillations replace the normal northward cold current off the coast of Peru with warmer water. Off Peru, fish catches drop—which usually happens around Christmas, an unwelcome present from the Christ child ("El Niño" in Spanish, hence the name of the phenomenon). But the effects are not limited to one locality. An El Niño event can change the weather around the globe. The storms, floods, and droughts associated with the 1982–83 El Niño event, probably the worst ever, caused over $8 trillion worth of damage from India to Tahiti to Bolivia.

The majority of humans live in the coastal zone. We rely heavily on food from the seas: In Asia, fish provides 40 percent of human protein intake. The oceans and their plant life play a major part in absorbing and converting carbon dioxide. As the United Nations Commission on Sustainable Development has noted, the oceans and seas provide vital resources to be used to ensure the well-being of present and future generations and their economic prosperity, to eradicate poverty and to ensure food security.

Our well-being therefore depends on the well-being of the oceans and seas. Yet at the same time we threaten them by many of our activities. There are seven major threats.

Over-fishing is a threat to the continued availability of a major food resource. Fisheries, when managed to be sustainable,

06 07 08 09 10 11 12 13 14 15

can secure global food security and incomes for both present and future generations. But we over-exploit many fish-stocks, and we have reached the sustainable limit for nearly all traditional stocks. We therefore begin to look for new stocks to exploit. The process of evolution means that many seamounts (submarine mountains that do not break the surface and become islands), like many islands, have species found nowhere else. These unique resources are now in turn threatened by over-fishing.

Shipping is essential to support world trade. Without it, the global economy could not function. Yet ships in poor condition are allowed to be used to carry cargoes, such as oil, that cause immense damage to the sea and the coasts when accidents inevitably happen.

Waste is still dumped in the sea on a massive scale. Some countries are not prepared to find a way to dispose in their own territory of the waste that they create on land, so they dump it at sea.

Waste from land-based activities is discharged through rivers and pipelines into the sea. Industrial waste brings hazardous substances into the marine environment that threaten the breeding of fish and shellfish and that can make them unusable as a source of human food. The nutrients contained in sewage produced by towns and cities can result in local over-enrichment that disrupts the natural ecosystem and can lead to the loss of oxygen in the sea and consequent fish deaths.

Coastal development frequently intensifies the human impact on the sea. Think of the tourist developments around the Mediterranean that add sewage, create erosion and consequent sedimentation on fish breeding grounds, destroy wetlands where wild animals breed, and disturb the ecosystem with intense human use of the coast. Worldwide, the pressure increases.

We look more and more to the oceans and seas as a source of minerals. The seabed is now a major source of oil and gas. More and more sand and gravel for construction is taken from the seabed, threatening spawning and nursery grounds for

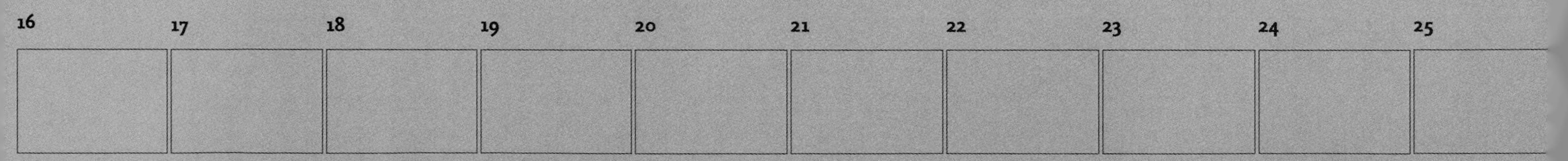

fish. Even metal mining is now a prospect. Extractive industries always have a heavy environmental impact, and those offshore have no less an effect, even though we may see less of it.

The changes in the world climate resulting from human activities also affect the oceans and seas. The increased ultraviolet radiation in some parts of the world, resulting from the hole in the ozone layer, affects fish breeding. Global warming will have an impact on the circulatory systems of the oceans, with effects difficult to foresee.

Yet the story is not all black. Over the last three decades, the international community has made major efforts to tackle these threats. Globally, the International Maritime Organization is trying to improve shipping standards. The U.N. Food and Agriculture Organization has developed action plans on fisheries. A global program of action on land-based pollution has been adopted, led by the U.N. Environment Programme. Around the world, the eighteen regional seas organizations are tackling the problems at a regional level.

These efforts need more political commitment to implement the agreements that have been reached. Commitments have been made. The task for governments and all stakeholders is now to deliver on them, so that the threats can be turned into springboards for achieving the sustainable use of oceans and seas to the benefit of the whole planet.

ALAN SIMCOCK
Executive Secretary
OSPAR Commission for the Protection of the Marine Environment of the North East Atlantic

Co-Chairman
UN Oceans Consultative Process, 2000, 2001, and 2002

26	27	28	29	30	31

MAY

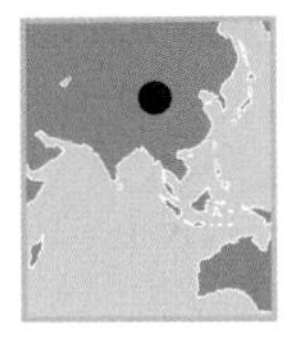

Mountains in the Gurvan Saïkhan national park in the Ömnögov (South Gobi), Mongolia (45°30' N, 107°00' E).

The Gurvan Saïkhan ("Three Beauties") national park owes its name to three groups of peaks, the "Beauties of the East, Center, and West" which tower over this ancient mountain range. Partly of volcanic origin, the range appeared between 550 million and 235 million years ago, like the Hercynian (that is, the Black Forest or central German) ranges in Europe or the Appalachians in the United States. These mountains are exceptionally rich in flora and fauna. Medicinal plants are plentiful, and two-thirds of the plant species endemic to Mongolia grow there. They are also home to fifty-two mammal species—eight of which are on the "red list" of Mongolia's threatened species, including the argali, the snow leopard, and the Siberian ibex—as well as 240 species of birds. Today, the chief threat to this fragile ecosystem is mining, for the Gobi is rich in coal, tungsten, copper, iron, gold, fluorite, molybdenum, anthracite, and semiprecious stones.

01

Flock of scarlet ibis, near Pedernales, Amacuro delta, Venezuela (9°57' N, 62°21' W).

From the Llanos region to the Amacuro delta at the mouth of the Orinoco River, more than a third of the area of Venezuela is made up of humid zones, the habitat of choice of the scarlet ibis (*Eudocimus ruber*). These waders nest in large colonies in mangroves and move no farther than a few miles to seek food. Carotene derived from the shrimp, crabs, and other crustaceans they eat helps create the characteristic pigmentation of the species. The scarlet ibis's feathers, at one time used by native populations to make coats and finery, are now a component in the manufacture of artificial flowers. This bird, sought after for its flesh as well as its feathers, is endangered today; fewer than 200,000 survive, in Central and South America.

MAY

Lakagigar Volcano Chain, Iceland (64°04' N, 18°15' W).

Lakagigar, in southern Iceland, still bears the scars of one of the most violent volcanic eruptions in recorded history. In 1783 two eruptive fissures, a total of 15 miles (25 km) long, opened up on both sides of Laki volcano, vomiting 4 cubic miles (15 km^3) of molten rock that engulfed 225 square miles (580 km^2) of land, the largest lava flow in human memory. A cloud of carbon dioxide, sulfur dioxide, and ash spread over the entire island and contaminated grazing land and surface waters. Three-fourths of the livestock was annihilated, and after a second eruption in 1785, a terrible famine decimated a fourth of the population (more than 10,000 people). The fissures of Lakagigar, crowned by 115 volcanic craters, are today closed up, and the streams of lava are covered by a thick carpet of moss. Iceland has more than 200 active volcanoes and has produced one-third of all discharges of lava occurring around the world in the course of the last 500 years.

03

MAY

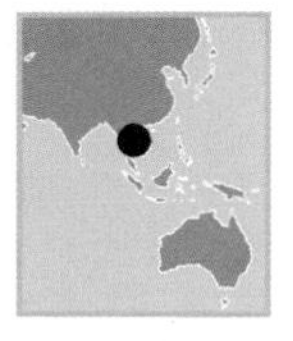

Water buffalo, Siem Reap province, Cambodia (13°22' N, 103°51' E).
Cambodia's central plain is the country's rice basket, and its myriad pools offer a haven for herds of water buffalo. Although these animals are chiefly used as beasts of burden in farming, they are nevertheless an important source of animal protein in Southeast Asia's poorest country. Meat consumption is low, however, and three-quarters of the population's daily calorie intake comes from rice. Malnutrition is causing serious concern: half of all children under five are suffering growth problems, and one woman in every three is malnourished. Lack of vitamin A, which is particularly severe in Cambodia, is a prime cause of the blindness that affects 500,000 children in the world every year.

04

MAY

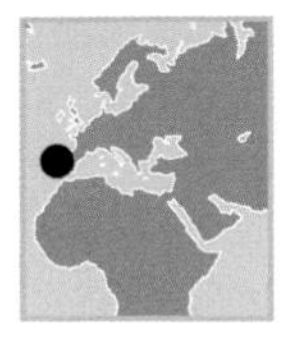

Palacio da Pena in the Serra de Sintra, Portugal (38°47' N, 9°22' W).

Gleaming, strange, and pretentious, the Pena Palace mocks the austere granite of the Serra de Sintra. North of Lisbon, in the heart of the exotic Pena Park, the palace's machicolations recall the papier-mâché décor of a theme park. But this astonishing mish-mash of architectural styles—Gothic, Renaissance, Moorish, and Manueline—was the brainchild of Ferdinand of Saxe-Coburg and Gotha, the husband of Queen Maria II, in the nineteenth century. Its construction, inspired by German Romanticism, was entrusted to Baron von Eschwege, a German architect, and it was completed in 1885, the year of Ferdinand's death. Other prestigious residences built along the same lines make the Serra de Sintra a center of European romantic architecture, whose parks and gardens have strongly influenced landscaping. Thanks to its cultural importance, the whole site was added to UNESCO's World Heritage list in 1995.

05

MAY

Fields near the town of Hammamet, Nabeul governorate, Tunisia (36°24' N, 10°37' E).

Northeastern Tunisia has a long tradition of irrigation and of planting crops so that they follow the contours of the land. Between 30 percent and 40 percent of the country's agricultural investment goes toward infrastructure for obtaining, transporting, and distributing water. In 30 years, the amount of irrigated land has quadrupled to total 380,000 hectares today, closely reflecting the growth in the country's population, which has doubled in 25 years. Agriculture uses 82 percent of Tunisia's water, but the exhaustion of water near the surface has led to a quest for supplies ever deeper underground. This increased pumping of ground water is a threat to farmland because it causes seawater to penetrate the aquifers, especially near coasts. Faced with a water crisis, Tunisia drew up a national water and soil conservation strategy in 1991-2000. About 20 percent of the planet's irrigated land has been rendered infertile by salinization, and 1.5 million hectares per year are affected by this phenomenon.

06

MAY

In the saddle, Tiergarten district, Berlin, Germany (52°30' N, 13°22' E).

Every year, more and more Berliners get on their bikes when the fine weather arrives, taking to the capital's extensive cycle path network. The bicycle seems to be the urban transport of the future for the twenty-first century, as cars travel ever more slowly through old cities. Traffic has become so heavy that a car in London or Paris today is no faster than a horse-drawn carriage was a century ago, and air pollution from transport alone is thought to be responsible for 3 million deaths worldwide each year. Economical, silent, and nonpolluting, a cyclist occupies a sixth of the road space taken up by a motorist, and a parked bicycle takes up a twentieth of a car parking space. The bicycle is on a roll: in the Netherlands, 30 percent of urban journeys are made this way, and even in mountainous Switzerland, 10 percent of the population pedals around town.

07

MAY

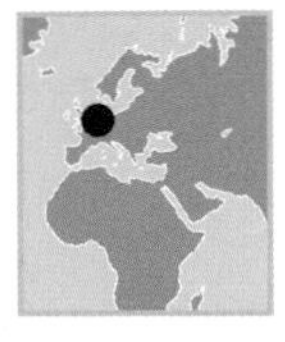

Fields of tulips near Lisse, near Amsterdam, Netherlands (52°15' N, 04°37' E).

In April and May of every year, Holland briefly dons a multicolored garb. Since the first flowering in 1594 of bulbs brought back from the Ottoman Empire by the Austrian ambassador, four centuries of selection have led to the development of more than 800 varieties of tulip. On more than 50,000 acres (20,000 hectares), half devoted to tulips and one-quarter to lilies, the Netherlands produce 65 percent of the world production of flowering bulbs (or some 10 billion bulbs) and 59 percent of the exports of cut flowers. Dutch agriculture, which employs 5 percent of the active population, is one of the world's most intensive and places the country third among world exporters of agricultural produce (after the United States and France). But chemical products have caused a deterioration in the water; Holland is thus beginning to use natural predators to protect its crops from illness and harmful insects, especially in the horticultural sector.

08

MAY

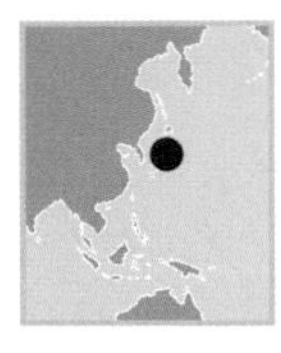

Cattle-raising near Fukuyama (east of Hiroshima), Honshu, Japan (34°31' N, 133°20' E).

Until World War II, most Japanese farmers devoted themselves chiefly to crops, only keeping two or three cows. Over the last 50 years, however, milk has become the country's second biggest agricultural product after rice. With limited land available, most of it in valleys and on the outskirts of town, Japan's farmers have concentrated on milk production and rendered it more intensive, to meet growing consumer demand. Between 1975 and 1990, the number of cows rose by 160 percent while the number of farms, which became increasingly specialized, declined continuously. A Japanese cow today produces about 1,850 gallons (7,000 liters) of milk per year, while a French cow produces "only" 1,438 gallons (5,450 liters). This trend is widespread in richer countries. It damages the rural economy, the diversity of food products, and the environment. Medium-sized farms are disappearing, and livestock farming is becoming separate from crops, breaking the natural cycle that returns to the soil the organic matter that animals took from it when feeding.

09

MAY

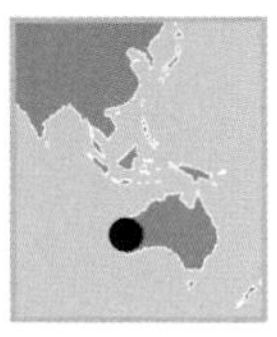

Cape Péron, Shark Bay, Australia (25°50' S, 113°51' E).

At the far western end of the Australian continent, the vast Shark Bay boasts exceptional biodiversity, which in 1991 earned it a place on UNESCO's World Heritage list. Its vast, rich beds of seaweed support a large population of dugong, a marine mammal also known as the sea cow, possibly because the females hold their young close while suckling them. On the beaches, astonishing colonies of stromatolite bacteria take the observer back 3.5 billion years. At Shark Bay, the earliest forms of life that appeared on earth, seen elsewhere only as fossils, flourish and are very much alive. It is estimated that existing species comprise only 5 percent of the total number of creatures that have ever lived on the Earth. Extinction is therefore a natural phenomenon; however, we are now living through one of the most massive extinctions that the planet has ever seen. The bottlenose dolphin, which still swims in the waters of the bay, is paying the price, as are 24 percent of existing mammal species.

10

MAY

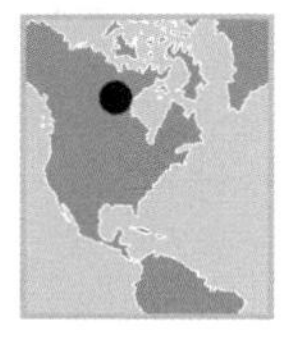

Landscape of Ice, Nunavut Territory, Canada (75°57' N, 92°28' W).

Nunavut is occupied by more than 20,000 Inuit, who represent 85 percent of the local population. The name means "our land" in the Inuit language, Inuktitut. The region was given the status of a territory in April 1999. This territory of archipelagos, water, and ice covers an area of 780,000 square miles (2 million km^2), reaching to within 125 miles (200 km) of the Arctic Circle. In the winter, when temperatures can go as low as –34° F (–37° C), the permanent ice floe at the center of the Arctic and the coastal ice floe formed by the freezing estuaries and bays link up, offering a landscape of continuous ice that can be traveled by dogsled and snowmobile. In the summer the ice floe breaks up, creating drifting platforms called packs. This seasonal release of the waters reopens the migratory routes of whales and other marine mammals and admits supply ships to the area. The Northwest Passage maritime route permits a link between Asia and Europe by way of the Nunavut Islands, providing an alternative to the Panama Canal.

11

MAY

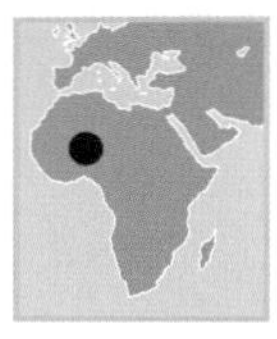

Peul village near Timbuktu, Mali (16°46' N, 3°00' W).

These brown cells bordered by hedges serve both as a place for growing crops and as livestock enclosures—and it is the presence of animals that explains the contrast between this dark-colored ground and the arid areas surrounding it. The animals fertilize the soil with their manure, which is spread on the plots while they lie fallow. Moreover, this natural fertilizer controls acidification and gives body to the soil. This system has been adopted only recently by the Peuls, as here in Mali. Originally nomadic, these shepherds became sedentary after the droughts of 1973–1975 and 1983–1985. The Peuls make up 14 percent of Mali's population; with the Moors, they were once the country's nomadic herders. Today the way of life of these peoples is tending to disappear as the government encourages the sedentary raising of livestock to increase yields. However, because its demands on the land are less intensive, nomadic herding is still the only viable way of using the arid lands of the Sahara and Sahel in the long term.

12

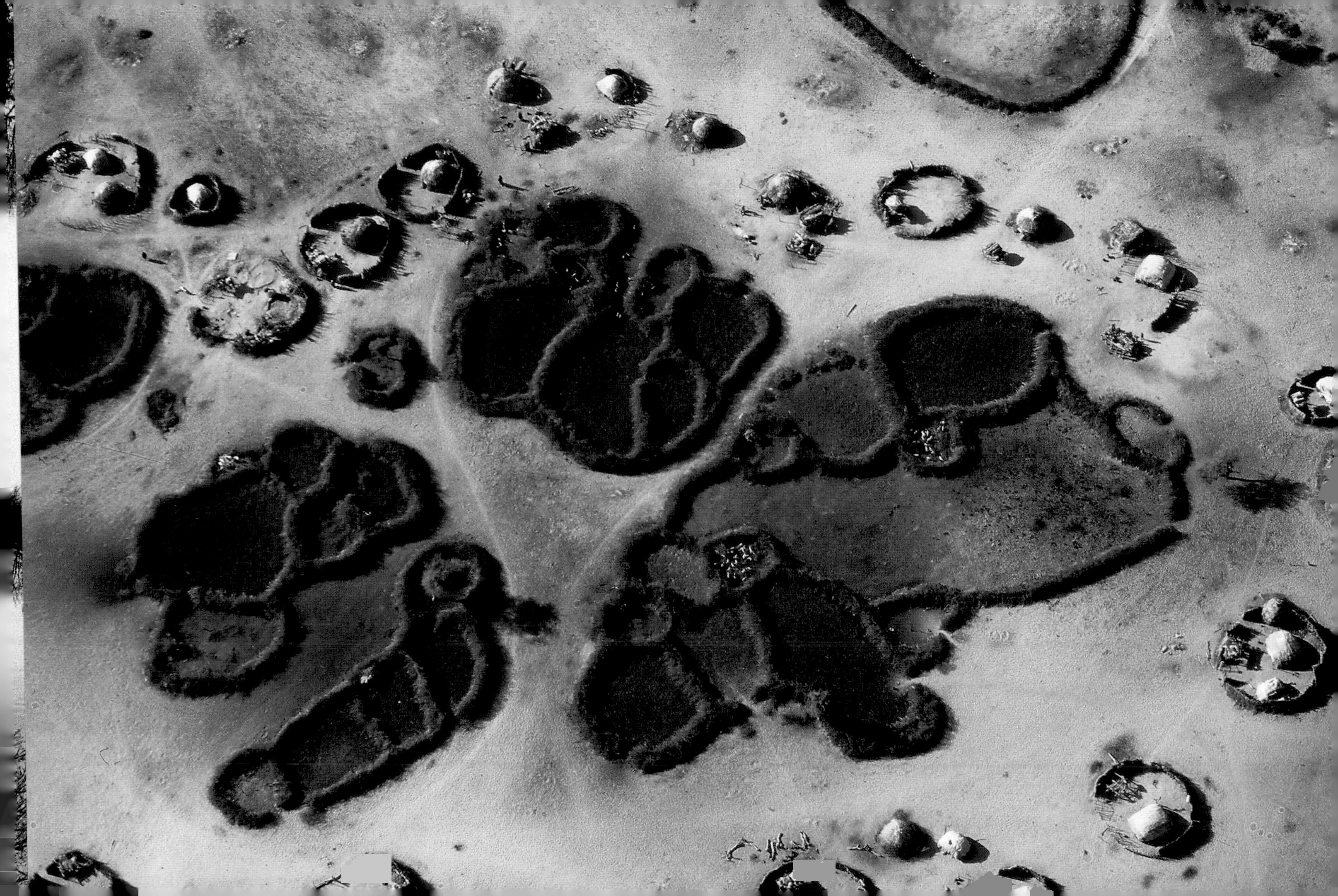

MAY

14

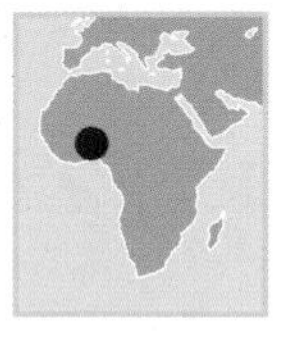

Crowd in Abengourou, Côte d'Ivoire (6°44' N, 3°29' W).

This colorful crowd, enthusiastically waving to the photographer, was photographed in Abengourou, in eastern Côte d'Ivoire. These children and adolescents remind us of the country's youthfulness; as in most of the African continent, 40 percent of the population is under 15 years of age. The country's birth rate is 5.1 children per woman, which is representative of the average for the continent (the world average is 2.8). Modernization and cultural and socioeconomic influences have gradually lowered the birth rate. The ravages of the AIDS epidemic in sub-Saharan Africa (home to 70 percent of the total 36.1 million people infected in the world) will have a severe impact on the region's demography: every day in Africa 6,000 people die of the AIDS virus and another 11,000 become infected.

MAY

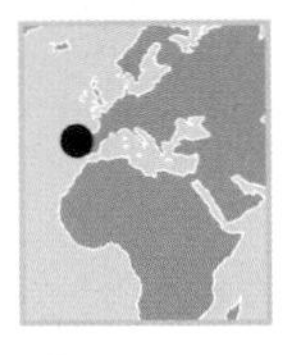

Cristo Rei monument overlooking the River Tagus, Lisbon, Portugal (38°43'N, 09°08' W).

Inspired by the statue of Christ the Redeemer that watches over Rio de Janeiro, this statue of Christ the King has stood facing Lisbon, on the Atlantic coast of the Iberian peninsula, since 1959. The artist Francisco Franco took a decade to make this monumental Christ, which is 95 feet (29 meters) tall. Standing 269 feet (82 meters) above the ground on a concrete pedestal on the banks of the Tagus estuary, it bears witness to the religious fervor of the Iberian peoples. Portugal was freed 50 years before its neighbor, Spain, from Moorish domination, and in that half-century it took the lead in exploration and conquest of the New World. Following in the wake of Henry the Navigator (1394–1460), Portuguese sailors were pioneers on all the world's seas.

15

"Love Parade" in Tiergarten Park, Berlin, Germany (52°31' N, 13°25' E).

In 1989 a Berlin disk jockey brought together 150 fans of electronic music for a modest party. Thirteen years later it has become the "Love Parade," an enormous festival that draws a million young people to dance to the rhythm of techno music. The Berlin Love Parade is already being imitated in Paris, Zurich, Geneva, and Newcastle. A Love Parade was planned for Moscow in 2001, but the mayor's office canceled it, issuing a statement that the festival would encourage debauchery and that homosexuality was immoral. The mayor may have confused the Love Parade with the Gay Pride Parade, for which he also refused permission. In many countries homosexuality remains a source of discrimination and violence.

MAY

17

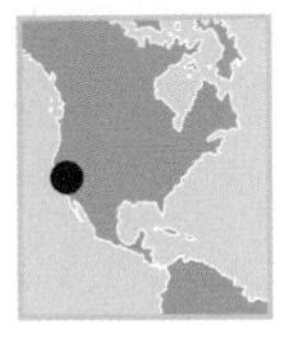

Watercourses around Soda Lake, California, United States (35°09' N, 116°04' W).

The beds of dried-up watercourses trace bright tree-like patterns on the arid plain of Carrizo. This 250,000-hectare plain, home to herds of deer and antelope, as well as the rare California condor, was declared a national monument on January 17, 2001, and is the last remnant of an ecosystem that once covered the whole of central California. Only 4 percent of the original grasslands remain, though their flora is still unusually rich. A total of 4,426 plant species (2,000 of them endemic) grow in California, more than in all of northeastern North America. This serious reduction in natural habitats is the result of the extraordinary economic and urban development of the region, a product of the "American dream." California's economy is one of the most dynamic in the world, and its agriculture provides half the food products consumed in the United States each year.

MAY

18

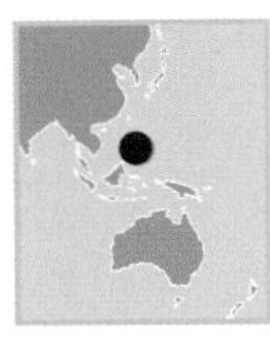

Discharge from the gold mine on the shore of Mindanao, Philippines (6°52' N, 126°03' E).

Exploitation of gold deposits on the island of Mindanao in the southern Philippines provides an important economic resource for the country, which now produces an average of 8 tons of gold per year. However, the refuse and sediments from the washing and sorting operations are discharged daily into the rivers and ocean. These discharges darken the waters and endanger marine flora and fauna both along the shore and out at sea, particularly the coral polyps that depend upon light for survival. Chemical products such as mercury and hydrochloric acid used for cleaning and refining the gold particles are also discharged into the water, adding their toxicity to the effects of this marine pollution. Damage from mining operations has also affected Hungary's Tisza River, which in January 2000 was contaminated by cyanide (used in the extraction process) escaping from a Romanian gold mine.

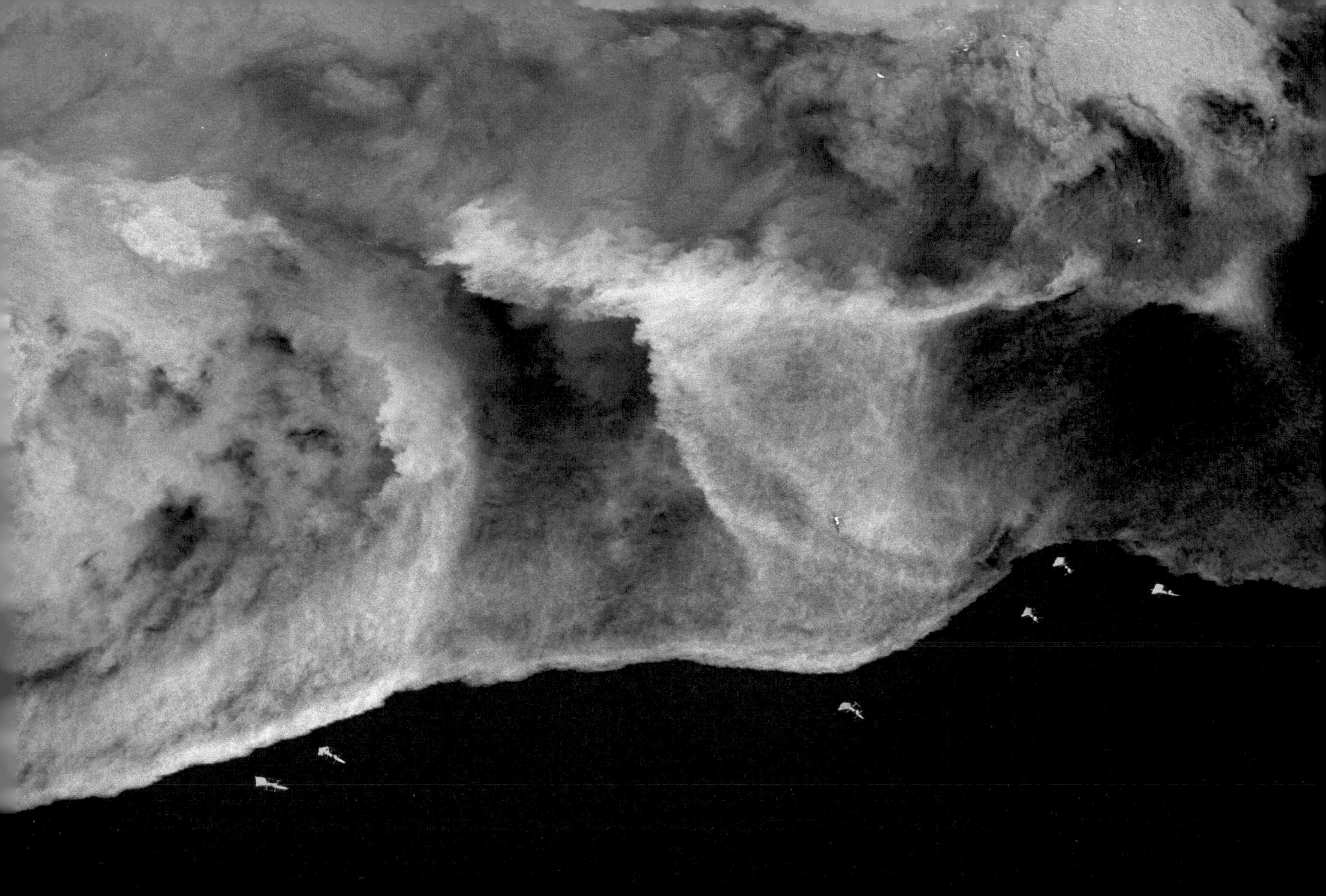

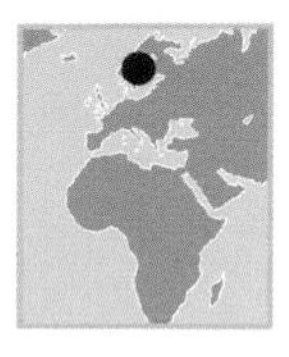

Stacks of wood at a wood pulp factory near Mörrum, Blekinge province, Sweden (56°11' N, 14°45' E).

The nine countries around the Baltic Sea have made efforts to reduce its pollution over the last 30 years. However, the state of this almost landlocked sea, tenuously linked to the North Sea by the Skagerrak and Kattegat straits, is still causing concern. A variety of chemicals—from wastewater, atmospheric pollution, and agricultural and industrial effluent—are building up in it. Now subject to regulation, the waste from many wood pulp factories, such as this one in Blekinge, southern Sweden, have been contaminating the region for a long time. The amount of paper used in the world has tripled since 1960, reaching a rate of 317 million metric tons a year. North America, Western Europe, and Japan—that is, 20 percent of the world's population—use 70 percent of this. If the whole of China, which uses 94 pounds (35 kilograms) of paper per person per year, rose to the same level as the United States (905 pounds, or 338 kilograms), present world output would not meet its needs. And if the average recycling rate of paper (43 percent) rose to that achieved by Germany (72 percent), a third of the wood now needed to meet world demand could be saved.

SVETRUCK

MAY

Deforestation in Amazonia, Mato Grosso do Norte, Brazil (12°38' S, 60°12' W). Almost 2 million hectares of the Amazon rainforest are cleared every year. This deforestation, which is speeding up, brings little benefit to local people. Its main function is to free up land for growing cheap cereals—especially soya—to feed livestock in rich countries. These cash crops for export generate foreign currency; in their quest for growth, big farming concerns do not hesitate to clear areas of forest, when they are not driving small farmers off the land and forcing them to retreat to forested areas. Worldwide, the expansion of farmland, the wood industry, and road building destroy almost 15.2 million hectares of natural tropical forest per year—the equivalent of the Netherlands, Belgium, Switzerland, and Denmark put together, or the whole of Florida.

20

Contaminated vehicles at Rassorva, Chernobyl region, Ukraine (51°20' N, 30°10' E).

The explosion of a reactor at the Chernobyl power station, Ukraine, in April of 1986 was the worst civilian nuclear disaster in history. A radioactive cloud burst from the reactor and contaminated Ukraine, Belarus, and neighboring parts of Russia before the wind carried it all over Europe. The 120 surrounding districts were evacuated, though belatedly. The precise number of victims remains unknown today, but it is estimated that several million people are suffering from illnesses linked to radiation, such as cancer and immune deficiencies. In December 2000, the power station's last reactor—which had remained in service to supply 9 percent of the country's electricity—was finally shut down in exchange for $2.3 billion in Western aid, which will allow two new nuclear facilities to be built. The nuclear industry has not yet resolved the problem of disposing of waste that will remain highly radioactive for many years. This waste, produced by 433 reactors in thirty-two countries, accumulates in storage centers.

MAY

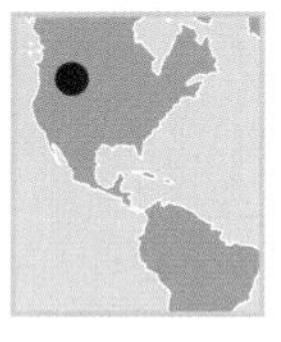

Banff National Park, Alberta, Canada (51°38' N, 116°22' W).

With its 2,563 square miles (6,641 square kilometers) of dramatic landscapes, Banff is the jewel of the Canadian Rockies. Created in 1885, Canada's first national park contains vast coniferous forests dotted with lakes and hot springs, lying at the foot of great mountains. The richness of its wildlife is one reason why the park was added to UNESCO's World Heritage list in 1984. However, this prestige has its price. Tourism has boomed, and 4.7 million visitors roamed the park's roads in 2000. Its animals have paid a heavy price for this "green tourism." For example, 90 percent of grizzly bear deaths occur close to human infrastructure and are caused by, among other things, collisions with careless drivers or encounters with walkers who are carrying firearms. On February 19, 2001, this paradox led to the passing of a law defending the "principle of ecological integrity" of Canada's national parks, aimed at reducing commercial and recreational development in these protected areas. This sacrifice is essential if the 86,644 square miles (224,466 square kilometers) of Canada's national parks are to be preserved for the future.

22

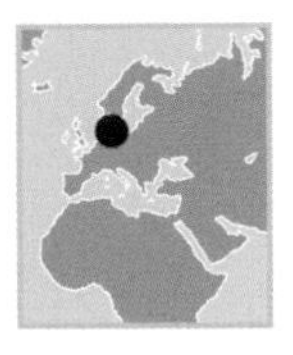

Power station at Hvidovre on the Baltic Sea, Denmark (55°39' N, 12°29' E). Completed two years ago, this plant on the Baltic coast southeast of Copenhagen produces energy from renewable sources such as wind, but also from fossil fuels such as oil and coal. Although these produce pollution, the plant's builder claims that it uses new technologies that reduce toxic emission by up to 80 percent. Power stations and cars are the chief sources of manmade air pollution. Children, the elderly, and those who are vulnerable to it are most at risk. According to the World Health Organization (WHO), almost 3 million people die every year from the effects of pollution. In Europe, half of these deaths are thought to be linked to vehicle emissions. Many countries, notably in South America, limit car use in large cities.

23

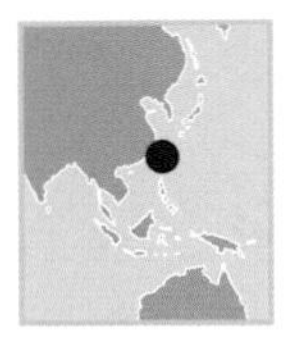

Fisherman near the Tungkang lagoon, Taiwan (22°26' N, 120°27' E).

This fisherman on Taiwan's western coast has chosen a spot on a breakwater that juts out into the sea like a promontory. The nearby city of Tungkang is the second biggest fishing port on an island that has a great love of seafood. Taiwan is well known for its shark catch of almost 50,000 metric tons per year, the fifth biggest in the world. The strait between the island and China is home to the biggest range of shark species on the planet. The Asian market is increasingly hungry for shark meat, and the world catch rose from 272,000 metric tons in 1950 to a record of 760,000 metric tons in 1996. Biologists have become alarmed at the increasing rarity of these marine predators: of the 100 species most often fished, twenty are thought to be in danger of extinction. Illegal fishing is believed to be the main threat to these selachians and on fish stocks in general. At the beginning of 2001, the FAO's fishing committee adopted an international action plan to try to eliminate this trade.

24

MAY

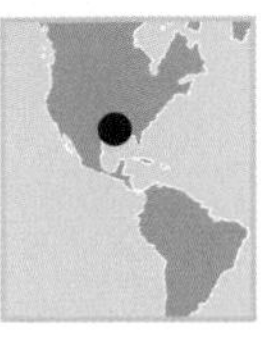

Flight of pelicans in Louisiana, United States (29°50' N, 90°13' W).

Several dozen pelicans form a compact band on this lake in the Mississippi Delta, south of New Orleans. With a wingspan of almost 10 feet (3 meters), these large birds—specifically, the Eastern brown pelican—are the state bird of Louisiana. Their impressive beak, with its pouch for holding fish to be fed to their young, has appeared on the official state flag since 1912. In North America, the pelican's distribution range shrank until the 1970s, but since then its numbers have increased. Today, with 100,000 pairs on the continent, the species is not in immediate danger of extinction. Another bird, familiar from its long association as fierce protector of the Star-Spangled Banner, has also been saved. The American eagle (or bald eagle) was common when it became the national emblem in 1782, but hunting, pesticides (which caused the birds' eggs to develop with too-thin shells), and habitat loss reduced it to a mere 417 pairs by 1963. Fortunately, protection campaigns saved this mascot, which now numbers almost 6,000 pairs.

25

MAY

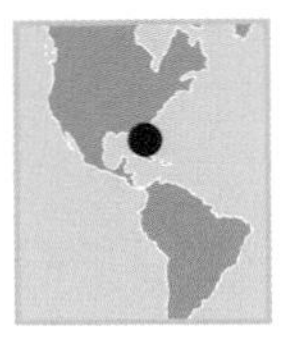

Tornado damage in Osceola County, Florida, United States (28°25' N, 81°20' W)

On February 22, 1998, a force-4 tornado (with winds of 185 to 250 miles per hour, or 300 to 400 km per hour) finished its course in Osceola County, after having devastated three other counties in central Florida. Several hundred homes were destroyed in its whirlwind, and 38 people were killed. This type of violent tornado, rare in Florida, is generally linked to the climatic phenomenon of El Niño, which causes strong meteorological disturbances all over the world about every five years. Major natural catastrophes are more frequent and devastating than ever before. Human activities have significantly disturbed natural sites, reducing their resistance and their ability to withstand the effects of extreme climatic events. People also aggravate these consequences by living in areas exposed to risks. The decade of the 1990s saw four times more natural catastrophes than occurred in the 1950s, and the economic losses thus caused in that decade totaled $608 billion, more than the costs for the four previous decades combined.

26

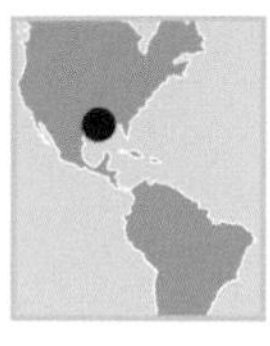

Mouth of the Mississippi river, Louisiana, United States (29°36' N, 89°49' W).

From its source in Lake Itasca, Minnesota, in the northern United States, the Mississippi river flows through the heart of the country, traveling 2,347 miles (3,780 km)—3,856 miles, or 6,210 km, if the Missouri river is included—to Louisiana, on the coast of the Gulf of Mexico, where it fans out in a vast delta. Known as "Old Muddy," the river washes down vast quantities of silt scoured from its enormous basin, which covers 1,243,692 square miles (3,222,000 km^2), or 41 percent of surface area of the United States. It deposits this load in the delta, south of New Orleans—a huge marshy plain measuring 248.4 by 124.2 miles (400 by 200 km), where various zones of silting bear witness to the river's changes of course over millennia. Brackish marshes and bayous, the backwaters of the Mississippi and its ancient oxbows, form an amphibious landscape suspended between land and sea, home to alligators and rich in bird life, that surrounds this boat. For geographers, "Mississippi Delta" merely denotes the place where the river flows into the sea. Usually, however, the name refers to a whole region—stretching from Memphis to Vicksburg at the confluence with the Yazoo river—that was the birthplace of the blues.

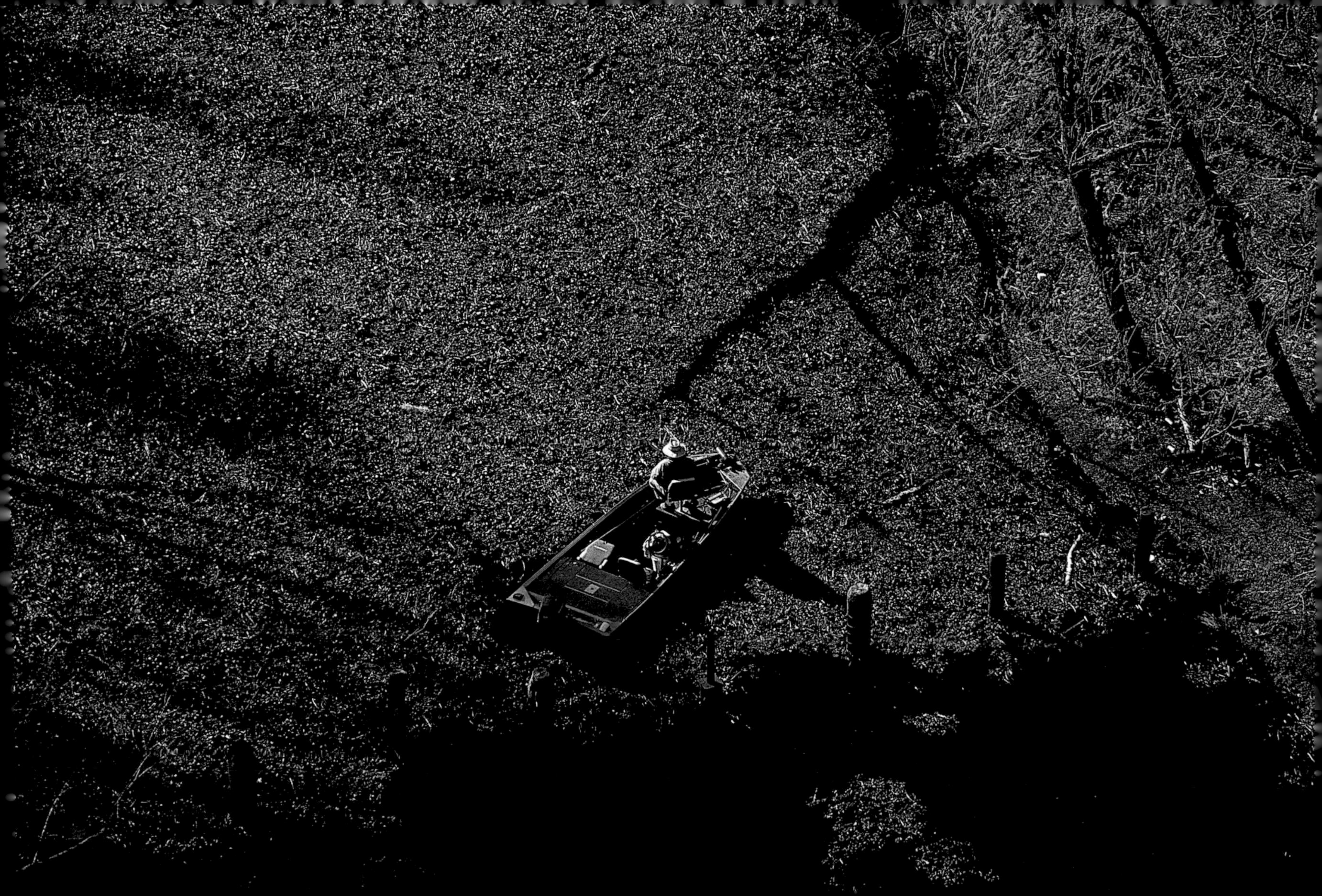

MAY

28

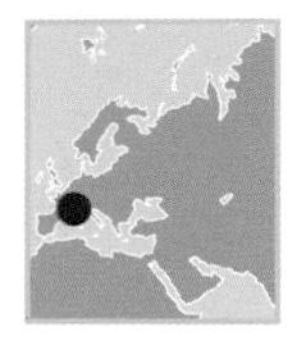

Beach at Saint-Raphaël, Côte d'Azur, France (43°25' N, 06°46' E).

This beach, crammed with sunbathers and with its concreted coastline, might belie the fact that France's Mediterranean coast is a rich environment. Land of the red and ochre volcanic massif of the Estérel, it is known as the Côte d'Azur (Azure Coast) because of the sparkling blue of its sea and sky. The simple fishing village of Saint-Raphaël has become a fashionable, world-famous resort and a holiday destination for many Europeans, chosen as much for its wide range of entertainments as for its sunny climate. Even back when it was part of the Roman empire, during the reign of Julius Caesar, rich Romans were already choosing this place—which they called Epula, or "Banquets"—to build their magnificent mansions. During the Middle Ages, the town tried to defend itself against the Saracen and Turkish invasions, the start of a reign of terror that was only brought to an end by William of Provence. Saint-Raphaël returned to fame in the nineteenth century, thanks to the many writers, artists, and musicians who lived there. Charles Gounod composed his opera *Roméo et Juliette* there in 1865.

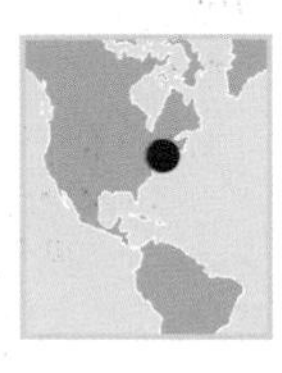

Financial district, Manhattan, New York, New York, United States (40°45' N, 73°59' W).

Four months after the terrorist attacks of September 11, 2001, destroyed the World Trade Center, a great empty space, entirely cleared of the debris of the twin towers, has been opened up in the heart of Lower Manhattan, the nerve center of New York. Despite its wounds, the city is determined to bounce back. Chicago and San Francisco have both been ravaged by fire in the past, and New York itself suffered two terrible fires, in 1776 and 1835. Each time the areas destroyed by the flames were rebuilt, as Manhattan's financial district will be. Soon after the attacks, a group of architects applied themselves to this task, with input from town planners, historians, and city authorities. The outcome of their deliberations will determine the new face of the site where the World Trade Center stood. The only thing that is certain is that reconstruction will take place, both to rehabilitate this part of the city and to pay homage to the thousands of Americans and others who died on the spot, which has become a sacred place.

MAY

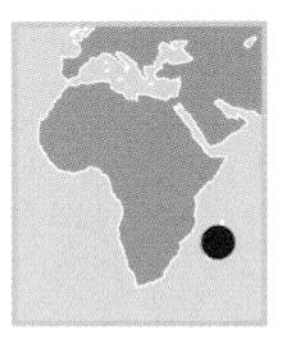

Traditional village north of Antananarivo, Madagascar (18°49' S, 47°32' E).

Not only is Madagascar one of the world's fifteen poorest countries, but in 1998 it had the lowest levels of health spending in Africa, at $15 per person. Despite this inadequacy, Madagascar is one of the few countries in sub-Saharan Africa that has largely escaped the HIV epidemic. This illness, now in its third decade, has taken on frightening proportions in the continent. In 2002, more than 39 million Africans were living with HIV, a number comprising 70 percent of the people infected in the world. In the same year, 2.4 million people in Africa died from it—compared with 8,000 in western Europe. In four southern African countries, the rate of HIV infection in adults is climbing at alarming rates: 38.8 percent in Botswana, 31 percent in Lesotho, 33.4 percent in Swaziland, and 33.7 percent in Zimbabwe. Loss of life has been so high that it has affected the continent's economy and led to food crises. The HIV phenomenon has not yet reached its peak, and the international community—governments, nongovernmental organizations, media, and international institutions—has yet to find a way of dealing with this disaster.

30

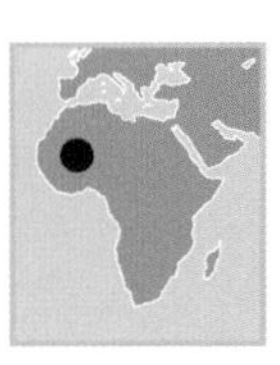

Dugout canoe on the Niger river near Timbuktu, Mali (16°39' N, 3°00' W).

As this small sample of humanity navigating the Niger river illustrates, the population of Mali is a mixture of many elements. The *Tamasheq* ("Tuaregs," in Mali), recognizable here by their immaculate white tunics, are only one of eleven ethnic groups in the country. However, their place in its history is unique. The Tuaregs were the first to revolt against the French colonial administration at the beginning of the twentieth century. In 1990, it was the turn of the state of Mali—or, more precisely, of General Moussa Traoré's military dictatorship—to experience Tuareg insubordination. From their stronghold in the mountains of the Sahara, the rebels inflicted heavy losses on the regular army, which took its revenge on civilians. It is estimated that 10,000 civilians were killed between 1990 and 1996 by the armies of Mali and neighboring Niger. In March 2002, Mali celebrated the sixth anniversary of the reintegration of the *Tamasheq*; however, this does not change the fact that the only Tuareg "country" is still the immense desert of the Sahara, which is cut up into eleven pieces by state borders.

FRESH WATER

JUNE

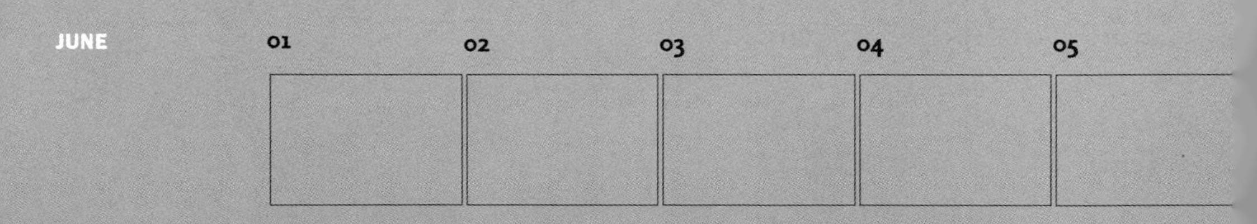

We call our planet "Earth," but the resource we most highly value and prize, and the one we notice first when it is absent, is fresh water. The presence or absence of fresh water has determined the course of civilizations, the wealth of nations, and the health of people. Humans can survive long periods with no solid food, but a scant few days without water. The history of human civilization is tightly connected with the history of our manipulation of the fresh water resources around us. The earliest agricultural communities formed where food could be grown with dependable rainfall and perennial rivers. Simple irrigation systems permitted more food to be grown and more mouths to be fed. As villages and towns expanded, local water supplies became inadequate, leading to advances in the engineering and science of water supply and waste disposal. The first industrial nations depended on waterpower to drive machines and multiply the effectiveness of human labor.

Water exists in vast quantities in the oceans, ice caps, aquifers, rivers, lakes, clouds, and even tissues of living matter, all circulating from one form to another through evaporation, precipitation, and runoff. Yet only a modest amount of the world's water is fresh—less than 3 percent of total global water supplies—and the majority of that is locked up in the ice caps and glaciers of Greenland and Antarctica and in deep groundwater aquifers. The remaining fresh water must serve to satisfy the needs of our natural ecosystems and our human society. As the world's population grows, competition for this water is also growing, leading to political, economic, and environmental frictions and tensions.

Today, we live in a hydraulic society. Our success in feeding 6 billion mouths depends on massive irrigation projects: less than 20 percent of the world's cropland is irrigated, but these lands produce 40 percent of the world's food. Overpopulated cities would wither without the complex arrangements of reservoirs, aqueducts, and wastewater systems we've built to serve them. A seemingly limitless supply of electricity is produced for voracious users by the power of falling water flow from massive dams on almost all of our major rivers. Thanks to improved sewer systems, cholera, typhoid, and other water-related diseases, once endemic throughout the world, have largely been conquered in the richer nations.

06 07 08 09 10 11 12 13 14 15

Yet there is a dark side to this picture. While many religions consider water to be a gift of God, we are treating it more and more like a commodity or a mineral, trading it from one place to another, mining it, and fighting over it. Despite some progress in efforts to reduce poverty, despite an ongoing electronic and information revolution now underway in the richer nations, half the world's population still makes do with water services inferior to those that were available to the ancient Greeks and Romans. More than a billion people lack access to safe and reliable drinking water; and nearly 2.5 billion people do not have adequate sanitation services. Preventable water-related diseases kill an estimated 10,000 to 20,000 children each day, and the latest evidence suggests that we are falling behind in efforts to solve these problems. New massive outbreaks of cholera still occur in the poorest regions of Latin America, Africa, and Asia. Tens of millions of people in Bangladesh and India are drinking water contaminated with unsafe levels of arsenic.

The effects of our water policies extend beyond the jeopardy they pose to human health. Entire cities and towns have been emptied to make way for the reservoirs behind dams. More than 20 percent of all freshwater fish species are now threatened or endangered because of human modifications to rivers and lakes. The Aral Sea is being destroyed by the use of water for growing cotton. Irrigation can degrade soil and water quality. In parts of India, China, the United States, the Persian Gulf, and elsewhere, groundwater aquifers are being used up faster than they are naturally replenished. An increasing number of major rivers, such as the Yellow in China, the Colorado in the United States and Mexico, the Jordan in western Asia, and the Nile in northern Africa, now disappear before they can reach the ocean because humans have taken all the water that used to flow in them. Disputes over shared water resources have led to local and even international tensions and violence.

Two paths lie before us. We can continue along the hard path and expand the construction of large infrastructure such as dams, aqueducts, reservoirs, and centralized water treatment plants. The hard-path water policies of the twentieth century brought tremendous benefits to hundreds of millions of people, but those benefits came at an economic, environmental, and societal cost we are only now beginning to reckon. And the hard path has failed to serve billions of people.

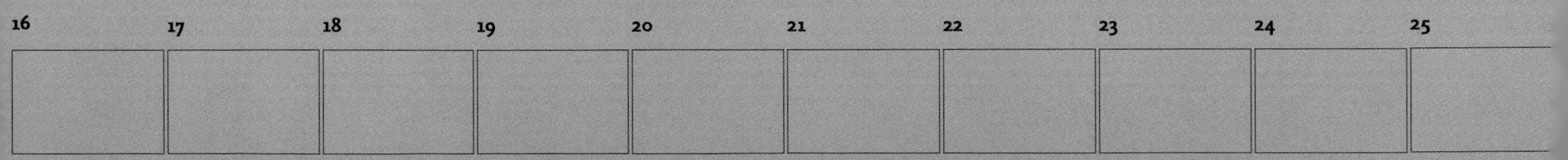

The alternative is a soft path, which taps the benefits of appropriate infrastructure but complements them with decentralized facilities, efficient technologies and policies, and the smart application of economics and human capital. The soft path seeks to improve the overall productivity of water use rather than finding endless sources of new supply. By becoming more efficient in our water use, we can produce more food, or steel, or computers—all with less and less water. The soft path requires that rather than merely supplying water, governments, local communities, and even private companies work together to meet water-related needs. It actively involves local voices and communities, efficient new technologies, and traditional approaches that draw on centuries of local knowledge and experience.

The water that flows through our landscape is a thing of beauty and power; it stimulates song and poetry and reverence. But it is more than that. Water is a vital and precious resource, necessary to the basic needs of humans and natural ecosystems. As we move into the new millennium, we must acknowledge that unless we protect our precious fresh water, we will have neither a healthy environment nor a healthy society. It is time for a new way of thinking about water. History shows that although access to safe and reliable fresh water cannot guarantee the survival of a civilization, a civilization most certainly cannot succeed without water.

DR. PETER H. GLEICK
President of the Pacific Institute for Studies in Development, Environment, and Security, Oakland, California

26	27	28	29	30

JUNE

"Lava truck" in a steelworks at San Felipe, Chile (32°45' S, 70°44' W).

As soon as the iron ore has melted with coke, trucks carry the liquid to a tank where it is mixed with oxygen and lime. The raw steel thus obtained is treated at the San Felipe plant in the Santiago region. This is where almost all of Chile's industry is concentrated, and the region is being suffocated through its by-products, for pollution is trapped by the surrounding mountains—the Andes to the east and the coastal range to the west. To improve the capital's air quality, the authorities implemented a plan in 1990 aimed at combating emissions of dust and greenhouse gases. However, the growing number of cars and the increase in industry may well cancel out these efforts. Worldwide, 3 million people die each year from the effects of air pollution, mostly in cities. This is likely to worsen, as 65 percent of the world's inhabitants will live in cities in 2050, compared with 47 percent today and only 35 percent 50 years ago.

01

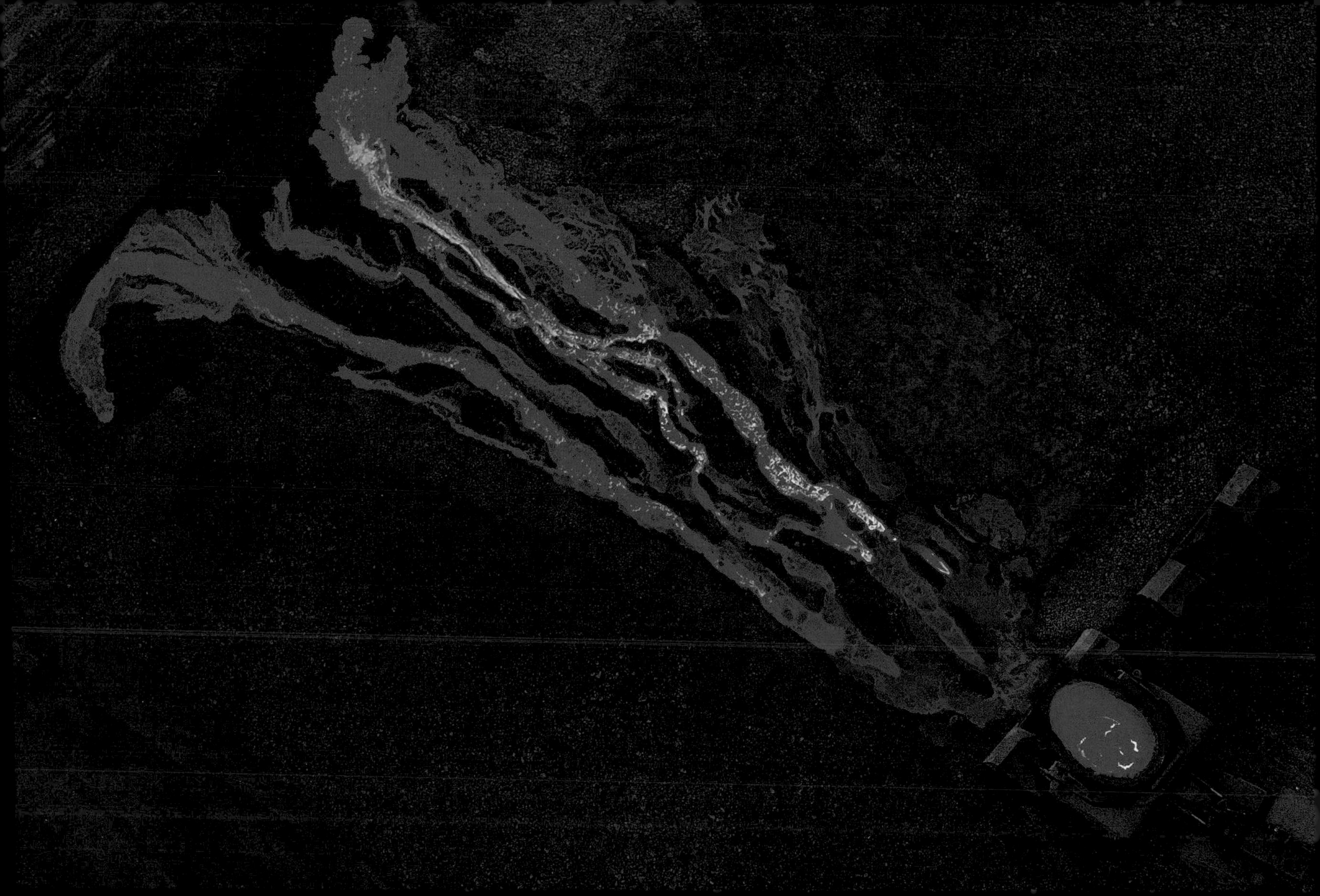

JUNE

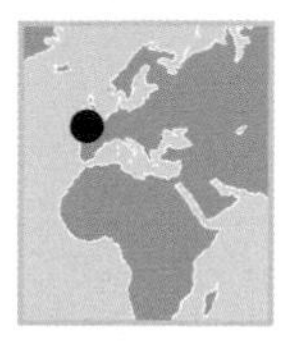

Catamaran in the Glénan archipelago, south coast of Finistère, France (47°44' N, 4°00' W).

South of the Brittany peninsula's tip, some 9 miles (15 km) off Concarneau, the five Glénan islands and their retinue of islets and rocks cluster together on a shallow seabed of pale granite, white sand, and marl, bathed by the Atlantic. This wildly beautiful place with its clear waters, no deeper than 6.5 feet (2 m) at low tide, attracts many yachts. It is also home to France's most famous sailing school, Les Glénans, founded in 1947, and its parade of trainees in summer. But its most distinguished inhabitant is the Glénan narcissus, which grows on the archipelago and nowhere else. However, the white corollas of this flower were almost decimated by the abundant gulls, thorn bushes, and ferns. The nature reserve on the island of Saint Nicolas was set up in 1974 to ensure the species' survival, and led to an increase in the number of flowering stems from 6,500 in 1985 to 120,000 in 1998. The fragile nature of the area also demands measures to minimize the impact of humans, such as ecological sanitation, gathering rainwater, keeping to footpaths, and installing solar panels and a wind turbine for local production of nonpolluting electricity.

02

JUNE

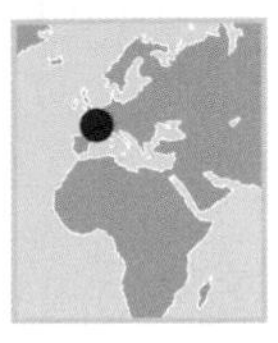

Vallée Blanche glacier at the foot of the Aiguille du Midi, Mont-Blanc Massif, Haute-Savoie, France (45°50' N, 6°53' E).

On the Vallée Blanche, in the Mont-Blanc Massif, skiers can hire the services of a guide and experience an unforgettable descent, nearly 12.5 miles (20 km) long. More than 80,000 people every year are seduced by this route, which starts at the Aiguille du Midi (accessible by cable car) at 12,598 feet (3,842 m), follows the Vallée Blanche, and continues along the Tacul and Mer de Glace glaciers. Depending on the snow conditions, it is sometimes possible to go as far as Chamonix. But, over the decades, the glacier has tended to recede. The front of the Mer de Glace, which has receded by 1.25 miles (2 km) since 1820 with a 32.9 °F (0.5 °C) rise in temperature, may yet settle at an altitude of 5,902 feet (1,800 m) in 2050, having lost another 1.86 miles (3 km) of its present length. Climatologists are forecasting a rise of between 37.4 and 42.8 °F (3 and 6 °C) over the coming century. The development of renewable energy sources and techniques of energy conservation, as well as reduced use of private cars—especially in cities—would mean a reduction in emissions of the gases that are causing climate change.

03

JUNE

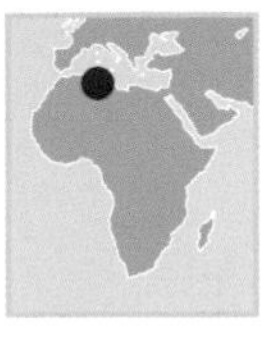

Kebili Oasis, Nefzaoua, Tunisia (33°42' N, 8°58' E).

Kebili is the main oasis of Nefzaoua, in Tunisia's far south. Surrounded by sand, this fertile area is irrigated, as are all oases, by ground water that breaks through the surface in plentiful springs. The tapping of ground water with motor-driven pumps has transformed this steppe on the edge of the desert into a modern agricultural region, with increasing numbers of irrigated areas. However, this method soon exhausted the water that lay close to the surface. Drilling then tapped deeper supplies; these, too, are now being exhausted. Soon this race to the depths of the Earth will necessarily come to an end. We have neglected to observe that this water is probably not a renewable resource. It is not so much that the desert advances as a result of human activity; it is that the grassland is being degraded. Abandoned areas are invaded by small sand dunes driven by the wind. Like the holes moths chew in clothing, these dunes gradually join up and form a single expanse, causing desertification. Natural and man-made causes join forces in the advance of the Sahara desert on its southern edge; in the Sahel, the same causes and effects can be observed. Worldwide, drought and desertification threaten more than 1 billion people in more than 110 countries.

04

JUNE

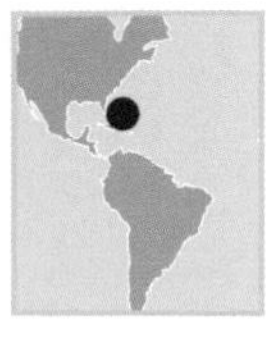

Islets and Seabeds, Exuma Cays, Bahamas (24°26' N, 76°44' W).

The archipelago of the Bahamas, which takes its name from the Spanish term *baja mar* ("shallows"), spreads out in an arc in the Atlantic Ocean, running 750 miles (1,200 km) with 5,405 square miles (14,000 km^2) of land above water level from Florida to Santo Domingo. It consists of more than 700 islands (29 of which are permanently inhabited), plus a few thousand rocky coral islands called cays. It is on the islands, specifically on Samana Cay, that Christopher Columbus first set foot on October 12, 1492, during his first voyage. During the sixteenth and seventeenth centuries the Bahamas were a center of piracy. They became a British possession in 1718, which they remained until their independence in 1973. Today the country is a "fiscal paradise": there is no income tax. It derives its resources from banking (20 percent of the GNP) but mostly tourism (60 percent of the GNP), which employs two out of three Bahamians. More than a thousand ships, or nearly 3 percent of the international commercial fleet, are registered with a Bahaman flag of convenience. The Bahamas have also become one of the centers for drug traffic (marijuana and cocaine) bound for the United States.

05

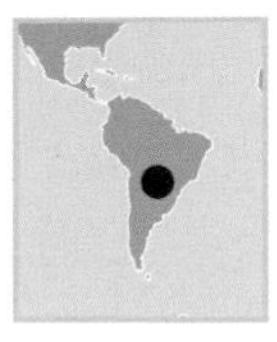

Fields in Misiones province, Argentina (27°00' S, 55°00' W).

At Argentina's northernmost tip, the province of Misiones is the cradle of *maté*, the ubiquitous drink of the Argentines. The *maté* herb originated in this region, where the guarani indians drank it as an infusion long before the Jesuit missions arrived in the sixteenth century. Today, plantations of this crop, whose leaves have tonic properties, follow the contours of the land. This mode of cultivation was made compulsory in 1953 to protect crops from erosion caused by the often torrential rains. Rural areas have been severely hit by the country's present economic crisis: 90 percent of small *maté* producers are particularly badly affected. By contrast, cultivation of GMOs has increased sharply, with the area under such crops rising thirty-fold between 1996 and 2001. Argentina now has 22 percent of the world's transgenic crops, which puts it in second place behind the United States.

JUNE

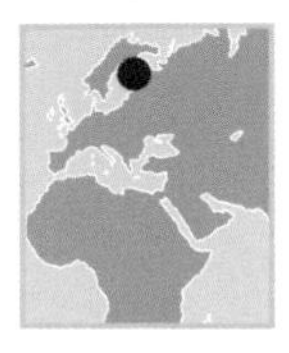

Ferry docking at Houtsala island, Turku archipelago, Finland (60°15' N, 21°50' E).

The Turku archipelago consists of 22,000 islands, scattered granite fragments that separate the Gulf of Bothnia from the Baltic Sea, southwest of Finland. In this biosphere reserve, it is essential to preserve traditional lifestyles, based on fishing and agriculture, and to place environmental constraints on new commercial activities such as tourism, leisure, and fish farming. In winter, the Baltic freezes, forcing food supply ships to use icebreakers. Transport by ship produces only 0.0125 percent of the greenhouse gases produced by air transport; the latter mode of transport is, however, growing at record rates (6 percent per year). Finland harvests timber from its coniferous and birch forests, which cover 70 percent of its area and yield more than two-thirds of its export revenue. The residue from the timber industry and the waste from logging are used for fuel, and these are an important source of renewable energy, meeting 20 percent of the country's energy needs and 10 percent of its electricity needs in 2000.

07

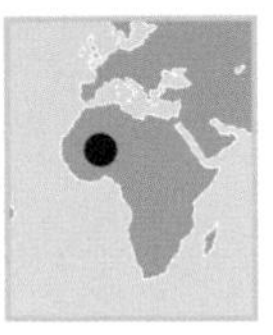

Enclosure in a village near Nara, Mali (15°10' N, 7°17' W).

Traditional African villages often consist of private enclosures, where families and their animals live in different types of hut. The biggest types accommodate a married couple or a mother with her young children. The others serve as kitchen, stable, or grain store. Built out of *banco* (a kind of brick made of clay mixed with straw and sand) and roofed with reeds, straw, or dried palm leaves, the huts do not withstand the weather for long. In less than a decade, they fall into ruin and are abandoned. The family then builds a new enclosure, after consulting local notables, that is, the "land chiefs" who, since rural land is collectively owned, are responsible for allocating plots.

JUNE

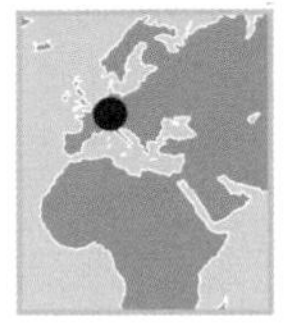

Landscape in Upper Franconia, Bavaria, Germany (50°12' N, 11°30' E).

The industrial region of Upper Franconia, in northern Bavaria, occupies a prime location in the center of Germany. Its economy is based on textiles (now facing tough competition from Eastern European and Asian production), porcelain, electronic goods, plastics, and machinery. Forest and farmland jostle for space and contribute to the appearance of the landscape. More than a third of Europe, or 304 million acres (123 million ha), is forest; it covers 30 percent of Germany, but only 9 percent of Ireland. Forest cover reached its lowest point during the eighteenth and nineteenth centuries, but during the twentieth century it increased at the expense of agricultural land. Incentives led to the replanting of 1.24 million acres (500,000 ha) in the European Union between 1993 and 1997. When the right tree species are planted in suitable locations, and the forest is properly managed, it plays an important role in controlling erosion, regulating land drainage, and combating global warming.

09

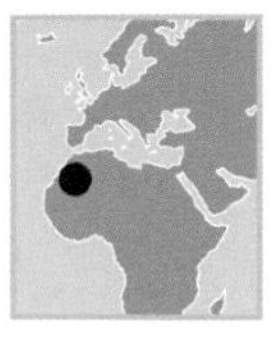

Salt marshes, Oualidia, Morocco (32°44' N, 9°08' W).

Oualidia is a small beach resort 108 miles (175 km) southwest of Casablanca. Its topography, climate, and geology allow the exploitation of the salt marshes, an activity that demands flat, nonporous ground, a climate that favors evaporation, and a lack of rain for a fairly long period during the year. Since salt is cheap and used throughout the world, it is used to combat certain deficiencies, especially of iodine and fluorine. Iodine deficiency affects 760 million people and is believed to be responsible for mental disorders in almost 50 million of these. Today, 70 percent of the world's salt is iodized, which has led, since 1990, to a reduction by more than half in the number of newborn children suffering from cretinism. Research into adding iron to salt is underway, but there is still a long way to go in the battle against vitamin and trace-element deficiencies, which affect 2 billion people.

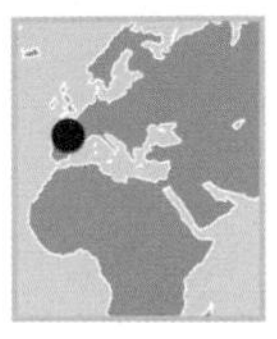

Cleaning up fuel oil leaked from the tanker *Prestige*, Biarritz, France (43°40' N, 1°35' E).

Three months after this Liberian-registered tanker was wrecked, France had already cleared up more than 13,500 tons of hydrocarbon waste from its shores. For this cleaning job, the most effective tools are scoops, modified dragnets, and drift nets. The victims of such ecological disasters must turn to the International Oil Pollution Compensation Fund (IOPCF–FIPOL). This intergovernmental organization, financed by the oil industry, provides up to 180 million euros ($204 million) for each such accident—a sum that is woefully inadequate in view of the total cost of such catastrophes, which runs into billions, and is even more laughable given the income that oil generates. In 2001, the French government levied 24 billion euros (about $27.6 billion) in oil taxes. This system, which some consider a polluters' charter, does nothing to encourage improvement in the tanker fleet, since the compensation amounts levied on ship owners are the same, regardless of the condition of the ships in service.

NOBILIS
BAYONNE

JUNE

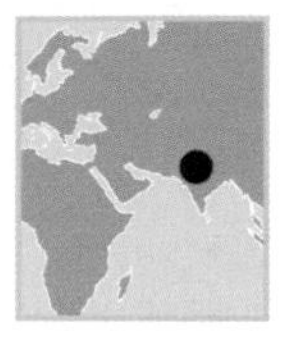

Old town of Jodhpur, Rajasthan, India (26°17' N, 73°02' E).

At the gates of the former kingdom of Marwar, the "land of death," Jodhpur's bluish tinge stands out defiantly against the great expanse of the Thar desert. The Brahmans are believed to have painted the houses blue to keep them cool within. Thus the former city of Jodhagarh took on the color of water, a most precious resource in this arid region. Lacking in water supply infrastructure, and daunted by the cost of providing it, many countries such as India are choosing to part-privatize this sector. As yet, private companies run only 5 percent of the world's water-supply grids, but the sector has already generated a market worth $200 billion. Privatization does not always ensure the protection of water resources, nor does it promise access to drinking water for the poor. In South Africa, it led to the country's worst cholera epidemic in 2001 and 2002. More users have had their supply cut off for nonpayment of bills, leading poorer people to seek water in polluted rivers and wells.

12

JUNE

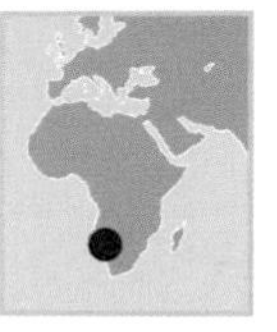

Beginning of the Namib desert, west of Gamsberg, region of Windhoek, Namibia (22°35' S, 17°02' E).

The road linking Windhoek, Namibia's capital, to the beach resort of Walvis Bay crosses the plateau of the Gamsberg, one of Namibia's highest mountains at 7,650 feet (2,334 m). The Namib desert, formed 100 million years ago, is believed to be the world's oldest. It consists largely of stony plains but also includes 13,260 square miles (34,000 km^2) of sand dunes, which are the highest on Earth (984 feet, or 300 m). The Namib is also the only place on Earth inhabited by the *Welwitschia mirabilis* plant, specimens of which are as old as 1,500 years. The fragile balance of this arid ecosystem is preserved in part by two national parks, which occupy a total of 26,000 square miles (66,400 km^2), one-fourth the total area of the Namib.

13

JUNE

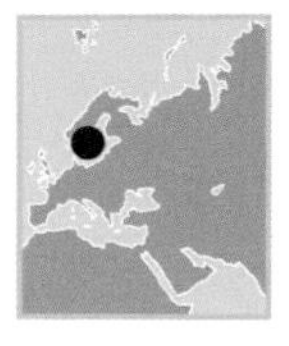

Storm at sea between Kalmar and the island of Öland, Sweden (56°39' N, 16°25' E).

There are 50,000 storms a day on the planet. They are caused by damp, unstable air that rises rapidly, cooling as it does so. Condensation occurs when the water-vapor saturation point is reached, forming a huge cumulonimbus cloud that can be 15.5 miles (25 km) across and 10 miles (16 km) high in the lower latitudes. Within this cloud, electrical charges produce lightning, followed by thunder and heavy rain. All over the world, storms cause serious damage to crops and buildings, disrupt air and land transport, interfere with communications systems, and kill hundreds of people and thousands of animals. Global warming is expected to cause an increase in the average amount of water vapor in the atmosphere and in the amount of rainfall. It is also expected to lead to more frequent extreme weather events such as storms, tornados, and cyclones.

14

JUNE

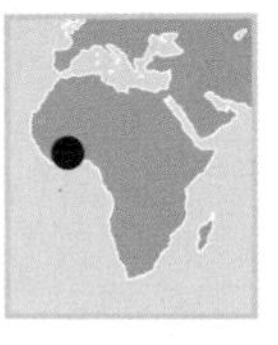

Picking pineapples, Abidjan, Republic of Côte d'Ivoire (5°19' N, 4°02' W).

Although Côte d'Ivoire is the biggest exporter of fresh pineapples to the European Union, the sector is suffering a crisis. Latin American countries are competing fiercely, and the EU is demanding higher standards in the traceability and level of chemical residues. Moreover, producers claim the Pineapple and Banana Trading Office is contributing to the continuing fall in prices, to the point that in 2003 they were lower than production costs. Worse still, since September 19, 2002, the crisis has been exacerbated by the country's civil war, not only because roads are closed but because farmland lies at the root of the conflict. Since 2001, Ivorians have angrily accused immigrants, chiefly from Mali and the Republic of Burkina Faso, who make up 30 percent of the population, of enriching themselves on their land. In the Bonoua district, immigrants have even been asked to stop growing pineapples, which are the area's chief resource.

15

JUNE

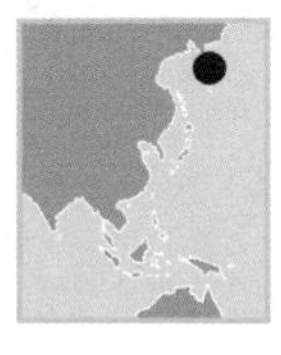

Uzon caldera, Kamchatka, Russia (51°00' N, 159°00' E).

The Uzon caldera is a vast crater formed after a volcanic eruption several thousand years ago. Its flat bottom, surrounded by enormous terraces 655 to 2,623 feet (200 to 800 m) high, covers about 39 square miles (100 km^2). In places where hot magma has reached the Earth's surface, activity that can continue for thousands of years after the eruption. The caldera offers a unique spectacle. The effects of the heat deep in the magma reservoir are clearly visible: lakes, marshes, small streams, and hot springs all constantly give off steam and contain high concentrations of boron, silica, and ammonia. Evaporation from the surface of the clay produces composite crystals and amorphous colored deposits. Bubbles of water vapor, filled with hydrogen sulfide, float on the water's surface, forming a film of sulfur. Thanks to its heated floor, hot springs, and enveloping vegetation, the caldera enjoys a unique microclimate that has produced an astonishingly rich biodiversity.

16

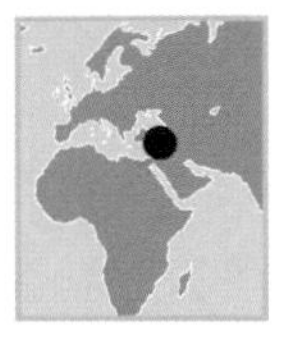

Flock of sheep near Kefraiya, Lebanon (33°39' N, 35°43' E).

These sheep form a curious arabesque design on the bare pastures of Mount Lebanon, the country's main mountain range. Once its slopes bore forests of centuries-old cedar, but these are now reduced to a few isolated shreds. Up to 60 percent of Lebanon's forests vanished between 1972 and 1994 as a result of civil war, urban expansion, and repeated forest fires. Deprived of vegetation, the ground became extremely vulnerable to erosion, especially from the heavy winter rains typical of the Mediterranean climate. Overgrazing has accentuated this process and made the land ever more barren—especially on Mount Lebanon, where the number of animals far exceeds the land's capacity to support them. Overexploitation of land is the chief cause of soil erosion in the world: it is estimated that 20 percent of pasture and common land is no longer productive.

JUNE

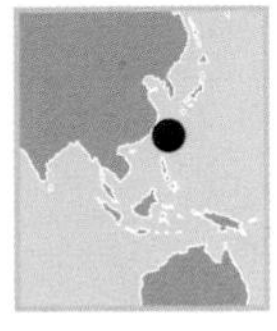

Aerator in a shrimp farm, Tungkang lagoon, Taiwan (22°26' N, 120°28' E).

The Tungkang lagoon, southwest of the island of Taiwan, is checkered with brackish pools devoted to farming seafood, particularly shrimp, which is highly profitable. Parting the white froth produced by the shrimp's waste, this aerator oxygenates the tanks where they are raised. Over the last 20 years, world production of shrimp has risen sharply, reaching 814,000 tons in 1999—more than sixteen times the figure for 1980, which was 50,000 tons. Asia, where the tiger shrimp is the predominant species, produces 80 percent of the world's output. Since shrimp need warm water to grow, shrimp farms are established in coastal regions in the tropics. Here, they often replace mangrove swamps—unique, fragile ecosystems that are a refuge and breeding ground for fish and crustaceans. Moreover, large amounts of waste from these intensive farms, as well as the antibiotics widely used, pollute the surrounding habitat. It is estimated that in certain areas, for every pound (or kilogram) of farmed shrimp, nearly half a pound (447 g) of wild fish and shrimps disappear. This is harmful to local people for whom shrimp farming, which is aimed at export markets, brings no benefits.

18

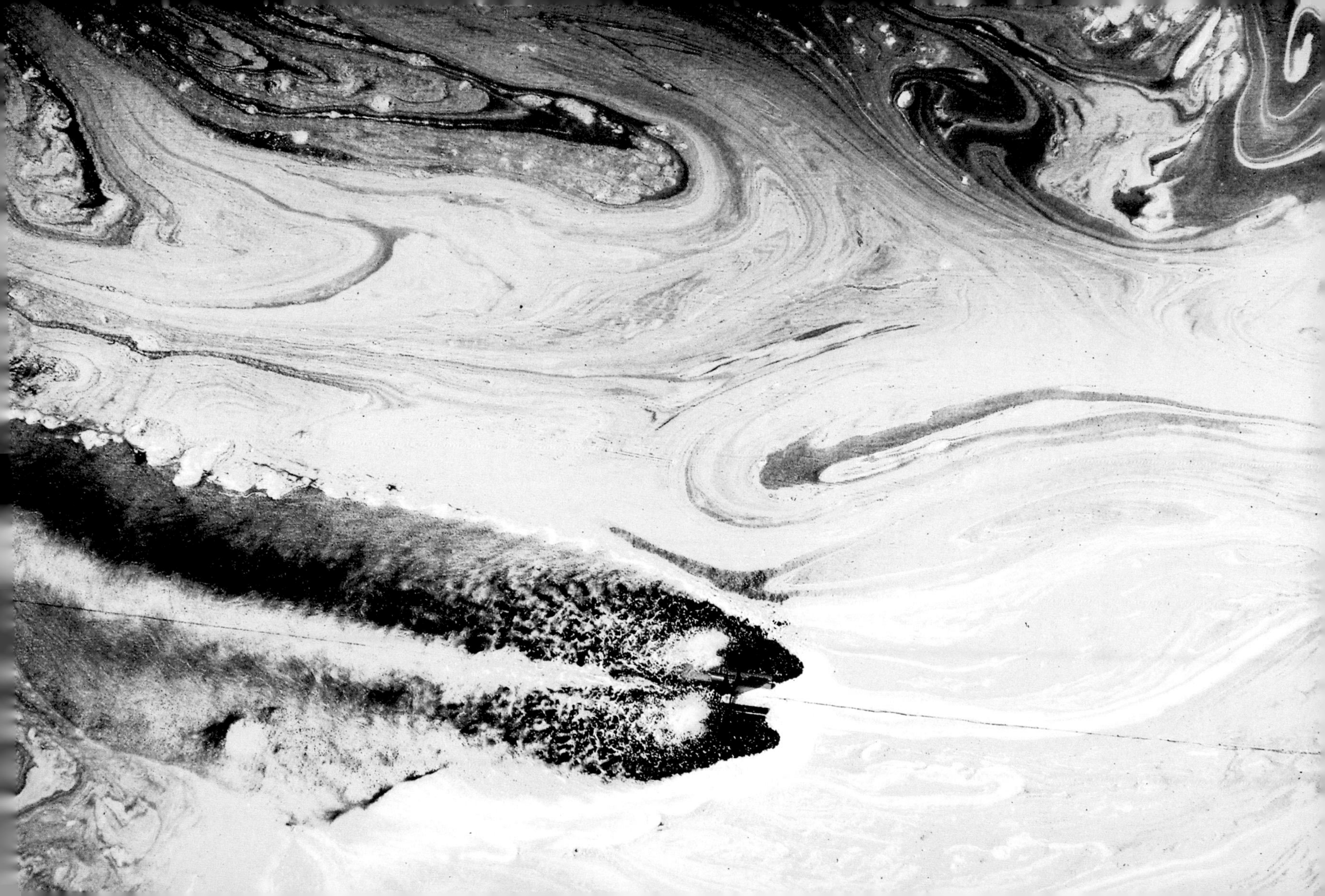

JUNE

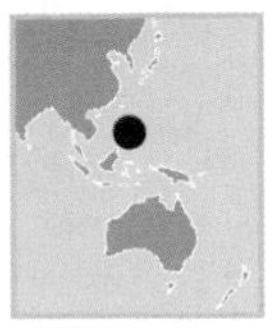

The Chocolate Hills, Bohol island, Philippines (9°35' N, 123°50' E).

In the central eastern part of the island of Bohol, the ground rises in curious rounded shapes, about 165 feet (50 m) high, covering an area of some 18.5 square miles (50 km^2). In the dry season, the tall grasses that cover them take on a brown color—hence the name Chocolate Hills. This unique landscape is the result of karstic erosion in a tropical setting. In limestone areas, many different kinds of relief are produced by the action of largely underground water, which dissolves calcium carbonate. The island has built tourist facilities with these astonishing hills as the central attraction. Foreign visitors, still few in number, for the most part treat the local natural resources with respect.

19

JUNE

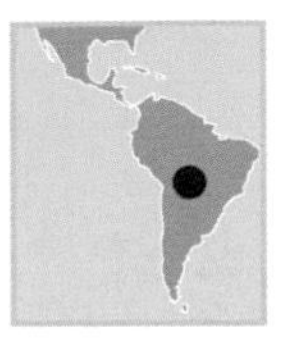

Chipaya settlement, Salar de Uyuni, Altiplano, Bolivia (19°26' S, 68°09' W). On the high plateaus of Bolivia, as elsewhere in the Andes cordillera, is the succession of vast salt plains that give the mountain range its other name, "Cordillera of Salt." Here, at an altitude of more than 13,000 feet (4,000 m), the world's biggest salt desert, the Salar de Uyuni, stretches over 4,416 square miles (12,000 km^2). On its edge, the Chipaya Indians have built their circular dwellings out of slabs of salt. The indians mine the precious rock salt, which is taken by truck to the market. Bolivia is the poorest country in Andean America: 70 percent of its people live below the poverty line. As part of the eight "Millennium Development Goals" set by the United Nations in 2000, Bolivia and 188 other signatory states have committed themselves to halving the number of people living on less than $1 per day by the year 2015. Today, these people number 1.2 billion worldwide.

20

JUNE

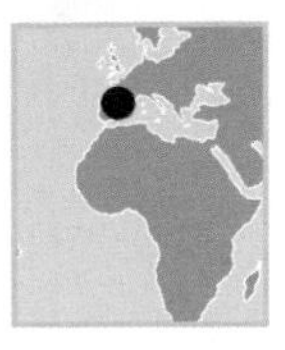

Unfinished nuclear power station, Armintza, Basque country, Spain (43°25' N, 2°54' W).

In 1984, Spain put a moratorium on the building of new nuclear power stations and on completing those under construction. Renewed in 1992, the moratorium left five plants, such as that of Armintza, in an unfinished state. The country has nine nuclear power stations, which meet 30 percent of its electricity needs. European countries remain divided over the nuclear question. France, which with fifty-nine has the most nuclear reactors, has committed itself to this energy source, which provides 72 percent of its electricity. Austria, Denmark, and Italy, on the other hand, have virtually abandoned it. In the United States, nuclear power is the primary source for 15 percent of electricity generated. Although nuclear energy allows emissions of greenhouse gas to be reduced while renewable energy sources are developed, it comes with a variety of other risks, such as those related to radioactive waste, nuclear accidents, and terrorist attack. Whatever policy is adopted, the uranium that powers nuclear reactors is not renewable. If we want to significantly reduce greenhouse gas emissions—which could triple by 2100 and quadruple by 2150 if nothing is done—we must reduce our energy consumption.

21

JUNE

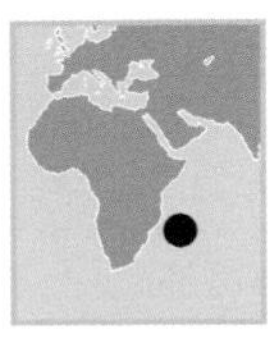

Fishermen in a lagoon of Sainte Marie Island, north of Toamasina, Madagascar (16°50' S, 49°55' E).

On the east coast of Madagascar, the island of Sainte Marie is a tropical paradise whose fishing villages pursue a centuries-old way of life. Fishing methods are traditional, using simple equipment and canoes made from hollowed-out tree trunks. Madagascar's marine resources are among the richest and most diverse in the western Indian Ocean. Fishing, 80 percent of which is for local consumption, is an essential source of food and employment for Malagasies. However, the destructive methods used by industrial fishing are devastating marine habitats—coral reefs, mangroves, marshes, and seaweed—and leading to an alarming drop in fish stocks. It has been estimated that 10 percent of Madagascar's coral reefs have already been wiped out. Unless methods change, 60 percent will have disappeared by 2025.

22

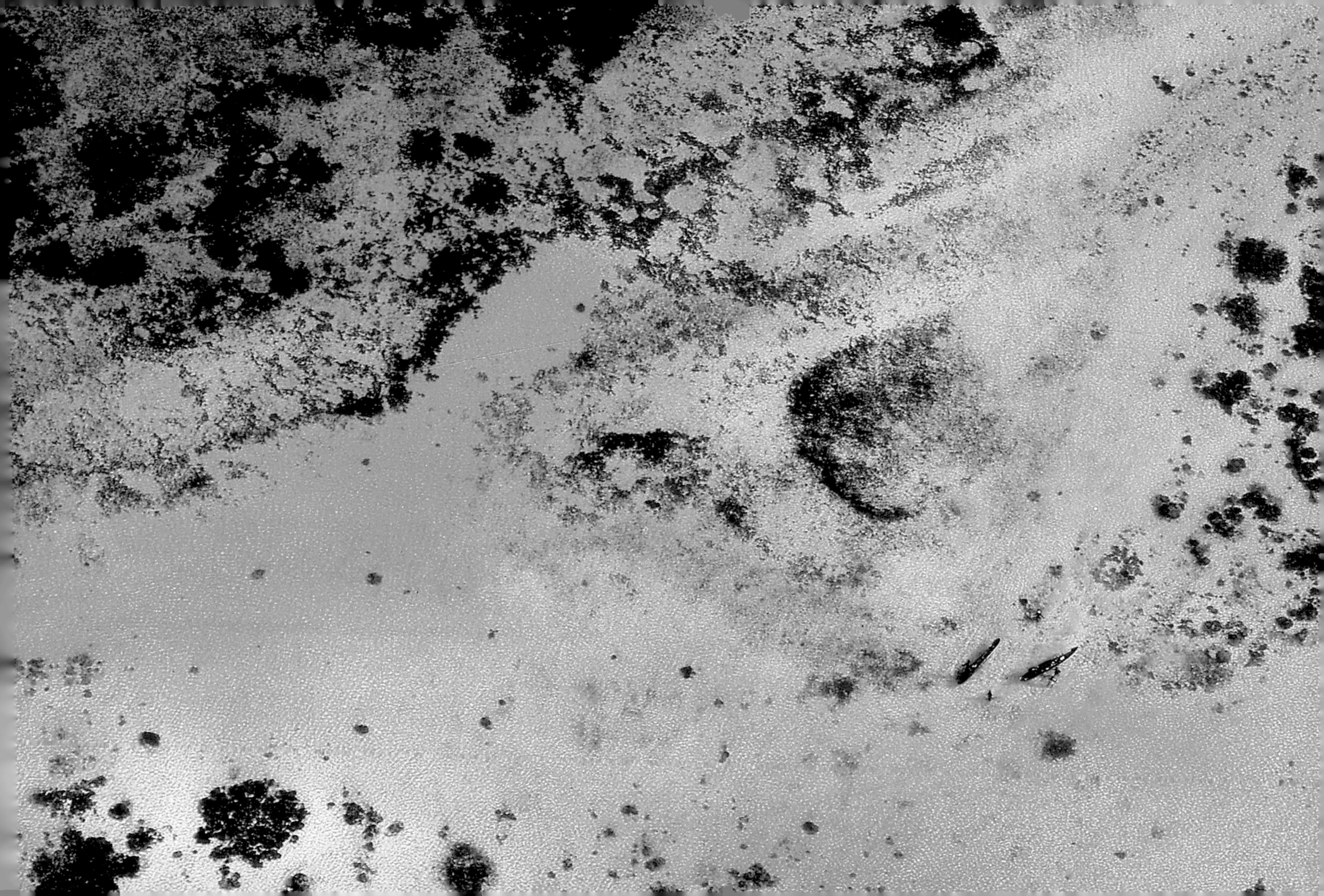

JUNE

Glacier flowing into the San Rafael lagoon, Chile (46°38' S, 73°60' W).

In the soft twilight, the ice takes on a bluish tinge. More than 30,000 years old, it cascades into the San Rafael lagoon, filling this marine lake that is linked by icebergs to the Pacific Ocean. This is the only glacier at this distance from the poles that reaches the sea. It flows slowly, with much cracking and creaking, from the mountains of Campo de Hielo Norte, where ice fields covering 1,620 square miles (4,200 km^2) are fed by the region's abundant rains (138 inches, or 350 cm, per year—six times that of London, England, and ninety-five times that of Riyadh, Saudi Arabia). Such an expanse is a considerable freshwater resource, but it is inaccessible: trapped as ice, it is only drip-fed into the surrounding rivers and lakes. Glaciers and permanent snows contain 70 percent of the world's fresh water. Most of the remaining 30 percent is polluted or inaccessible, which is why, although water might seem to be plentiful, a third of the world's population has only limited access to it.

23

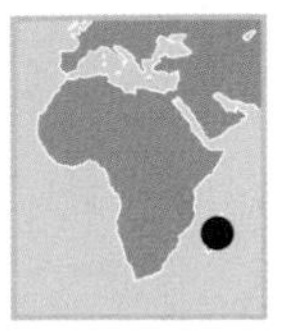

Working in the rice paddies on the shores of Lake Itasy, Antananarivo region, Madagascar (18°55' S, 47°31' E).

Over the last two centuries, the region around Lake Itasy has been given over to rice cultivation, which is controlled by big landowners. The transition from mixed farming to irrigated monoculture has caused malaria to spread in Madagascar's high plateaus. Rice's growing season coincides with that of the reproduction of *Anopheles funestus*, a species of mosquito that is an efficient carrier of the disease. Malaria kills at least 1 million people every year, most of them in poor countries. Since the 1950s, the World Health Organization (WHO) has endeavored to eradicate the disease, but it has not been able to secure enough funds for research and treatment. To try to correct this imbalance, the WHO created a world health fund in 2001. In 1992, more than 90 percent of world spending on medical research was devoted to just 10 percent of the illnesses that affect the planet.

JUNE

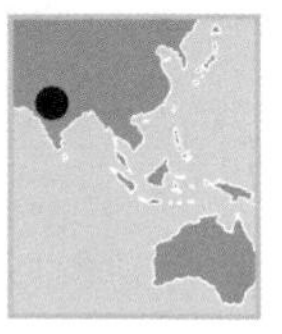

Fortified city of Jaisalmer, Rajasthan state, India (26°55' N, 70°54' E).

Rajputana (now Rajasthan) in northwest India, the "land of the sons of kings," once encompassed about twenty princely states. Founded in 1156 by Rao Jaisal, a ruling rajput who established the Bhatti clan, the fortress of Jaisalmer was strategically placed on the route used by the spice caravans traveling from central Asia to India. The citadel's crenellated ramparts recall the attacks and endless sieges of the conflicts between the Bhattis and the Muslims of the sultanate of Delhi, when the rajputs and their wives demonstrated their bravery by preferring *jauhar* (collective self-immolation by fire) to surrender. In the sixteenth century, the *maharawals* of Jaisalmer resisted the Mogul offensives and then accepted imperial rule. The city prospered, and the gilded sandstone facades of the sumptuous *haveli*, merchants' houses in the lower town, were decorated with golden lattices, balconies with small, finely carved columns, and delicate tracery in stone. The development of sea trade routes in the nineteenth century led to the decline of Jaisalmer, which now is but a timeless mirage on the edge of the Thar desert.

25

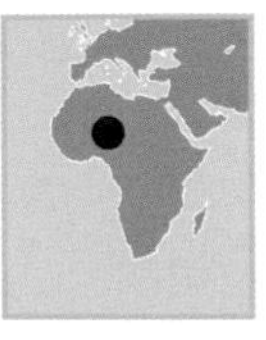

Pirogue on the Niger river in the Gao region, Mali (16°12' N, 0°01' W).

The Niger river, which has its source in the massif of Fouta Djallon in Guinea, is the third-longest (2,600 miles, or 4,184 km) river in Africa, after the Nile and the Zaire. Crossing through Mali for a length of 1,050 miles (1,000 km), it forms a large loop that rises to the southern border of the Sahara, watering major centers such as Timbuktu and Gao. The short rainy season stimulates the regeneration of aquatic plants, through which pass pirogues, a customary means of travel, transport, and trade among people living near the river. The Niger, subject to seasonal flooding, also irrigates nearly 2,000 square miles (5,000 km^2) of land used for rice paddies and market gardens. It is the chief source of water for approximately 80 percent of Mali's population, who live by farming and herding.

Chichen Itza castle, Yucatán province, Mexico (20°71' N, 88°53' W).

The capital of the Itzas, a northern Toltec people who invaded the Mayan empire in the tenth century, Chichen Itza is a remarkable testimony to the union of the two cultures. The castle's perfect pyramid, comprising ninety-one steps on each of its four sides and topped by a platform, represents the 365 days of the Mayan calendar. At the vernal and autumnal equinoxes, the interplay of light and shade forms the figure of a snake—an important element in Toltec art—on the pyramid's north face. Although today the majority of Mexico's inhabitants are of mixed race, the Indian population, descended partly from the Mayas, makes up a tenth of the total and is the largest in the Americas. It is also one of the poorest. Average earnings are as little as a quarter of the national average. Mortality is 40 percent higher than among people who live in the capital, and a third of Indian children do not go to school. Since 1994, groups in the province of Chiapas, in the south of the Yucatán peninsula, have been demanding improvements in Indians' living conditions, as well as recognition of their distinct culture within the country.

JUNE

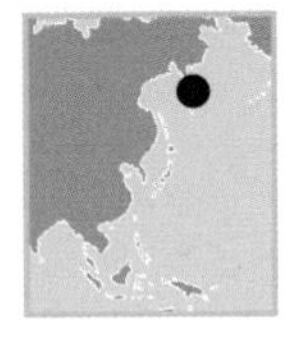

Karymsky volcano erupting, Kamchatka, Russia (54°05' N, 159°43' E).

A mountainous peninsula of volcanic origin at Siberia's far eastern end, Kamchatka is a place apart in the Russian Federation. It is remote from the capital—by more than 3,700 miles (6,000 km)—and the Russian authorities have done little to encourage its development since the breakup of the Soviet Union. Yet Kamchatka plays a part in Russian economic life, thanks to its forest and agricultural resources, the development of its coastal towns, and its fisheries. The population is concentrated in the towns and consists largely of Russians, who mingle with older residents such as the Kamchadales. These nomads, also known as Itelmens, have retained their traditional way of life and live chiefly from fishing. Only about 18,000 are thought to remain of a people who were once the most numerous on the peninsula.

28

JUNE

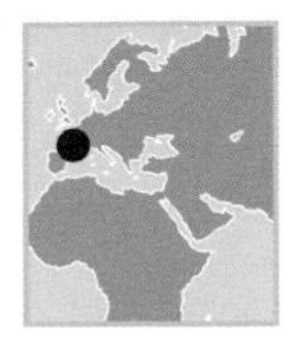

AZF factory chimney after an explosion, Toulouse, France (43°36' N, 1°27' E). On September 21, 2001, a deadly blaze raged at the Grande Paroisse AZF factory, a subsidiary of the Total-Fina-Elf group, in a southern district of Toulouse. It killed thirty people and wounded more than 3,000. The effects of this nitrate explosion were felt all over the city, to a distance of several miles (or kilometers), and ten times farther than the so-called security zones. The disaster reignited the debate in the risks attached to industrial installations in urban areas. The precautionary principle clearly had an application in this instance. Both government and industry now give priority to reducing risks at their source, rethinking urban planning and land use, and also keeping citizens informed. On March 7, 2003, the French Chamber of Deputies adopted a draft law on technological and natural risks. One of its main points was the adoption of plans to prevent technological risk by limiting urban development near France's 672 high-risk establishments.

29

AZ

30

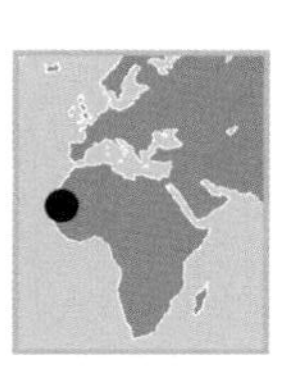

Dromedary caravan in the dunes, near Nouakchott, Mauritania (18°09' N, 15°29' W).

The Sahara, the world's largest sand desert, covers 3.5 million square miles (9 million km^2)—equivalent to the area of the United States—spread over eleven countries. Mauritania, which lies on its western border, is three-quarters desert and is thus particularly vulnerable to the phenomenon of desertification. Excessive grazing, harvesting of firewood, and agricultural expansion are gradually destroying soil-retaining vegetation on the perimeters of the great dune ranges. This facilitates the advance of sand, which today endangers cities, including the capital, Nouakchott. In arid and semiarid zones (which make up two-thirds of the continent of Africa), fragile arable lands deteriorate rapidly if farming and other exploitation become too intensive. In the past half-century, 65 percent of arable lands in Africa have suffered degradation, resulting in a drop in agricultural yield. In this vicious cycle, so difficult to break, poverty is both a cause and a consequence of the depletion of arable soil and the decline in agricultural productivity.

DIVERSE AGRICULTURE: THE STAKES OF WELL-CONTROLLED GLOBALIZATION

JULY

01	02	03	04	05

Seen from above, the planet's agriculture is as varied as its landscapes. We can see how the works of human civilizations have left their mark on nature. We can discern the limitations and artifices of peasants (often true gardeners) and herders, and guess at the history of their settlements and migrations. From the peasants who cling to the precipitous slopes of Cape Verde's volcanoes, to the rice paddies carved into Indonesian hillsides, to the great wetland plains of Latin America's southern cone, the human footprint on ecosystems is astonishingly varied. Seen from ground level, however, from the point of view of the global economy, agriculture is ever more integrated. International trade in its products is growing; homogenized markets are obliterating diversity; and technology is reducing the range of farming practices.

Integration is not the only threat to the variety of farming systems that human beings have adapted to differing ecosystems. We face two additional challenges in this century: the pressure of a growing population, and increased demand for food.

Landscapes already show the signs of pressure from population growth: deforestation, for instance, has reduced the world's forest cover by 20 percent since the dawn of agriculture. This process has speeded up dramatically in the last 20 years. Every year, 25 million acres (10 million ha) of mostly tropical forest are erased from the map, chiefly because of demand for more farmland.

As the demand for food grows, it brings a growing demand for water for agriculture. Of all the water gathered, 70 percent is destined for use in farming. Since the green revolution, increases in production have depended far more on developments in irrigation than on technological progress. Soil is being contaminated with salt. Land is abandoned when ground water supplies are exhausted; rivers and soil are polluted through a combination of irrigation and farm chemicals. The Aral Sea is drying up, and soil is being poisoned. All these demonstrate the ecological disaster that economic pressure can bring upon farming systems.

The globalization of food, too, is sweeping away these farming systems. The seas of plastic sheeting that allow fruit and vegetables to be grown year round have transformed the landscape of whole regions. The spread of certain species and varieties, and their adaptation to large-scale agriculture to meet growing worldwide demand, have sometimes triggered unforeseen and violent changes in ecosystems.

An example would be the so-called "invasive" species, which displace local species, often irreversibly. Globalization of technology leads to the spread of plant varieties. These are poorly diversified and have a narrow genetic base: while whole regions may have barely a few dozen varieties of wheat or rice, a Mexican peasant's plot may still contain up to ten varieties of maize.

The workings of international markets also have a part to play in maintaining the diversity of farming practices. Produce prices over the long term have been falling by rates ranging from 1 to 1.5 percent per year, depending on the sector. This is due to increased productivity and technological advances. But the distribution of this progress is highly unequal among different markets. While an American or European farmer might produce 500 tons of grain, an Andean or Sudanese farmer might produce only 1 ton. While due partly to natural conditions, this difference is chiefly the result of how much capital is invested in machinery and the resources consumed: energy, fertilizers, water, and pesticides. Small, labor-intensive farms have to compete with those that are larger and capital intensive. This does not mean, however, that modern agriculture is always the most efficient: European and American farmers produce high yields but are generously subsidized by their governments. Grants for farmers, the unequal nature of competition, and especially subsidized exports by developed countries are among the factors that damage peasant farmers. Between them, developed countries spend $365 billion per year on agriculture—or more than the annual income of the roughly 900 million people living below the poverty line in rural areas. Without these distortions in the market, and if they were provided with a minimum of infrastructure, such as roads, access to loans, and technology, peasants could produce far more for the world's markets, and thus earn more.

16 17 18 19 20 21 22 23 24 25

Agriculture is vital to developing countries' economies. In rich countries, it accounts for a minimal percentage of the gross domestic product—2 percent in the United States, and 4 percent in Europe—in the poorest countries agriculture is responsible for up to 35 percent. In Africa, it provides 70 percent of all jobs. Protecting the diversity of peasant agriculture is thus one of the best ways of combating rural poverty and feeding the hungry, and this protection can be achieved by modifying the way international markets work. Fair trade practices are one way of recognizing the unique contribution of peasant farmers to the common good and of guaranteeing more economic fairness.

Preserving agricultural diversity can also help protect the planet's environment. If peasant farmers are not reduced to overexploiting their resources of water or soil in an attempt to survive, they can have the freedom and ability, on the contrary, to protect biodiversity through their farming. Properly controlled, globalization can protect this heritage of cultures, knowledge, and resources; poorly controlled, it will obliterate it.

LAURENCE TUBIANA
Director, IDDRI (Institute of Sustainable Development and International Relations, of the French Ministry of Ecology and Sustainable Development)

01

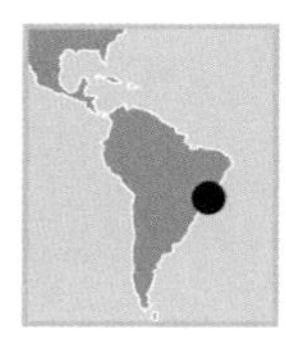

Favelas in Rio de Janeiro, Brazil (22°55' S, 43°15' W).

Nearly one-fourth of the 10 million cariocas—residents of Rio de Janeiro—live in the city's 500 shantytowns, known as *favelas*, which have grown rapidly since the turn of the twentieth century and are wracked by crime. Primarily perched on hillsides, these poor, under-equipped neighborhoods regularly experience fatal landslides during heavy rains. Downhill from the favelas, the comfortable middle classes of the city (18 percent of cariocas) occupy the residential districts along the oceanfront. This social contrast marks all of Brazil, where 10 percent of the population controls the majority of the wealth while nearly half of the country lives below the poverty level. As a result of urban growth, approximately 25 million people in Brazil, and 600 million in the world, inhabit the slums of great metropolitan areas, where over-population and poor conditions threaten their health and their lives.

JULY

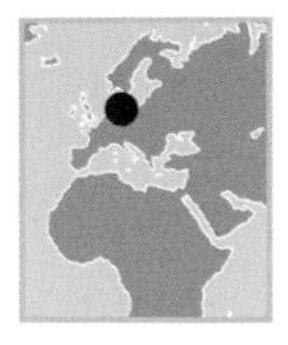

Floating house in the Christianshavn (Christiana) district, Denmark (55°40' N, 12°35' E).

Deep in Copenhagen's canal network, the old barracks of Christianshavn are unique in Europe. In 1971, hippies occupied its disused buildings and declared the district a "free city." Now, after years of clashes, the authorities regard Christiana as a "social experiment." Its 2,000 inhabitants have their own architectural regulations. They allow cannabis but firmly banish hard drugs and live in a wild environment where the bicycle reigns. The district, which was built on pile foundations at the beginning of the seventeenth century, links the canals of the port of Copenhagen to the island of Amager. Over the last 30 years, it has acquired houses and buildings of original design that bear witness to the creativity of its inhabitants. This floating house uses natural materials, such as wood and metal, and takes advantage of the sun's energy, thanks to large south-facing bay windows.

02

JULY

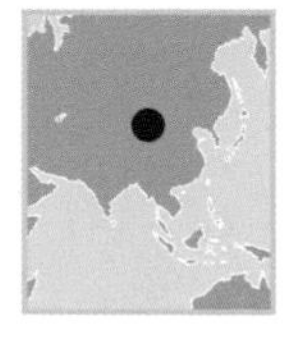

Gurvan Saïkhan Gobi, Mongolia (43°50' N, 103°30' E).

The *gobi*s of Mongolia—great stony basins where water is scarce and the steppe becomes meager, salty, and arid—cover a third of Mongolia. They are expanding as a result of a long series of droughts in this part of the world. Continental regions such as central Asia are particularly vulnerable to the enhanced greenhouse effect and the increase of 33.8 to 42.8 °F (1 to 6 °C) in average temperatures expected during this century. Although carbon dioxide, the main greenhouse gas, is mostly emitted in North America, Europe, and Russia, the speed with which currents move around the atmosphere means it is very evenly distributed throughout the world—a reminder that pollution knows no boundaries. International treaties on climate change commit the industrialized countries that have signed them to cutting average carbon dioxide emissions by 5 percent from 1990 levels.

03

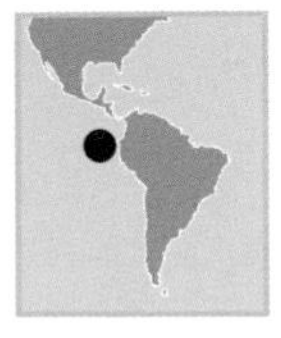

Tortuga Island, Galápagos archipelago, Ecuador (1°00' S, 90°52' W).

The slender shape of Tortuga Island rises from the water like the back of a gigantic sea monster. This young crater, born within the last 1 million years, bears witness to the volcanic activity on the Galápagos archipelago. Even today, the 128 islands and islets, whose biodiversity fascinated Darwin, are being reshaped by eruptions and earthquakes, particularly in their western part. The sea around this bare land contains an astonishingly rich variety of life because it lies at the crossroads where the warm waters of the Gulf of Panama meet the cold Humboldt Current. Sea lions and sharks from warm seas, penguins, and marine iguanas can all be found here, protected by a marine biological reserve that covers 30,880 square miles (80,000 km^2). Since the end of the 1990s, despite the permanent presence of guards, marine resources on the islands have been plundered by fishermen, some of whom come from Southeast Asia to gather sharks' fins by the thousands.

04

JULY

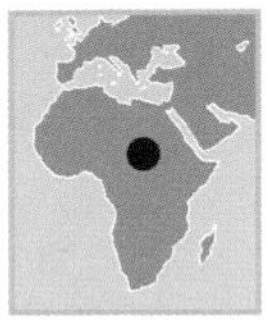

Laundry drying by the Chari river near N'Djamena, Chad (12°07' N, 15°03' E). Carpets, hangings, and other brightly colored materials brighten the many sandbanks of the Chari near Chad's capital, N'Djamena. The Chari river is the main tributary to Lake Chad, whose surface has literally melted away in the last 30 years, shrinking from 9,650 to 965 square miles (25,000 to 2,500 km^2). The river's waters, which are used for laundry, personal hygiene, and kitchen use, may well suffer from these competing demands, and their quality could deteriorate. Sources of fresh water are rare in this part of the Sahel, which has suffered repeated droughts; only 27 percent of Chad's population has access to drinking water. This is the third most critical situation in the world, after Afghanistan and Ethiopia. Ground water supplies are also threatened by the planned Chad–Cameroon oil pipeline, which is supported by the World Bank. Any oil leaks could contaminate rivers and wells. In 1990, thirteen African countries suffered water supply problems or shortages. By 2025, this number could double.

05

JULY

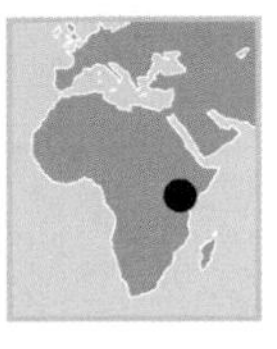

Small agricultural plots near Kisii, Kenya (0°40' S, 34°46' E).

The hills of the town of Kisii, in southwestern Kenya, are exceptionally fertile. The Gusii, Bantu farmers who cultivate this land, quickly became relatively prosperous from tea, coffee, and pyrethrum, a natural insecticide extracted from geraniums. Unlike the big farming companies in the north, the Gusii and their neighbors are small operators. As a result of their high birth rate and intensive methods, their land has been divided up into smaller and smaller plots. Today, Kenya is the world's third-biggest tea producer, after India and Sri Lanka. Tensions have surfaced since August 1997 between the Gusii and Masai peoples; some Gusii, whose livelihood depends on the fields they rent from the Masai, have been barred from them. Africa is regularly torn by interethnic strife: the conflict that erupted in Rwanda in 1994 is thought to have killed almost 500,000 civilians.

06

JULY

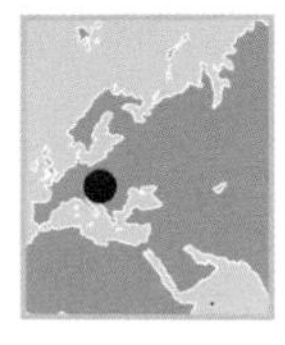

Episcopal cathedral of Székesfehérvár surrounded by modern buildings, Hungary (47°12' N, 18°25' E).

Between Budapest and Lake Balaton, Székesfehérvár and its contrasting architecture bear witness to the greatest eras of Hungarian history. A thousand years old, and capital of the Hungarian kingdom for five centuries, the city became a bishop's see in 1777 and was enriched by the ecclesiastical buildings in the purest Hungarian Baroque style. The Episcopal cathedral is one of the few buildings of the period that survive. Modern apartment blocks, built when Hungary was part of the Soviet bloc, stand where once there were ancient city walls. Since the fall of Communism, Hungary, like the rest of central Europe, has witnessed a strong resurgence of Christian faith, and people can now visit churches freely. The Christian church suffered 50 years of religious persecution in central Europe. During World War II, 3,000 priests were interned in the concentration camp at Dachau. Under Communism, the faith was not officially forbidden in all countries but nevertheless went underground. Some monks and nuns were imprisoned and even assassinated.

07

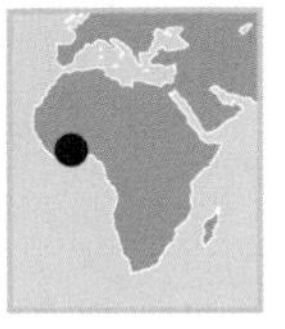

Farmers near the Buyo dam, Republic of Côte d'Ivoire (5°47' N, 7°05' W).

The village of Buyo has been transformed since a hydroelectric dam was built there in 1980. Due to irrigation of the surrounding land, it has become part of Côte d'Ivoire's "coffee belt," attracting a million farmers. The newcomers have lost no time in cutting down patches of forest so that they could build their houses and plant their coffee. Now they are suffering from poor quality drinking water because, to increase yields, they spray their crops with pesticides such as DDT, lindane, aldrine, and heptachloride, so toxic that many countries have forbidden or limited their use. These pollutants, which are associated with some cancers and development abnormalities, are concentrated in the lake at Buyo. Moreover, since the area lacks sanitation plants, waterborne diseases such as malaria and dysentery are increasing. Poverty makes it all the harder for these people to obtain treatment: the country has suffered from falling coffee prices since 1996, and a civil war has been raging since September 2002.

Fisherman, Tunisia (34°15' N, 11°00' E).

Towed along in his little boat by his motorized colleagues, this man is one of the 6,200 small-scale fishermen in the Gulf of Gabès. All kinds of fishing vessels can be found on the Mediterranean. Depending on their size and specialization, they use anything from traps to trawl nets, and sometimes even drift nets. The last are still used in North Africa, Turkey, and Albania but have been banned in the European Union since January 2002; their sharp meshes wound the dolphins that get accidentally caught up in them. Some countries go further in their efforts to protect these marine mammals, suggesting a ban on plastic grocery bags, which dolphins mistake for jellyfish and swallow, with fatal results. Ireland has already reduced their use by 90 percent since March 2002, thanks to a fifteen-cent tax on each bag handed out in shops. Taiwan, Spain, and Corsica have followed Ireland's example. Consumers are in favor, too, since plastic bags produce air pollution by releasing toxic chemicals when incinerated with garbage.

JULY

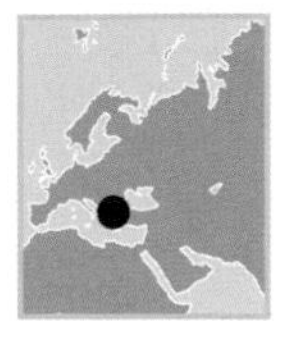

Acropolis, Athens, Greece (37°58' N, 23°43' E).

On a hilltop of little more than 7 acres (3 ha), the genius of Greek civilization built one of the most remarkable collections of buildings in history, which today is the official emblem of UNESCO. Four masterpieces of classical Greek art—the Parthenon, the Propylaea, the Erechtheum, and the temple of Athena Nike—affirm, by their great size and magnificence, the power of the Athenian democracy under Pericles. Now restored under a European Union program, the building vividly illustrates the cultural influence of Greek civilization, the cradle of the European identity. Zeus was the first to seduce Europa, and he sired her children. More than ever before, Europe is an idea, a shared heritage of values—perhaps even a myth—more than a geographical entity. On April 16, 2003, a highly symbolic event bore witness to this. During the Greek presidency of the European Union, a treaty welcoming ten new member states was signed. The ceremony was held at the foot of the Acropolis.

10

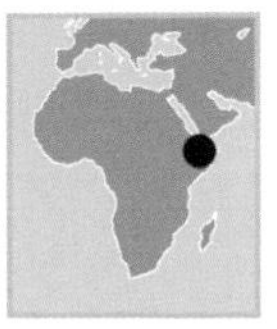

Volcanic landscape between Lake Ghoubbet El Kharâb and Lake Assal, Djibouti (11°41' N, 42°20' E).

The region between Lake Assal and Lake Ghoubbet is an extraordinary one where, little by little, an Eritrean ocean is in the process of being born. Here, at the southeastern end of the Assal rift, the Earth's crust has been torn apart, gradually detaching Africa from the Arabian peninsula. The crust has stretched, become thinner, and broken up into many faults. As a result, it has also sunk—Lake Assal lies 515 feet (157 m) below sea level. Lava has flowed out from the depths of the Earth and, on reaching the surface, has cooled, forming the volcanic cones in these black "fields" which will, one day, lie beneath an ocean of the future. Here, the Earth has grown new skin and is forming a new crust. However, since the planet's surface as a whole cannot be stretched, the old surfaces elsewhere must disappear, thousands of miles of them sinking beneath the continents to rejoin the deeper mantle of the Earth from which they came.

11

JULY

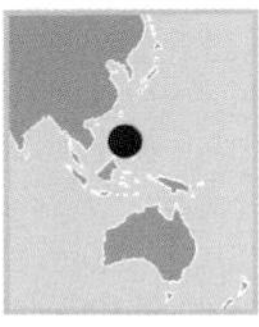

Houses in the marshes near Malolos, Luzon island, Philippines (16°00' N, 121°00' E).

On Luzon, the biggest and northernmost of the islands of the Philippines, resources and poverty sit side by side. The central plains of the Bucalan region are a veritable rice bowl, yet 40 percent of Filipinos live on less than $1 a day. Those who live in rural areas often set up house in the heart of the damp rice paddies. Their houses are mostly built out of bamboo with roofs of curved tiles. Despite an ambitious plan aimed at combating poverty, in rural areas this has increased in recent years. Conditions are made worse by the fact that people living in these fragile shacks are in danger from the frequent floods. An average of thirty storms hits the Philippines every year.

12

JULY

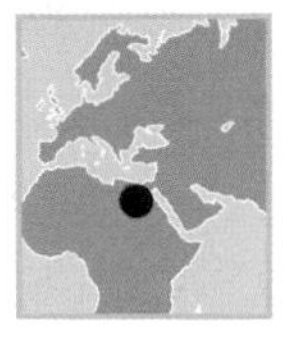

Adrere Amellal "Ecolodge" hotel, Siwa oasis, Egypt (29°12' N, 25°31' E). Barely distinguishable from the surrounding landscape of salt lakes at the oasis of Siwa at the foot of the "white mountain," the Ecolodge is a luxury hotel—even though guests have neither electricity nor air conditioning—in the arid Egyptian desert. To cope with the extreme environment, the building—which was completed in 1997—takes advantage of local techniques that have been developed over 2,500 years: walls of salt-bearing rock extracted from the beds of lakes burned by the sun, and roof, insulation, ventilation, plaster, furniture, and accessories designed and made by 150 local craftsmen using the oasis' own resources. Economizing in the transport of materials and people minimized energy consumption, while the use of local human resources revived the region's economy. Since the Ecolodge was built, almost 600 people make a living from the revival of interest in the architecture of Siwa.

13

JULY

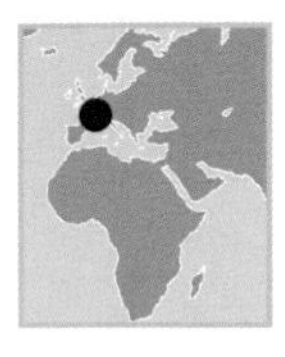

The Puy de Dôme, Auvergne volcano range, Puy-de-Dôme, France (45°47' N, 2°57' E).

In the heart of the Massif Central, the eighty or so extinct volcanoes of the Puys range may not look recent, but they were formed barely 15,000 years ago. Seeming to rise in order to look across the Rhône valley to the Alps on the other side, they rise above a series of plateaus, rich in hot springs, which overlook the plain of the Rhône. This landscape did not happen by chance. It has its origins in the depths of the Earth's crust, which is greatly thickened beneath the Alps and drags the rocks of the surrounding regions downward. Faults are formed, and blocks subside, gradually giving shape to plateaus, hillsides, and plains where the Earth's crust has become thinner and heat is very close to the surface. This is the reason for the series of plateaus that stretch from northern Italy to Slovakia. The Alps are therefore surrounded by areas with high potential for geothermal energy, and every year many studies seek to discover how this resource can be tapped.

14

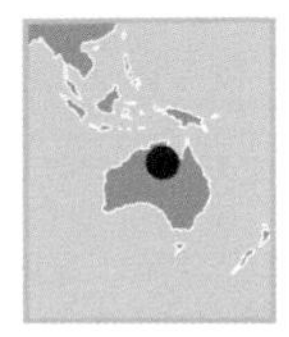

Uranium mine in Kakadu National Park, Northern Territory, Australia (12°41' S, 132°53' E).

Kakadu National Park is a rich source of uranium, making up 10 percent of the world's resources. It is divided into three plots in aboriginal territory, Ranger, Jabiluka, and Koongarra, which are enclosed in the protected park (a UNESCO World Heritage Site since 1981) but are statutorily excluded. The plan to open a mine on Jabiluka has caused controversy concerning the pollution risks, and the Mirrar aborigines, traditional proprietors of these sacred lands, are stringently opposed, mobilizing international public opinion. Only Ranger is authorized for mining. In this waste zone, large sprinklers water the marsh banks to increase evaporation and reduce the risk of dust build-up, leaving sulfate deposits. Australia also has two other large deposits of uranium, and in 2000 the country produced more than 20 percent of the uranium extracted in the world (34,400 tons). Uranium provides fuel for the world's nuclear sites, divided primarily between the United States, France, and Japan.

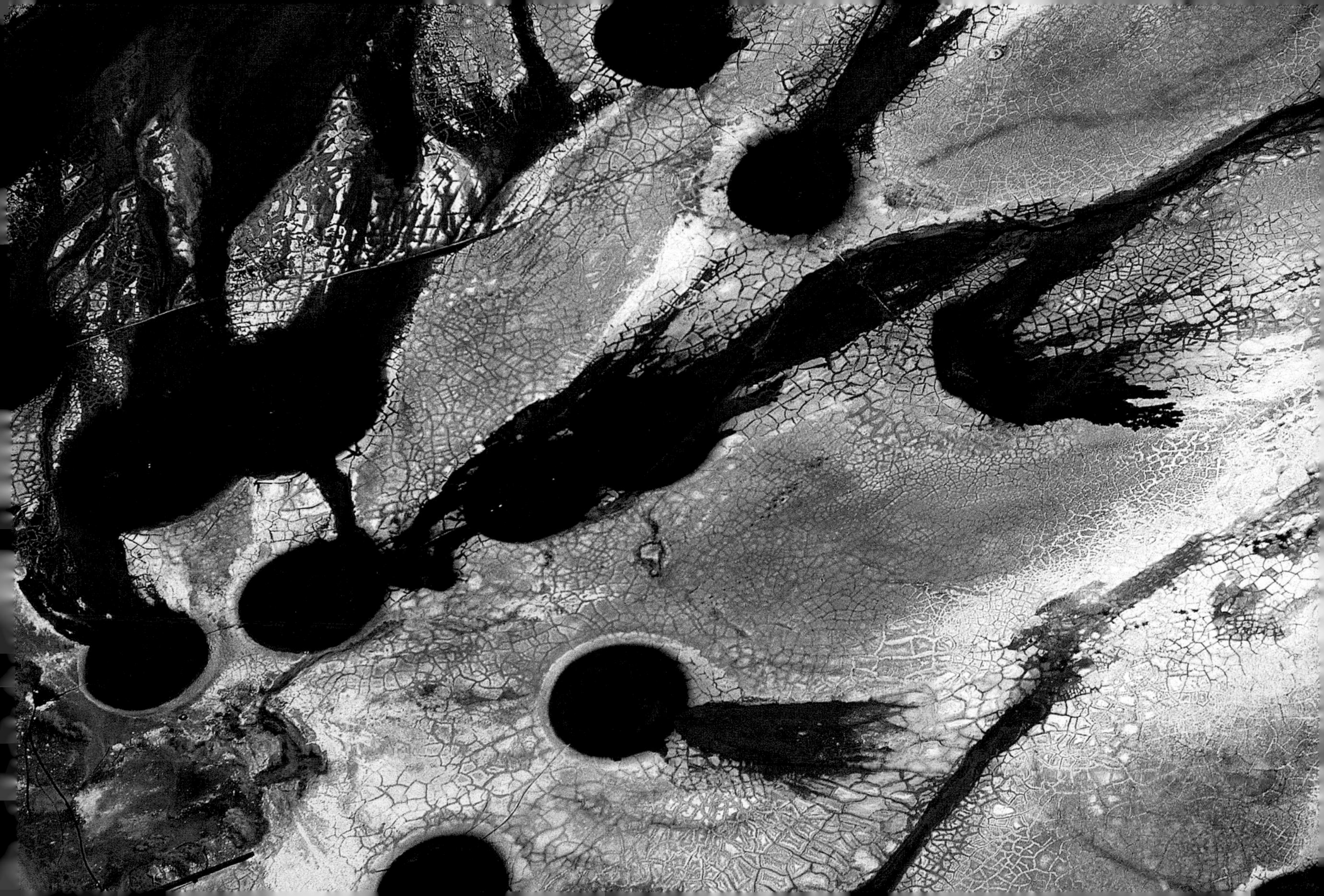

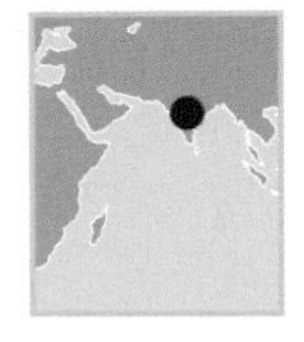

Varanasi, the Ghats: ritual bathing in the Ganges, Uttar Pradesh, India (25°20' N, 83°00' E).

The word *"ghats"* refers both to the vast plateaus that stretch from the Himalayas to the river Ganges and to the steps on the banks of the river itself. The *ghats* of the holy city of Varanasi are visited by Hindu pilgrims for purification, worship, or cremation of their dead. Leading a virtuous life and carrying out their *dharma* (duty) increases their chances of being reincarnated in a higher caste. India's 2,000-year-old caste system uses the accident of birth to dictate the place an individual occupies in society. Almost 170 million Indians—or one in six—are "untouchables," or *dalits,* who are excluded from the four main castes: Brahmins (priests and teachers); Ksatriyas (warriors and rulers); Vaisyas (farmers, merchants, artisans); and Sudras (servants and laborers). Although India's constitution forbids discrimination based on caste, these *dalits* have no access to land, live in separate districts, and are forced to accept the most menial jobs and violations of their basic rights. In 2001, UNESCO estimated that two-thirds of *dalits* were illiterate, and that only 7 percent had access to clean drinking water, electricity, and sanitation.

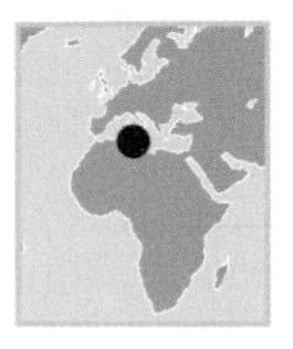

Fishing by traditional methods off the Gulf of Gabès, Tunisia (34°40' N, 11°10' E).

An arrow points the way to where the fishermen lie. The fish follow it, little suspecting that this path, marked by palm leaves, leads them straight to one of the cage traps known as *charfia*. The concessions determining where such devices are placed have been handed down from generation to generation since the seventeenth century; this has the effect of limiting the number of fishermen and the size of their catch. The installations also keep commercial fishing fleets at bay, thus inadvertently acting as nature preserves. The Tunisian government gives the fishermen legal protection and financial aid. Although 20 years ago Tunisia, like other countries, sought to boost fishing, today it promotes sustainable and responsible fishing. In an attempt to protect its fish stocks, and therefore its fishermen, the country follows the code drawn up by the United Nations' Food and Agriculture Organization (FAO). Worldwide, 36 million people who live by their catch are threatened by overfishing.

17

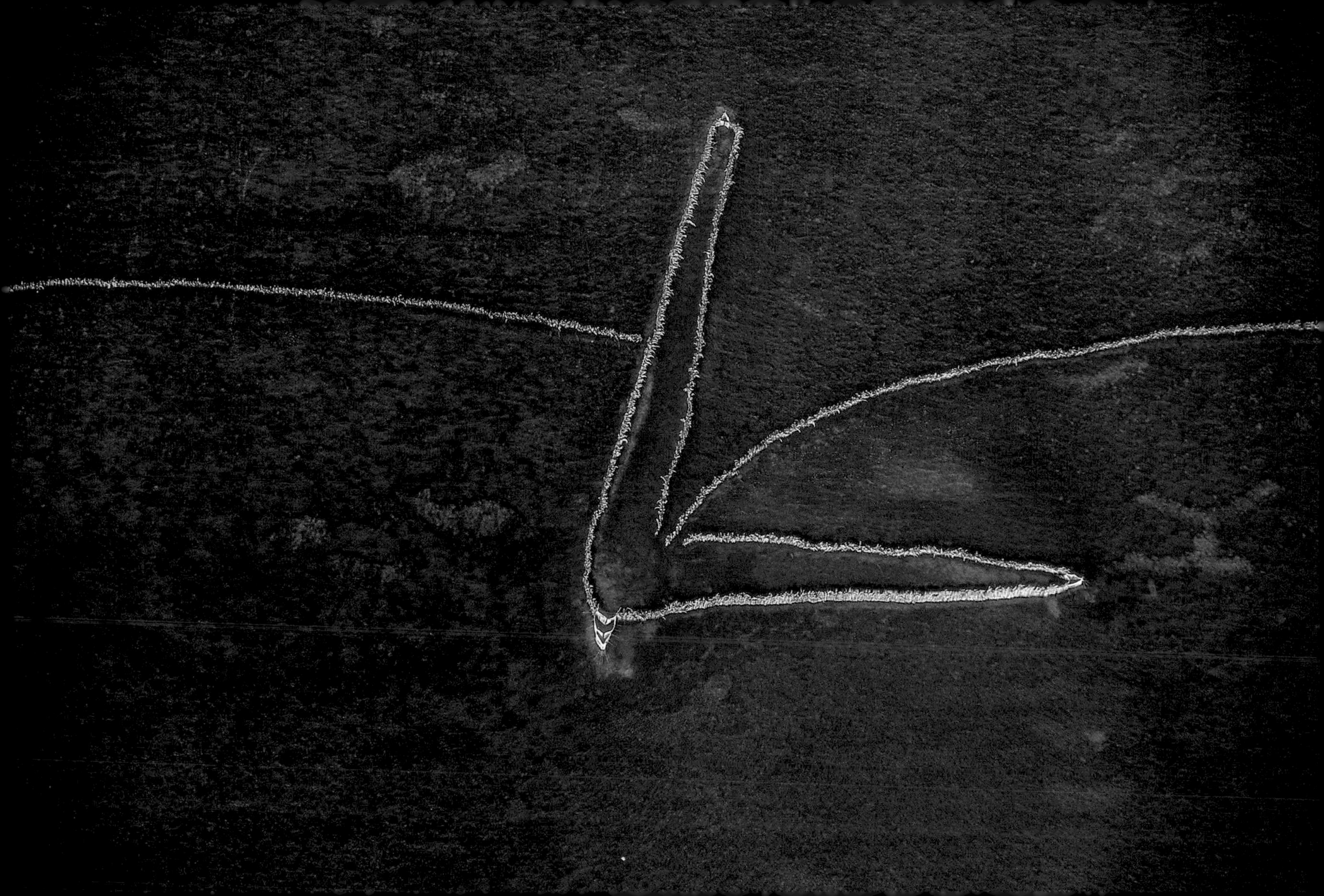

JULY

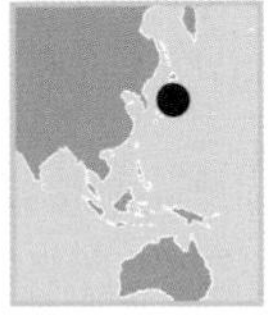

Freeway interchange near the port of Yokohama, Honshu, Japan (35°42' N, 139°46' E).
Since it was linked to Tokyo by a railway line in 1872, the small fishing port of Yokohama has been growing continuously; today it is Japan's main international port and second-biggest city after the capital. The freeways that encircle it symbolize a type of economic development largely built around road transport, as in many industrialized countries. This dominant model has led to an increase in freeways all over the world. The number of vehicles has risen to almost 800 million, most of these in developed countries: 29 percent are in the United States alone, and just 2.4 percent in Africa. The level of ownership is also unequal: there are 790 vehicles per 1,000 inhabitants in the United States, but just eight in India. Despite the pollution and congestion in cities, the number of cars continues to grow relentlessly. Transport is the chief emitter of greenhouse gases, and the sheer number of users renders the measures for controlling it complex. Although emissions from industry have fallen since the Rio Earth Summit in 1992, emissions from transport have risen by 75 percent.

18

JULY

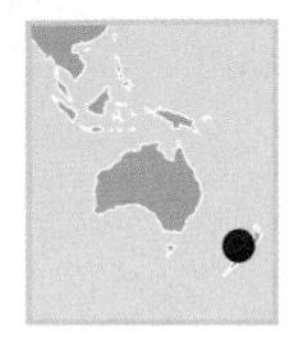

Humpback whale off Kaikoura, New Zealand (42°25' S, 173°43' E).

A majestic humpback whale moves peacefully among the dolphins in the dark waters near the peninsula of Kaikoura. This part of the east coast of New Zealand's North Island is one of the world's meccas for whale watching, being frequented by migrating humpback whales, many dolphin species, and above all by sperm whales. Watching these great cetaceans is increasingly fashionable among tourists. In 1998, this type of ecotourism attracted more than 9 million people from across the globe, and generated more than $300 million in income. While this craze may have increased public awareness of the whales' plight, it now constitutes a new threat to these gigantic creatures. Every year, some are killed in collisions with boats laden with sightseers. Formerly persecuted by hunters, two of the eleven whale species—the enormous blue whale and the Basque whale—are close to extinction.

19

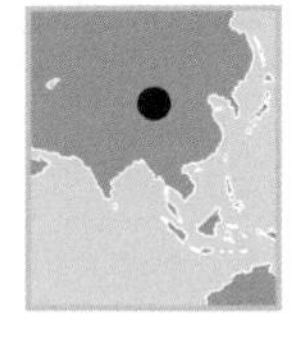

Goats amid the shadows in the Khustain Nuruu Reserve,

Mongolia (45°50' N, 106°50' E).

How many immaculate Kashmir goats still brighten the arid steppes of the Mongolian *gobi*s? One in every three animals raised in Mongolia is a goat, thanks to these animals' hardiness and their precious wool. However, they are disappearing, as are the nomads themselves. The price of wool is falling and, although Mongolia's industry has developed over recent years and tried to adapt to Western tastes, it faces competition from China. This leaves little room for the processing of the luxurious Kashmir wool within the country. With the help of government grants, the nomads are trying to increase the size of their herds, but the climate—which brings droughts and bitter winters, with temperatures plunging to –76 °F (–60 °C) in February 2000—has led to desertification of the steppe, killing millions of animals since that year. Some World Bank experts believe nomadism here is doomed, but despite their difficulties, neither Mongolia's government nor its population agree.

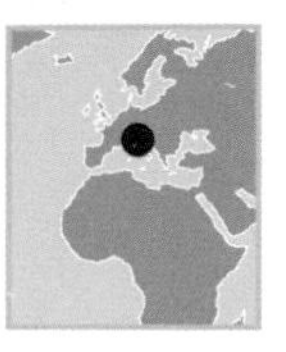

General view of Venice, Veneto, Italy (45°35' N, 12°34' E).

Venice is not an island, but an archipelago of 118 islands that are separated by 160 canals spanned by more than 400 bridges. The Grand Canal—Venice's main artery—is lined with the city's most beautiful buildings. A hundred or so Renaissance and Baroque palaces were built on the banks of the Grand Canal by rich Venetian merchants. They bear witness to the importance these traders gained when Venice opened up to the outside world. From AD 1000, the city dominated the Adriatic. It extended this dominance to the entire Mediterranean, establishing a number of trading posts, until the end of the seventeenth century, when trade by land supplanted trade by sea, eclipsing Venice on the international commercial scene. Today, the eclipse risks becoming total. "La Serenissima" could vanish under the waves, a victim of the floods that have increased as a result of canal widening, the sinking of the ground on which Venice is built, and the rise in sea level (0.24 inches, or 6 millimeters per year).

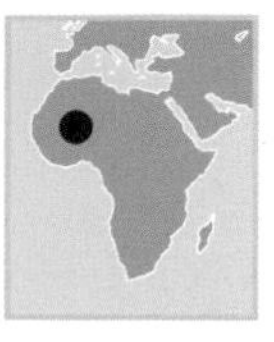

Mosque in the village of Kwa, Mopti region, Mali (14°29' N, 4°01' W).

The towns on the banks of the Niger river in Mali are celebrated for their fortresslike mosques. In 1325, Mansa Musa, celebrated ruler of Mali, made a pilgrimage to Mecca. Having converted to Islam, he became the first to order the building of imposing, conspicuous mosques—in contrast to the places of worship of traditional African religion, which were sited away from the gaze of nonbelievers. The "Sudanese" type of mosque soon became the nerve center of villages and towns in Mali, a country that is now 90 percent Muslim. The unfired bricks and fine clay rendering allowed a bold architectural style and the flexible use of shapes, which bristled with rows of posts to strengthen fragile structures. Original as they are, these mosques retain the traditional square plan of the mosque, with a courtyard and *qibla*, the eastern wall against which stands the *mirhab*, the chair that faces Mecca. The finest example of this architecture is in Djenné, not far from the port of Mopti on the Niger, which was added to UNESCO's World Heritage list in 1988.

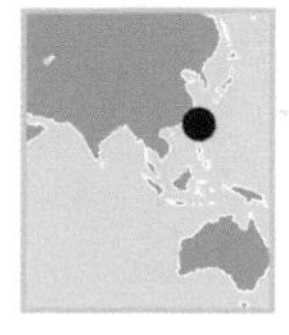

Military exercises at Makung on the island of Penghu, Taiwan (23°34' N, 119°34' E).

Taiwan is an island in a complex situation. Having declared its independence in 1949 after the Communist victory founded the People's Republic of China, Taiwan formed its own government and army. However, for the last 30 years, both the United Nations and the United States—which supplies arms to the Taiwanese forces—have officially recognized only one representative of China: Beijing. Although the country is equipped with the most sophisticated weaponry and defended by an army 300,000 strong, Taipei, the Taiwanese capital, lives in constant fear of military operations by Beijing that would force the island to form "one China" with the mainland. Nevertheless, from both sides of the Formosa Strait, economic ties are being woven ever more closely. The economic recession suffered by the "little dragon" since 2001 has driven many businessmen to relocate their operations to the mainland, thus taking the first steps in a process of economic integration that Beijing would like to have a political dimension.

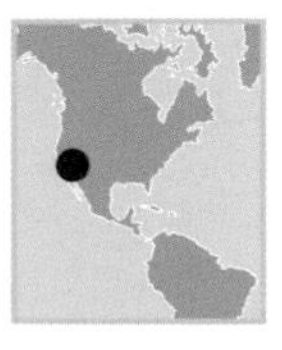

Oilfields near Bakersfield, California, United States (35°22' N, 119°01' W).

Californian oil is heavy and viscous, like tar. Before it can be pumped, it must be warmed and rendered more fluid by water vapor injected into the oil well—a process that depletes the region's already scarce water resources. The technique is expensive, but the United States cannot do without these oil reserves for, while it is the world's second-biggest oil producer after Saudi Arabia, it is also the world's biggest oil importer. In a wider sense, all developed countries depend on this hydrocarbon, especially for transport and for the plastics industry. Today, the amount of oil that remains in the ground worldwide is roughly the same as the amount we have already burned or made into other products. Oil will become increasingly difficult to extract—that is, ever more scarce and expensive. For this reason it is essential to diversify our energy sources, promoting the development of renewable forms such as wind, solar, and geothermal energy.

24

JULY

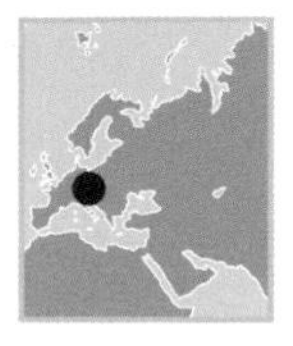

Lake Velence, Hungary (46°53' N, 16°39' E).

Hungary has about 1,200 natural and artificial lakes, the biggest of which, Lake Balaton, lies in the heart of Transdanubia. Lake Velence is in the same region, 30 or so miles (50 km) from Budapest; largely covered with reeds, it is popular with fishermen as well as with board sailors. Lakes are a type of wetland, a category of habitat whose importance has been gradually recognized after massive drainage operations to reclaim land. Wetlands are a natural filter that preserve our supplies of drinking water. They also act as reservoirs that protect inhabited areas from floods and as buffer zones shielding certain ecosys-tems from some human pollution. Human activity—especially farming and building—have seriously damaged freshwater systems, and they are responsible for the loss of about 50 percent of the world's wetlands during the twentieth century. Since 1971, the Ramsar Convention has protected certain wetlands, but only 8 percent are scheduled, and 84 percent are still under threat.

25

JULY

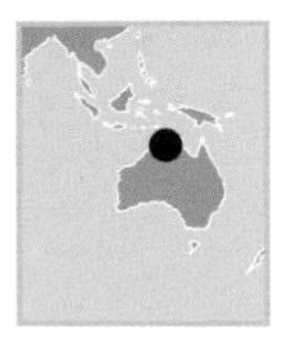

Bungle Bungle National Park, Halls Creek, Kimberley, Australia (17°27' S, 128°35' E).

In the heart of Bungle Bungle National Park (called Purnululu by the aborigines), in western Australia, stands a series of sandy columns and domes approximately 330 feet (100 m) high. This labyrinth of gorges covers 310 square miles (770 km^2). The rocks are made up of solidified sediment from the erosion of former mountains, fissured and raised from the force of motion of the Earth's crust. Their tigerlike orange-and-black appearance is the result of alternating layers of silica and lichen. Known for centuries by the aborigines, this site was revealed to the public only in 1982 and was declared a national park in 1987.

26

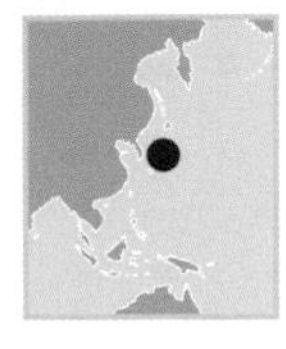

Shinjuku district of Tokyo, Japan (35°42' N, 139°46' E).

In 1868 Edo, originally a fishing village built in the middle of a swamp, became Tokyo, the capital of the East. The city was devastated by an earthquake in 1923 and by bombing in 1945, both times to be reborn from the ashes. Extending over 43 miles (70 km) and holding a population of 28 million, the megalopolis of Tokyo (including surrounding areas such as Yokohama, Kawasaki, and Chiba) is today the largest metropolitan region in the world. It was not built according to an inclusive urban design and thus contains several centers, from which radiate different districts. Shinjuku, the business district, is predominantly made up of an impressive group of administrative buildings, including the city hall, a 798-foot-high (243 m) structure that was modeled after the cathedral of Notre Dame in Paris. In 1800 only London had more than 1 million inhabitants; today 326 urban areas have reached that number, including 180 in developing countries and 16 megalopolises that have more than 10 million. Urbanization has led to a tripling of the population living in cities since 1950.

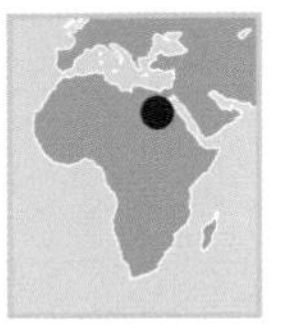

Islands at Siwa oasis, Egypt (29°12' N, 25°31' E).

The oasis of Siwa was internationally famous long before Alexander the Great visited it in 331 BC, for it was the site of one of the biggest temples of Amen, the most powerful Egyptian god of the time. Its oracle was so renowned that the illustrious Macedonian conqueror crossed the Egyptian desert to consult it. According to legend, the priests confirmed his divine nature—a verdict that was convenient, as he had just been crowned pharaoh. During the Roman occupation, the temple of Amen fell into disuse, but the oasis remained. Its springs, famous for their medicinal properties, quenched the thirst of many a caravan on the way from the Mediterranean coast to central Africa. Today, many Egyptian, Roman, and medieval ruins survive in this area, which the Egyptian government would like to see added to UNESCO's World Heritage list. This would earn it a grant toward the preservation of the archaeological site, which would no longer be solely Egypt's responsibility but the shared responsibility of the organization's 175 member countries.

Agricultural landscape near Quito, Ecuador (0°13' S, 78°30' W).

Ecuador was freed from Spanish rule in 1822 but retains a heavy colonial burden, as the place of Indians in agriculture shows. This has not changed much since independence, despite the abolition in 1964 of the *huasipungo* system, which obliged indigenous people to serve big Creole landowners, the *hacendados*. Many Indians are still employed on the *haciendas* on the Pacific coast, whose owners still have title to the best land, where bananas, cocoa, and coffee are grown for export. Others try to survive on a few unproductive plots in the Andes, which they have inherited as a result of agricultural reforms. There is a clear social divide between Creoles and indigenous people, not just in agriculture but in all other sections of society. The economic crisis in the country, as well as natural disasters (such as volcanic eruptions and *el Niño*), periodically act as a reminder of the depth of this divide, as the poorest people are always most vulnerable to them.

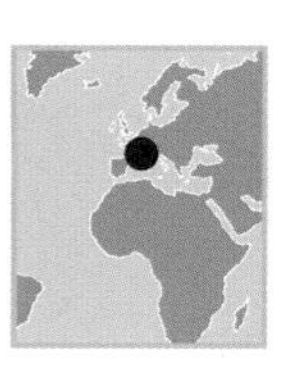

The 2001 Formula 1 Grand Prix,
Principality of Monaco (43°42' N, 7°23' E).

Every year, at the end of May, Monaco is gripped by the excitement of the Grand Prix weekend. The first Grand Prix was held on April 14, 1929, and it was won by the Bugatti 35B driven by "Williams," in a time of 3 hours, 56 minutes, and at an average speed of about 50 miles (80 km) per hour. Since then, the race's 100 laps have been reduced to seventy-eight and the distance, originally 197.5 miles (318 km), to 163 miles (262.6 km). Now the race lasts 1 hour, 47 minutes. Leaning on the parapets of the Ermanno Palace, hundreds of spectators cheer the racing cars that roar past the foot of their perch at 161.5 miles (260 km) per hour. The building looks over the Sainte-Dévote bend and offers an unsurpassed view of the celebrated race, which spectators pay 2,000 to 3,000 euros ($2,300 to $3,500) to watch—a drop in the ocean compared with the astronomical salaries commanded by the drivers. Michael Schumacher (four-time world Formula 1 champion and five-time winner at Monaco) is the world's highest-paid sportsman, with annual earnings estimated at 65 million euros (about $74.75 million). Two Americans complete the top three: the golfer Tiger Woods (62 million euros, or $71.3 million) and the boxer Mike Tyson (56 million euros, or $64.4 million).

San Andreas fault, Carrizo Plain, California, United States (35°08' N, 119°40' W).

More than 100 geological faults cut across California. The best known is the San Andreas fault, which is 621 miles (1,000 km) long and a continuation of the East Pacific ridge onto the North American continent. From the northern end of the Gulf of California in Mexico, the fault runs the length of California, reaching as far as north as San Francisco. The sides of the fault are pushed by their respective tectonic plates at different speeds. As a result, instead of pushing against each other evenly, the sides slip, creating a movement differential from one side to the other of about 2.16 inches (5.5 cm) per year. At this rate, in some 10 million years' time, Los Angeles will be next door to San Francisco—if either still exists. The accumulated tension between the rocks is enormous; if this were ever fully released, the results could be disastrous for the region and its dense population. The 1906 earthquake destroyed two-thirds of the city of San Francisco.

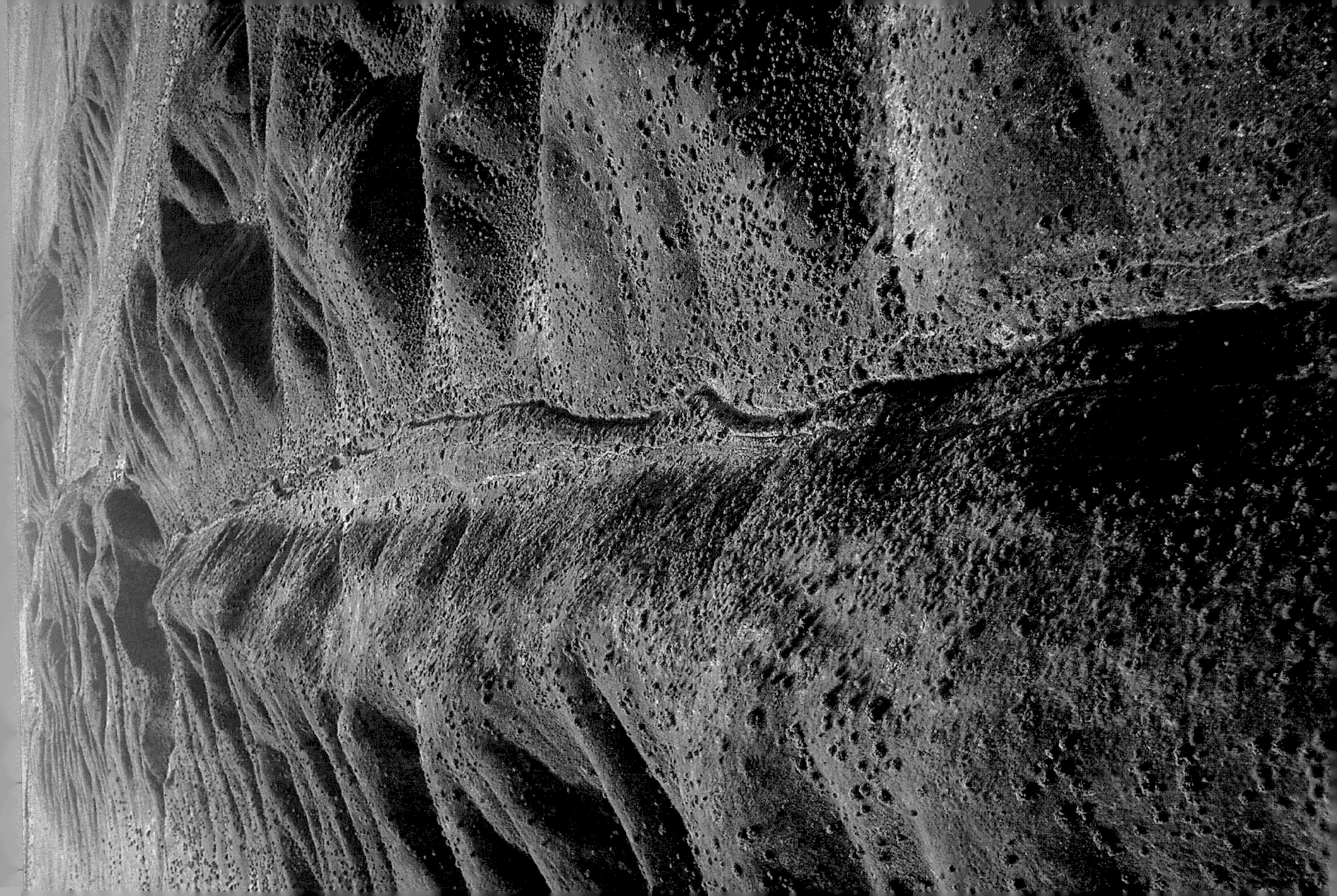

AUGUST

01

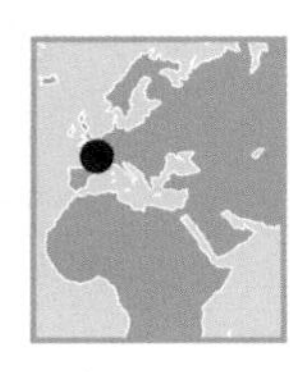

Airbus delivery area, Toulouse, Haute-Garonne, France (43°38' N, 1°22' E).

Airbus's factory in Hamburg, Germany, and its counterpart in Toulouse, southwest France, delivered 325 aircraft in 2001. Between them, they share the assembly of the European constructor's five families of aircraft, soon to be joined by the superjumbo A380, which will carry 555 passengers and come into service in 2006. Air transport is growing at 6 percent. It produces the highest emissions of greenhouse gases, such as carbon dioxide (CO_2), which contribute to global warming. Passenger aircraft emit twice as much greenhouse gas per passenger mile as cars, and six times as much as trains. Freight aircraft produce six times as much CO_2 as trucks and eighty times as much as ships and trains. To reduce the risk to the planet's climate, world emissions of CO_2 would need to be halved or cut to a third of their present levels. This would involve, for each of the world's 6 billion inhabitants, an annual quota of just 0.5 tons of carbon emitted into the atmosphere. At present, the average American emits 6 tons of carbon per year, a European 2 tons, and an Indian just 0.3 tons. Taking a single transatlantic air journey would be enough for an individual to use up this quota.

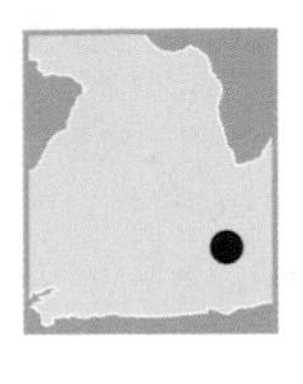

Scientific party on the way to Concordia European base (Dome C), Antarctic (South Pole) (75°00' S, 124°00' E).

In 1959, the Treaty of Washington brought an end to claims of sovereignty over the southern continent, reserving it for peaceful activities and scientific research. Analysis of air bubbles trapped in the ice has revealed 400,000 years of climatic history. Vostok, the name of the Russian base in the heart of the Antarctic, is also the name of the lake that was discovered under the glacier immediately beneath it in 1996. It is 1640 feet (500 m) deep and as big as Lake Ontario. When the party that found it came within 400 feet (120 m) of the lake's surface—which lies beneath a 2.3-mile-thick (3,750 m) layer of ice—they stopped, and discussed how best to reach the water without the risk of contaminating or irreversibly disturbing this environment, untouched for millions of years. The Madrid Protocol forbids the exploitation of mineral resources, requires all waste to be returned to its country of origin, and demands that protected areas and wildlife be respected. It demonstrates the desire of all nations to preserve this virtually virgin land and to be aware of the possible effects of all actions. Antarctic tourists (who already number 15,000 per year) must also respect it.

Iguazu waterfalls, Misiones province, Argentina (25°41' S, 54°26' W). On the border between Argentina and Brazil, the 230-foot (70-m) Iguazu falls form a semicircle 1.67 miles (2,700 m) across, which attracts 1.5 million tourists every year. On the Argentine side, the falls are part of the Iguazu National Park, which was added to UNESCO's World Heritage list in 1984. The park is home to 44 percent of the country's animal species and is one of the best-preserved remnants of South America's Atlantic forest. This unique subtropical forest ecosystem runs along Brazil's ocean fringe and stretches slightly inland in Paraguay and Argentina. It is regarded as one of the five highest-priority areas for the preservation of the world's biodiversity, a "hot spot" where 20,000 types of plant live—8,000 of them endemic—as well as 1,668 species of land vertebrates, of which more than a third are found nowhere else. This wealth of biodiversity has been granted a stay of execution, for deforestation and urban development have already destroyed 90 percent of the original habitat.

AUGUST

04

Islands in Upper Lough Erne, Northern Ireland (Ulster), United Kingdom (54°23' N, 7°30' W).

The River Erne leaves Lough Gowna, in the Republic of Ireland (Eire), and flows 65 miles (105 km) before reaching Donegal Bay in the northwest of the island. Soon after crossing the Ulster border, it widens out to form Lough Erne, which is 50 miles (80 km) long. Strewn with 154 islands and beloved of fishermen, the lake is divided into two: Lower Lough Erne and Upper Lough Erne. Along its banks is a patchwork of small woods and fields bordered by hedges, with lanes running between them—a landscape that was extremely common in the eighteenth century, especially on Atlantic coasts exposed to the wind. Despite their vital role in maintaining ecological balance, combating erosion, and as a habitat for many plant and animal species, hedges have disappeared from the landscape in some parts of Europe because they are an obstacle to intensive agriculture. In France, 217,350 miles (350,000 km) of these linear forests were torn up between 1960 and 1990 as land was reorganized and new urban areas built. Paradoxically, many species have taken refuge in towns, far from the chemicals spread over large areas under cultivation: 260 species of butterflies have been recorded in a park in central Munich.

AUGUST

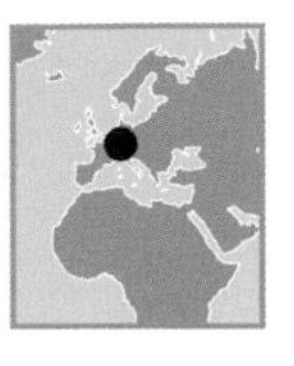

Janschwalde power station near Peitz, Brandenburg, Germany (51°51' N, 14°29' E).

Janschwalde, on the border of the former East Germany, is one of ten power stations supplied by Europe's biggest lignite deposit, which straddles Germany, Poland, and the Czech Republic. Lignite is a form of coal; it has the dubious distinction of being the fossil fuel that produces the most carbon dioxide when burned. Coal is by far the most abundant fossil fuel in the Earth's crust, and it still drives most of the world's power stations. It thus provides more than 50 percent of the world's electrical energy. In 1999, 38 percent of carbon dioxide emissions were due to power generation, making this sector the chief producer of greenhouse gases and putting it far ahead of transport, which in the same year was responsible for 24 percent of carbon dioxide emissions.

05

AUGUST

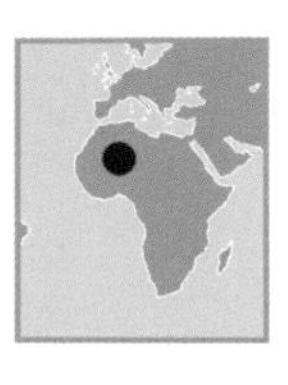

Village of Araouane, north of Timbuktu, Mali (18°54' N, 3°33' W).

In the Saharan portion of Mali, 168 miles (270 km) north of Timbuktu, the village of Araouane stands on the great caravan route, once heavily traveled, linking the north of the country with Mauritania. Araouane's numerous wells, which contributed to its ancient prosperity, still attract nomad campers to its periphery. Little by little, however, its fortlike houses, in which the absence of windows testifies to the permanent struggle against heat and sand, are being swallowed up by the sand dunes driven by the winds, which are erasing the village. The Sahara is the largest warm desert on Earth, extending for 3.5 million square miles (9 million km^2), in eleven African countries. It consists of not only sand but also *reg*s (gravel plains from which wind has eroded the sand), large stony plateaus (*tassili*s and *hamada*s), and high mountain ranges (Ahaggar, Aïr, Tibesti); these ranges take up 20 percent of its area. Scattered throughout this hostile, rigorous environment, the towns and villages of the Sahara contain 1.5 million people.

06

AUGUST

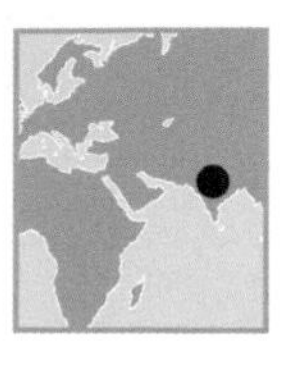

Lake Palace and City Palace, Udaipur, Rajasthan, India (24°35' N, 73°41' E).

Founded in 728, the city of Udaipur reached its golden age when the Maharajah Udai Singh II made it the capital of Mewar in 1567. Mewar, a fertile plain in southeastern Rajasthan, is separated from Marwar, the "land of death," by the Precambrian Aravalli range, which runs from north to south for 435 miles (700 km) and divides Rajasthan in two. One half benefits from the influence of the ocean while the other, which is arid, receives only 7.86 inches (200 mm) of rain per year. In 1746, Jagat Singh II built the Lake Palace at Udaipur, a jewel of marble on a small island, which was the royal family's summer residence. Since India achieved independence, the palace has been converted into a hotel. This magnificent building makes use of the interplay between water and marble, its facades being reflected in the water, and water in turn running through the building via a string of fountains, ornamental pools, and hanging gardens. Thus the maharajahs succeeded in making the mirage of the floating palaces of the Thar desert into a reality. It is regrettable, however, that access to the palace is restricted to customers of the hotel and restaurant.

07

AUGUST

Landscape of brightly colored fields near Sarraud, Vaucluse, France (44°01' N, 05°24' E).

On the Vaucluse plateau, an arid limestone upland in the east of the *département,* lavender fields blossom in the Mediterranean summer heat. Cultivation of fine lavender began about 1920; the crop was distilled to produce an essential oil for perfume. Now, however, it faces competition from lavandin and synthetic products. By 1992, annual production had dropped to 25 tons (a sixth of production totals in 1960). This decline is all the more worrying in that lavender cultivation, which makes use of arid land, supports rural communities in mountainous areas where agriculture is in decline. A program to relaunch and modernize this activity was started in 1994. In 2000, 9,884 acres (4,000 ha) produced 65 tons of essential oil (70 percent of world output), and a further 1,235 acres (500 ha) produced flowers and bouquets. The perfumed, purple carpets that are strewn over the landscapes of Haute Provence are also a considerable asset for tourism in southeast France. In less than a century, the evolution of rural life has given this little flower an important role in developing the local economy.

08

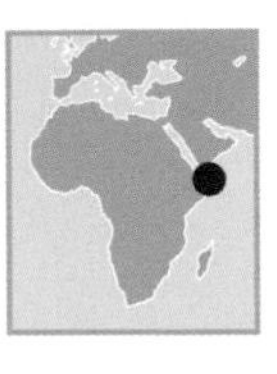

Terraced fields near Sheikh Abdal, Somaliland (9°59' N, 44°48' E).

The self-proclaimed Republic of Somaliland and the Somali Democratic Republic (Somalia) are victims of malnutrition. Only 2 percent of their total land area is cultivated, as in these terraces where melons and tomatoes grow. But the food shortage is mainly due to the frequent droughts as well as to the political instability of the Somali state, which is torn by armed conflict among the different clans. The countries of the Horn of Africa—Somalia, Ethiopia, and Eritrea—all face similar problems. In 2002, 12 million of their people were suffering famine. Droughts and civil war mask another, less obvious aggravating factor: the import of cheap food—including the food brought in as emergency aid. Competition from countries in which farm subsidies bring down export prices hits the sales of local producers and acts as a brake on agricultural development. According to the World Trade Organization, rich countries spent $57 billion on development aid in 2001, but during the same period handed out more than $350 billion to their own farmers.

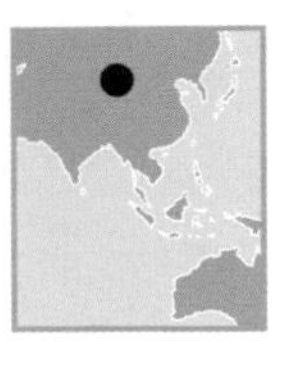

Yurts on the outskirts of Ulaanbaatar (Ulan Bator), Mongolia (47°55' N, 106°53' E).

Three times as big as France, yet with only 2.4 million inhabitants, Mongolia has 1.4 herders. Ulaanbaatar, the capital, is set in the heart of the steppes, southwest of the Hentii mountain range. A quarter of the Mongolian population is concentrated there, compared with an eighth ten years ago. Tens of thousands of nomadic herders have settled in the city, driven off the land where their herds once grazed by desertification. Often making a precarious living, they pitch their yurts—circular tents made from white wool stretched over a wooden framework—which are the traditional nomad family dwelling. This exodus from the land has gathered momentum since 2000, when an exceptionally severe drought, followed by one of the bitterest winters of the century, affected 45 percent of the population and killed 2.4 million head of cattle. Natural disasters have become far more frequent over the last two decades; the poorer the country involved, the more severely affected are its people. Natural disasters claim almost fifty times as many victims in a poor country as in a country such as the United States.

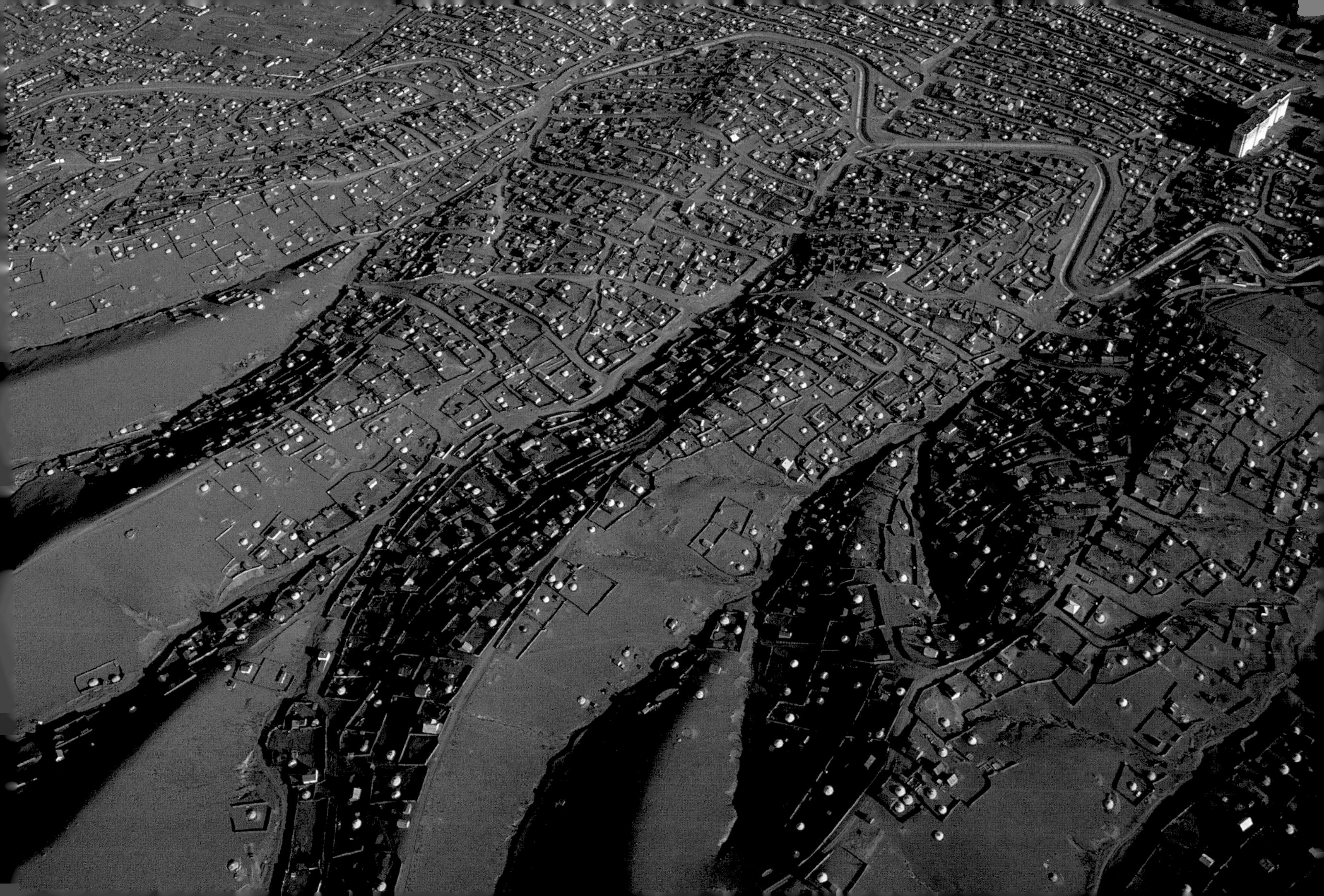

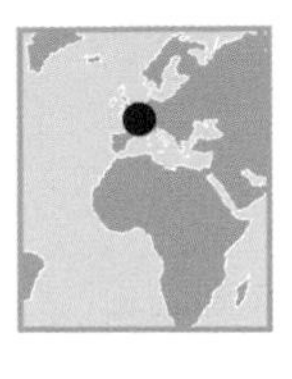

Loisinord ski slope on a slagheap at Nœux-les-Mines, France (50°29' N, 02°40' E).

It's all there: skiers, ski lifts, and the cold. Only the snow is missing. But snow is rare anyway in the Nord-Pas-de-Calais region; it is replaced here by a green turf that is watered constantly to keep it slippery. At 1,050 feet (320 m) in length, Loisinord is Europe's longest but by no means only synthetic ski slope. The United Kingdom has 110 of them, also built on the slag heaps produced by coal mining. The inhabitants of coal fields have clearly decided that rather than simply tolerating these mountains of waste, they should use them. By transforming the giant heaps into ski slopes, paraglider runways, amphitheaters, and even bird preserves, they have created tourist attractions in areas that only a few years ago were deserted by tourists. Thus, the Nord-Pas-de-Calais has become France's sixth-most-popular tourist region. Tourism has even become the second-biggest employer, breathing new life into an area that suffered severe unemployment when the mines were closed.

AUGUST

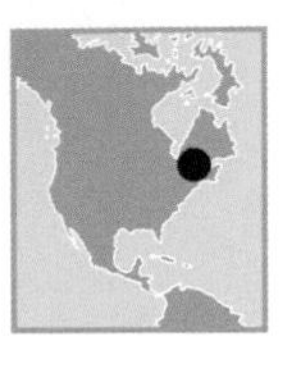

Timber raft on the St. Maurice river, Québec, Canada (46°21' N, 72°31' W).

In this valley, one of the most heavily industrialized in Québec, the sight of a vast train of floating logs belongs to the past. As a result of pressure from riparian owners and ecologists, the floating of timber down the river was prohibited in 1993. This way of transporting logs could have caused lasting damage to the river's ecosystem, altering the riverbed by eroding the banks and causing debris to accumulate on the bottom. The excess organic matter would have contributed to depletion of oxygen levels in the water. Today, the recent increase and concentration of pig farming presents a similar threat to Québec's rivers, because large quantities of liquid manure runoff produces nitrogen and phosphorus, which cause eutrophication of aquatic habitats. This phenomenon, due an excess of nutrients in water, causes plants to grow rapidly; when these die and decay, they deplete the oxygen supply, asphyxiating the habitat. According to the World Water Commission, 250 of the planet's 500 largest rivers are severely polluted.

12

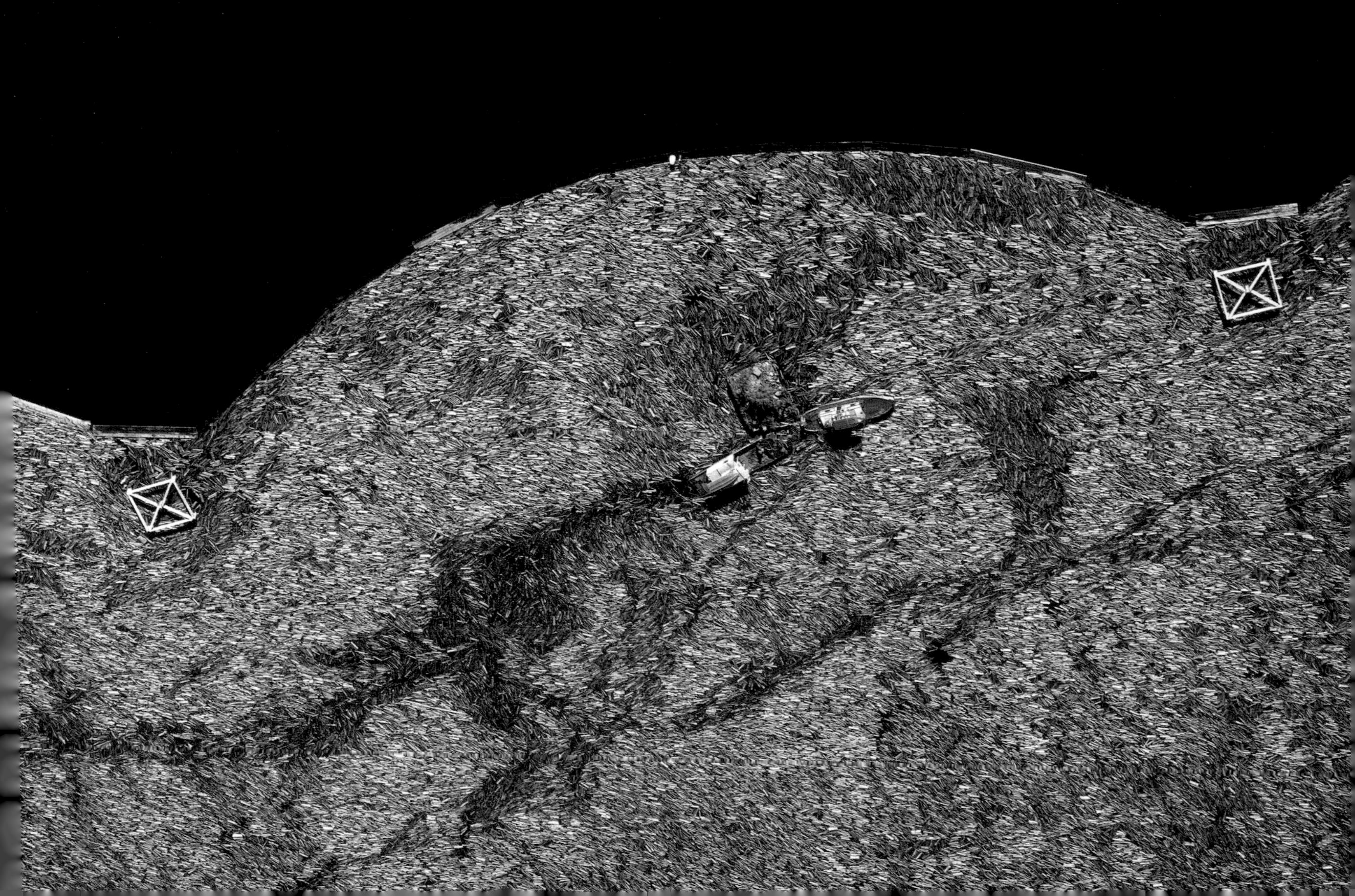

AUGUST

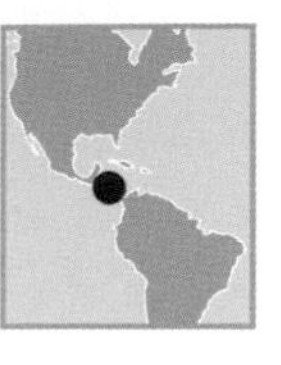

House destroyed by flooding of the Ulua river following Hurricane Mitch, San Pedro Sula, Honduras (15°27' N, 88°02' W).

Hurricane Mitch sprang up south of Jamaica and reached its climax on October 26, 1998, with gusts of 180 miles (288 km) per hour, four days before it hit the coast of Central America. For two days, Honduras was swept by devastating winds, torrential rain, and mudslides that razed entire towns and killed several thousand people, leaving behind more than a million victims and causing at least $58 million worth of damage. Hydrometeorological disasters (caused by water and meteorological conditions) have become increasingly frequent in recent decades. They have been especially devastating in developing countries. In just ten years, between the 1980s and the 1990s, the number of people increased by almost 1.5 who were affected by natural disasters.

13

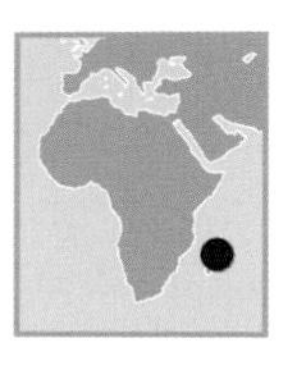

The Mahajilo river crossing the eroded plateaus east of Miandrivazo, Madagascar (19°31' S, 45°28' E).

As if scarred by great claw marks, these plateaus bear deep grooves scooped out by flowing rainwater. These ravines, or *lavaka*s, carry laterite, a red sediment scoured from the hills by erosion, to the river. There are no longer any trees to retain the loose earth, for the forest has disappeared, cleared by slash-and-burn farming and overgrazing. Although these practices are now forbidden, they have increased in recent decades as a result of the doubling of the country's population over the last 30 years. Farmers suffer the consequences. Since the eroded areas are no longer fertile, the land that can be farmed has been reduced to 5 percent of the island's total surface. They are sometimes obliged to work in the damaged areas, when they are not burning more forest to gain further space. Almost 5 billion acres (2 billion ha) of land are degraded in the world; in 30 percent of cases, deforestation is to blame.

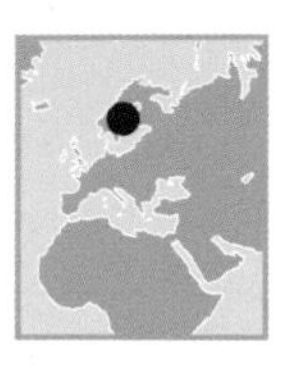

Islet off Göteborg, Sweden (57°38' N, 11°46' E).

Sweden's coasts are dotted around with more than 150,000 islands, like these granite rocks that rise from the sea west of Göteborg, the country's second-biggest city. They are havens of peace for yachtsmen, who flock by the dozens to the smallest rock. Sweden, where almost one household in five owns a boat, is one of the few countries where you can see a boat traffic jam. With gross domestic product per capita among the highest in Europe, and a record education budget, Swedes favor the personal development of the individual and family life, preferably in the open air. On the other hand, Sweden has Europe's highest rate of work lost due to sick days, a fact that reflects not just the quality of social provision but also a dehumanization of working conditions. In 2003, the government announced that it was giving priority to tackling this absenteeism—which costs three times as much as the entire education budget.

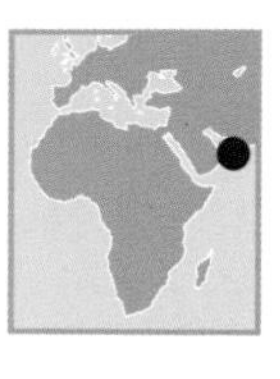

Terraced fields in Jebel Akhdar, Oman (23°30' N, 56°56' E).

The Sultanate of Oman contains some rugged mountains. In the north of the country, the sheer precipices of the Jebel Akhdar, which rise to an altitude of almost 10,000 feet (3,035 m), are scoured by sudden, short-lived torrents called *wadis*. Despite the scanty rainfall, these mountains have for centuries been brightened by a mosaic of terraced fields. A complex, 2,000-year-old water distribution system called *falaj* allows dates, lemons, and apricots to be grown. *Falaj* is generally fed by ground water, often taken from a depth of a hundred feet or so (several tens of meters), which flows by gravity through manmade underground passages over several miles before surfacing near villages. A hierarchy governs the water's use: first comes cooking, followed by men's washing, then the "women's fountain," and finally irrigation of farmed plots. Since 1990, intensive farming methods have reduced ground water supplies and increased their salinity.

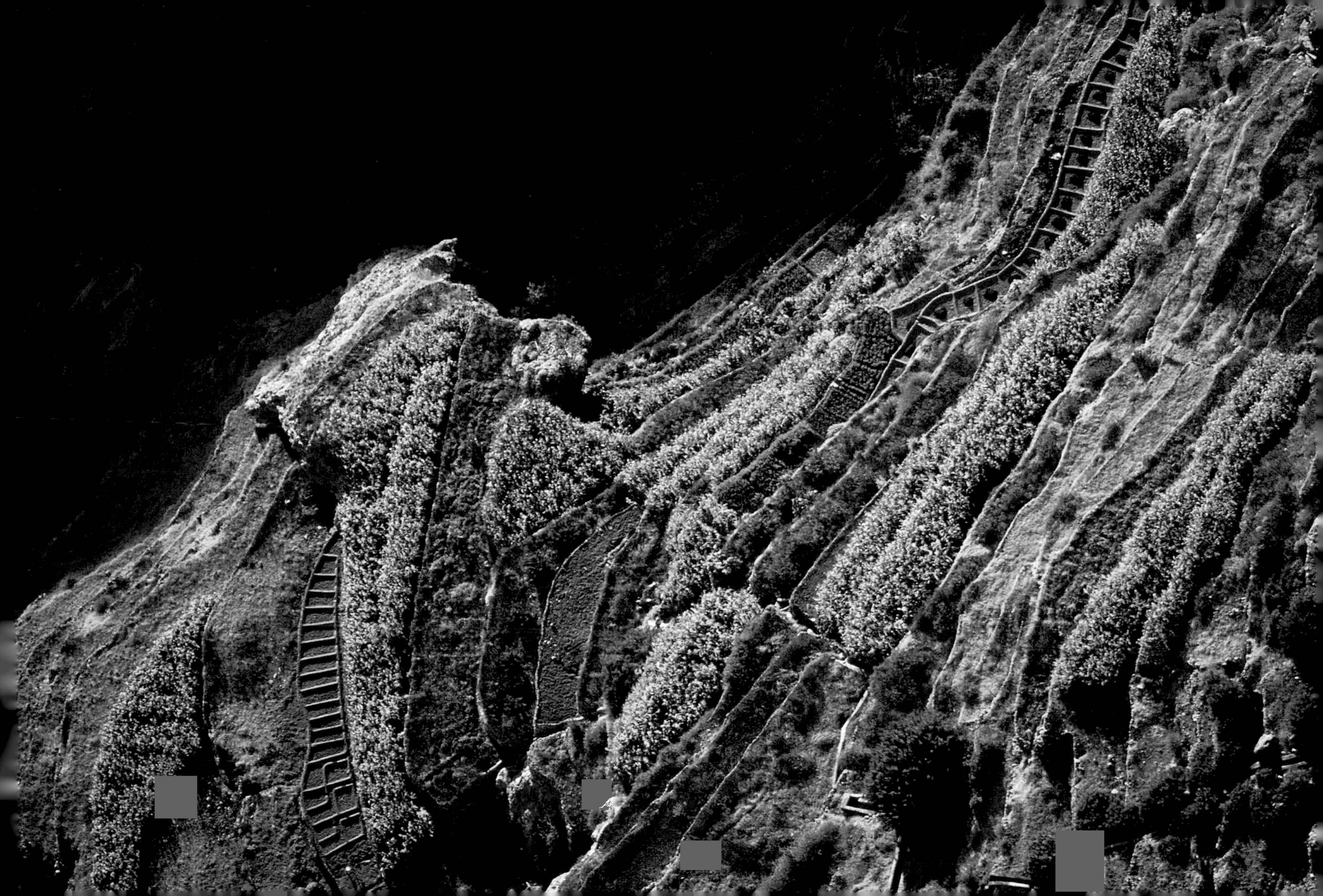

AUGUST

17

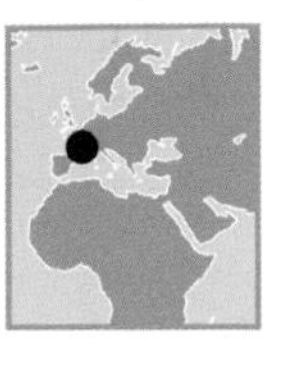

Encampment near Vallot mountain refuge on Mont Blanc, Haute-Savoie, France (45°49' N, 6°51' E).

The impregnable Mont Blanc was finally summited in 1786. This victory, for which a substantial reward had been offered, was achieved by the guide Jacques Balmat and a doctor from Chamonix named Gabriel Paccard. A century later, some 3,000 people had attempted the ascent and more than half, including sixty-seven women, had succeeded. Today, Europe's highest peak attracts 3,000 climbers every year. They make their approach via the Mont Blanc du Tacul and Mont Maudit route of 13,110-foot (4,000-m) summits, or they climb via the Dôme du Goûter. Every day during the high season, 300 to 400 people crowd this glamorous peak. This heavy traffic, which is the chief economic resource of the valley below, threatens the site itself. A cleanup operation that lasted from 1999 to 2002 removed almost 10 tons of trash from the mountain. Extreme environments, such as high mountains, are highly fragile. They are sensitive to the slightest disturbance and easily damaged by mass tourism.

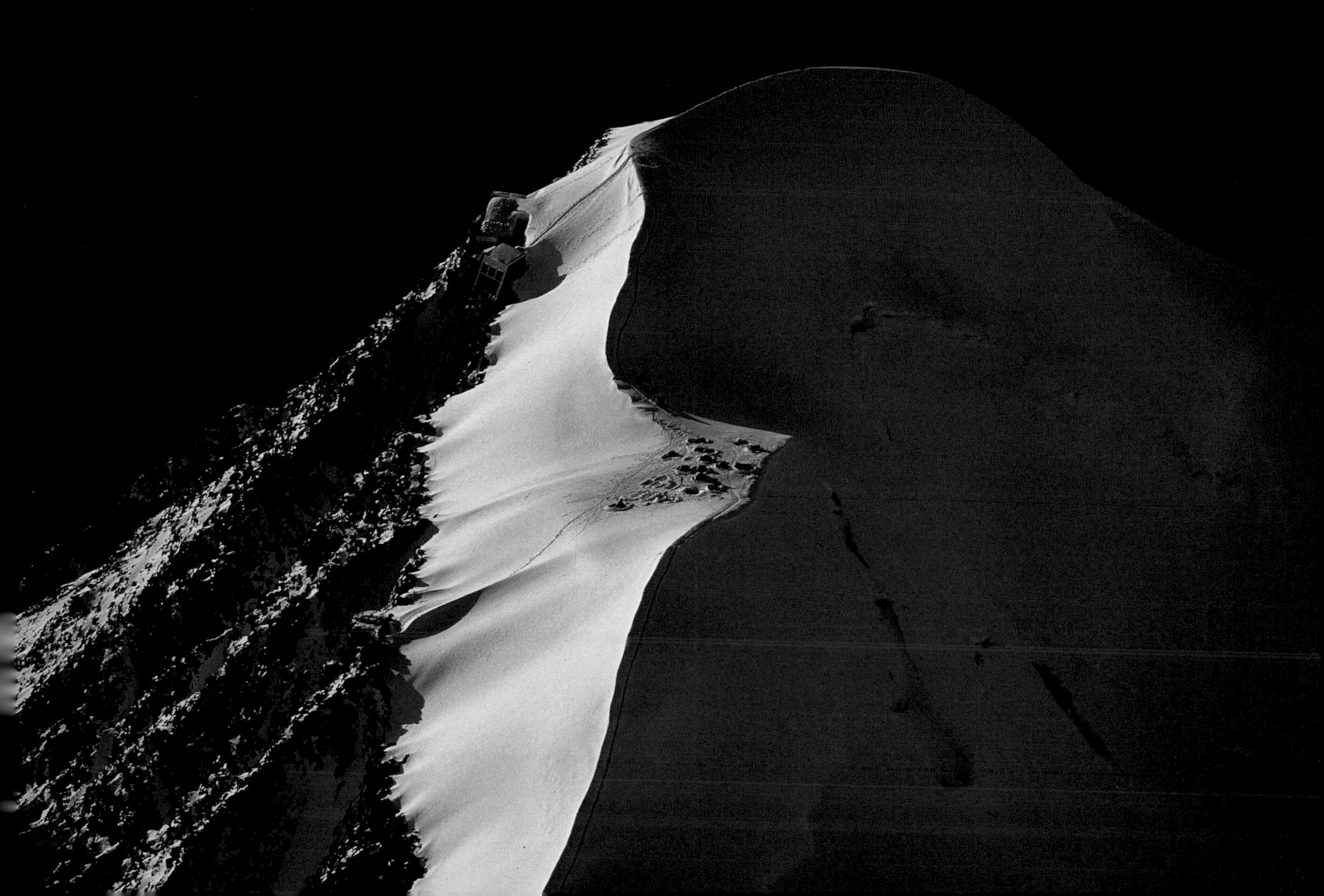

AUGUST

18

Oil platform on Lake Maracaibo, Venezuela (9°50' N, 71°37' W).

This great expanse of oil derricks indicates how much the discovery of oil in 1920 transformed Venezuela's economy. Since then, the country has lived to the tune of the price of a barrel of oil. In 1972, Venezuela found Eldorado; soaring oil prices put the country at the top of the Latin American league in terms of earnings per capita and economic growth. But the price fall of 1983 took the Venezuelans by surprise and showed how much they had come to depend on black gold. This bonanza gave rise to a form of capitalism founded on speculation, rather than on the enterprise and the desire to create wealth. Moreover, despite government efforts, the rural economy never managed to recover. A sizeable middle class certainly sprang up, but oil wealth, which was unequally distributed, has done nothing to reduce inequality: a third of the population cannot afford to buy basic foodstuffs.

H

Church of the Transfiguration, Kizhi Island, Karelia, Russia (62°00' N, 35°15' E).

At the heart of the northern islands in Lake Onega, in southern Karelia, lies a civilization based on wood. The Kidzi *pogost*, an architectural masterpiece, has been on UNESCO's World Heritage list since 1990. One of its jewels is the Church of the Transfiguration, with its twenty-two domes clad in silver-covered aspen wood tiles, which was built 1714. This monument demonstrates the virtuosity of the carpenters of the period, who were able to produce complex shapes and harmonious curves without using a single nail. However, the church's preservation faces serious problems: it is exposed to wide temperature variations as well as fungus, and it must also adapt to unaccustomed constraints in the form of a recently installed metal framework. Despite UNESCO's wise decision to protect it and to employ experts, who have worked on it since 1992, there have been disagreements over how the church should be restored.

AUGUST

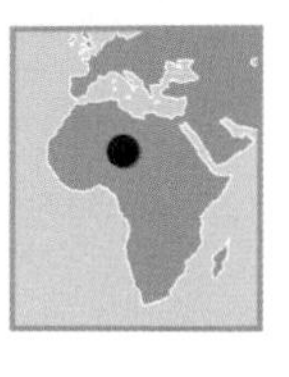

Dromedary caravans near Fachi, Ténéré desert, Niger (18°14' N, 11°40' E).

For decades the Tuareg have traded salt by driving dromedary caravans over the 485 miles (785 km) between the city of Agadez and the Bilma salt marshes. The dromedaries, connected in single file, travel in convoys at a rate of 25 miles (40 km) per day, despite temperatures reaching 114.8° F (46° C) in the shade and loads of nearly 220 pounds (100 kg) per animal. Fachi, the only major town on the Azalaï (salt caravan) route, is an indispensable stop. Caravans, at one time made up of as many as 20,000 dromedaries, generally are limited today to 100 animals; they are gradually being replaced by trucks, each of which can carry as much merchandise as 250 dromedaries. The number of motor vehicles on the planet has risen from 40 million in 1945 to 680 million today. Although the number remains low in developing countries, this is where automobile use is growing the fastest. If the entire world had the same vehicular use as the United States, the total number of vehicles on Earth would be 3 billion.

20

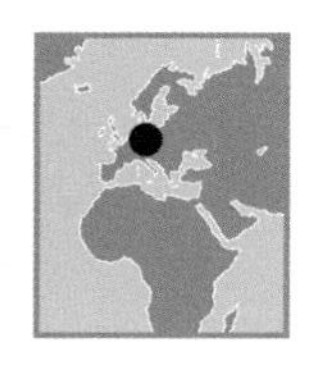

Johannes mine (Grube Johannes), Bitterfeld industrial area, north of Leipzig, Saxony-Anhalt, Germany (51°37' N, 12°20' E).
The old quarry of Grube Johannes, first worked in 1843, has been a dump for industrial waste since the 1930s, and until 1990 the effluent from a factory that produced cellulose fibers was poured into it. Nicknamed Silbersee (Silver Lake), this stricken place symbolizes both the failure of the former East Germany's economic and environmental policies and the considerable efforts made by the reunited Germany to remedy the serious pollution it inherited with its eastern addition. Here, nature itself is harnessed to repair the damage done by humans. The green bags, which form a sealing layer on the surface of the ground, are biological filters. They contain bark chippings, on which microorganisms live that transform the fetid hydrogen sulfide released from the industrial sludge into hydrogen sulfate, which is odorless.

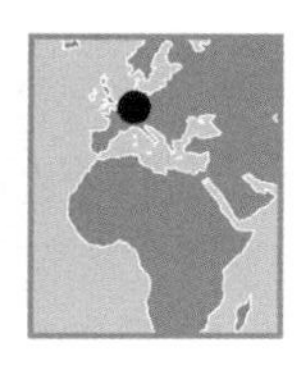

Jewish Museum, Berlin, Germany (52°30' N, 13°25' E).

The ground floor of this museum is totally empty. Its director, Michael Blumenthal, the son of a German Jew who was interned in Buchenwald, intended to fill it with the memory of the 6 million Jews killed by the Third Reich. The American architect Daniel Libeskind designed the building in the shape of a broken Star of David in order to convey, via this chaotic shape, the horror of the most painful memory in the European consciousness. Berlin's Jewish Museum was opened on September 9, 2001, and traces 1,700 years of Jewish history in Germany, including an emotional commemoration of the Holocaust, the biggest act of genocide of the twentieth century. Other acts of genocide include the mass killing of Armenians by the Turks from 1915 to 1916 (1.2 million dead), of Cambodians by the Khmer Rouge regime between 1975 and 1979 (1.7 million), and of Hutus by the Tutsis in Rwanda in 1994 (500,000). Under international law, genocide has been a crime since 1948.

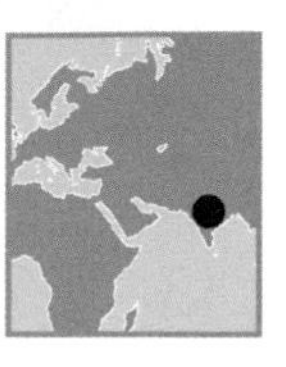

Brickyard east of Agra, Uttar Pradesh, India (27°04' N, 78°53' E).

Numerous brickyards have set up shop in the vicinity of Agra, a metropolis of 1.2 million in Uttar Pradesh, a state that comprises one-sixth of India's population. These small enterprises provide work in a region hard hit by unemployment, a problem throughout the country. In 1999 India ranked 144th in the world in gross domestic product per capita (adjusted accorded to buying power). Production of these terra-cotta bricks is particularly intended for urban centers; the rural population generally lives in housing made of adobe, a raw, claylike earth that is lower in cost but vulnerable to bad weather. The strong urban growth of the Agra area, where the population has increased by 50 percent in the past twenty years, points to a prosperous future for local producers of construction materials.

AUGUST

24

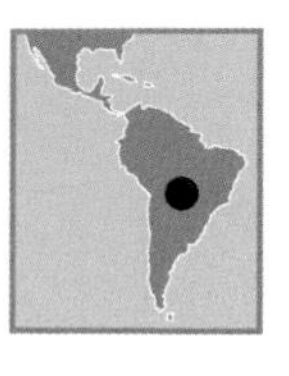

***Fazenda* (ranch) surrounded by the waters of the Rio Vermelho, Pantanal, Mato Grosso state, Brazil (17°00' S, 56°54' W).**

The vast sedimentary depression of the Pantanal is shared by Paraguay, Bolivia, and the two Brazilian states of Mato Grosso and Mato Grosso do Sul. This 54,040-square-mile (140,000-km^2) region is prone to flooding—*pantano*, the root of Pantanal, means marsh. It is wedged between highlands from which countless tributaries of the Paraguay river flow down. Its flatness, combined with its high rainfall (almost 50 inches, or 1,250 mm, per year) make it subject to recurring floods: the prairies with their grass cover, highly suited to the raising of livestock in the dry season, are submerged from November to March. They then become vast lakes—*baias*—dotted with small islands, called *cordilheiras,* where the animals gather. This exceptionally rich habitat, home to more than 270 bird species, is vulnerable to the pollution and silting of rivers that forest clearance and mining cause. More than 500,000 wild animals vanish from the Pantanal each year, either exported to menageries in rich countries or illegally hunted. The skin of the *jacaré*, or cayman, is highly prized. These animals are now farmed, and a proportion of all the young hatched in captivity are released into their natural habitat.

AUGUST

25

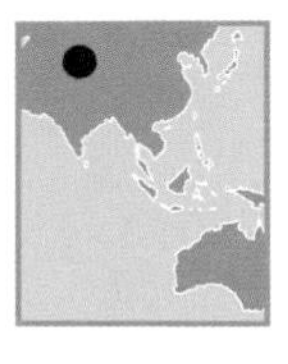

Rice field north of Pokhara, Nepal (28°14' N, 83°59' E).

The Himalayan mountain chain runs north of Nepal, separating it from its giant neighbor, China. The mountains crown Nepal with a string of eight peaks—out of a world total of fourteen—higher than 26,232 feet (8,000 m). The economy is based on agriculture, which employs 80 percent of the working population and accounts for 41 percent of the gross domestic product of one of the world's poorest countries. Generations of farmers have tamed the mountainsides and prevented erosion by cutting terraces. Rice paddies thus rise in tiers as high as 9,800 feet (3,000 m) above sea level, covering 45 percent of Nepal's cultivated land. Rice is the staple food of 3 billion Asians, including 25 million Nepalese. It is also used to make *chang* (a rice-based fermented drink) and *rashki* (an alcoholic drink made from rice or grain). The world consumes 400 million tons of rice a year, but the quantities differ among the continents: more than 0.1 tons per person in Asia, 0.04 to 0.06 tons in Africa and Latin America, and just 0.005 tons per person in Europe. Asia's population is growing at the rate of 50 million rice-eaters per year, Africa's at 5 million a year, and Latin America's at 2.5 million a year.

AUGUST

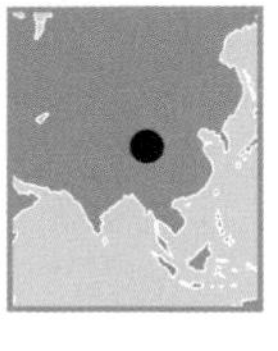

Nomads on the Züün Saikhan, Ömnögov Gobi, Mongolia (45°32' N, 107°00' E).

Mongolia, which is known as "Outer" to distinguish it from its Chinese counterpart, Inner Mongolia, is wedged between Russia and China. It covers an area three times as big as France, and lies at an average altitude of 5,180 feet (1,580 m). This land, largely peopled by Turko-Mongol nomads, produced the greatest conquerors in history, against whom China built its Great Wall beginning in the fifth and fourth centuries BC. In the thirteenth century, the Mongol prince Genghis Khan conquered the biggest empire in history, which stretched at its height from the Pacific to the banks of the Volga, and from Cambodia to Iran. It was broken up and divided between Turks, Russians, and Manchus. What is now Mongolia was completely under Manchu control from 1696 to 1911, when part of it—Outer Mongolia—became independent. At the end of the twentieth century, after seventy years of Communist rule, the difficult task of economic restructuring worked together with one of the century's worst droughts to take a third of the Mongolian population below the poverty line.

26

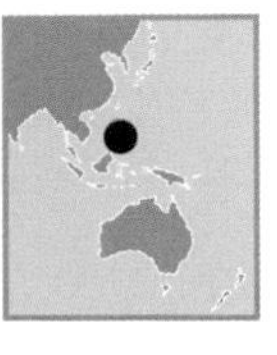

Village near Panducan, Philippines (6°15' N, 120°36' E).

The Panducan region, in the Pangutaran group of islands, is part of the Sulu archipelago. The islands are home to the Tausug, "people of the sea currents," who number 400,000. Formerly smugglers, the Tausug now live from trade and fishing. They live in small hamlets of bamboo houses on stilts, scattered along the coasts fringed with coral. The Philippines are home to 9 percent of the world's coral reefs. These reefs have the greatest biological diversity, but they are also the most endangered. The practice of fishing using cyanide or explosives has had devastating effects on the coral reefs and the marine fauna that depend on them. They are also choked by the sedimentary deposits that loosen from ground erosion. This damage is a grave concern to the many communities that depend on the health of the coral reefs for their livelihood.

27

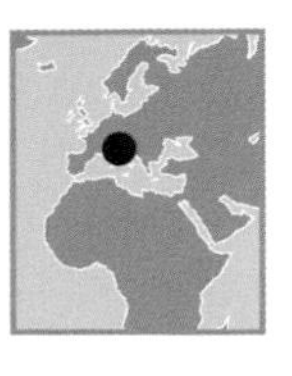

Marble quarry, Carrara, Apuan Alps, Italy (44°05' N, 10°06' E).

At little more than 6 miles (10 km) from the Tuscan coast, the Apuan Alps can be approached at a height of 2,625 feet (800 m) via the mountains of Carrara, which are such a dazzling white that they seem to gleam under a permanent layer of snow. Here, the famous white marble of Carrara has been quarried since Roman times. Its exceptional purity made it the most sought-after marble for sculpting statues, and it was prized not just by Michelangelo but by all Renaissance Europe, for from the fifteenth century on, Carrara exported beyond the Mediterranean to Britain and the Netherlands. Carrara's marble producers thus attribute the robustness of their 2,000-year-old local tradition to the fact that it opened up to international markets. Less widely used in sculpture today, Carrara marble is now used—as in Roman times—to build and dress prestigious public buildings, such as the arch of La Défense in Paris.

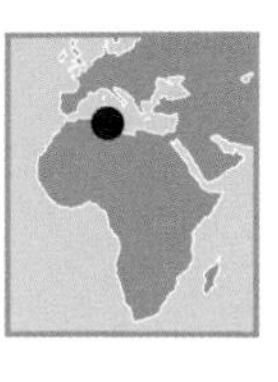

Oil river and delta in the desert, Tunisia (34°00' N, 9°00' E).

Black gold normally lies trapped between rock strata and does not form rivers on the surface of our planet—unless, that is, it has leaked from a pipeline, as here in the middle of the Tunisian desert. There are 360,180 miles (580,000 km) of pipelines carrying hydrocarbons around the world—a length equivalent to fourteen times the circumference of the Earth. Leaks are caused by wear, natural disasters, or sabotage, and they contaminate ground water and fertile land. In the countries of the Persian Gulf, oil extraction causes frequent leaks of hydrocarbons, of which 10 percent are thought to end up in the marine ecosystem. In January and February 1991, during the Gulf War, 396 million gallons (1.5 billion l) poured into the sea. On August 4, 1990, as a result of negligence, 17,000 cubic yards (13,000 m^3) of hydrocarbons oozed into the water table near a refinery in Seine-Maritime, France. Here, emergency measures made a rapid clean-up of the environment possible, but more remote regions lack the means to deal with such situations.

29

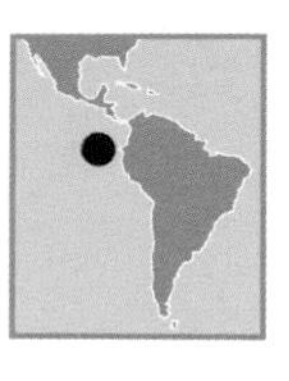

Volcanoes on the Galápagos archipelago, Ecuador (0°20' S, 90°35' W).

The nineteen islands of volcanic origin that constitute the Galápagos archipelago emerged from the Pacific between 3 and 5 million years ago. Despite their lunar landscape, they are exceptionally rich biologically. In particular, they are home to the world's biggest colony of marine iguanas, as well as the giant—or Galápagos—tortoise, which gave the archipelago its name. Visitors who sail there are entranced; Charles Darwin, however, was inspired to develop his theory of the evolution of species. The Galápagos were designated as a national park in 1959, and in 1978, UNESCO added the islands to its World Heritage list. However, nothing has prevented the increase in human settlement, the introduction of exotic species, or a boom in tourism (though this has been strictly controlled since 1998) from endangering this natural laboratory of evolution. The archipelago miraculously escaped being polluted by some 600 tons of fuel oil that leaked from the tanker *Jessica*, shipwrecked in January 2001—but other coastlines were not as lucky.

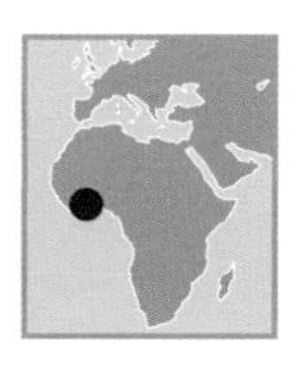

Commercial forest northeast of Yamoussoukro, Republic of Côte d'Ivoire (6°50' N, 5°15' W).

Côte d'Ivoire's forests have long been its "green gold" and principal source of wealth. "The Ivorian miracle" that was held up as an example for many years was the result of intensive exploitation of forests. These covered 29 million acres (12 million ha) in 1960, but by 1994 were down to 5 million acres (2 million ha). Alarmed, the Ivorian government decided to replant its forests, ration wood consumption, and educate the population so that it could diversify its fuel sources and stop lighting bush fires to clear land or renew young grass growth for flocks to graze on. Export of logs was prohibited from 1995, and producers concentrated on processing, which now accounts for 80 percent of the sector's turnover. Nevertheless, the wood industry is suffering from the high cost of production and of spare parts, which have to be imported. Only 20 percent of forestry operations are in good financial health.

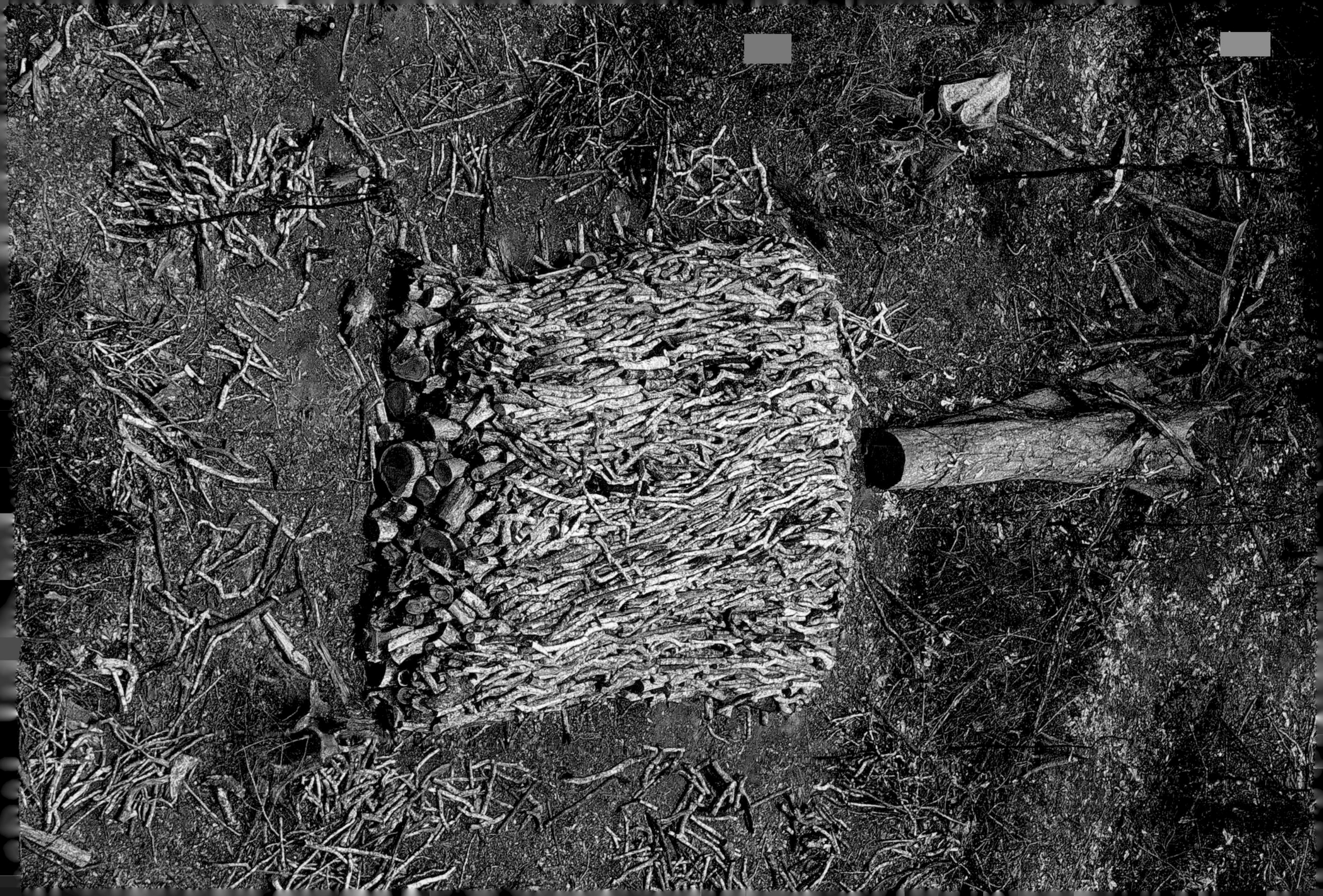

TRANSPORT

SEPTEMBER

01	02	03	04	05

Transport has a major impact on everyone's lives. Lack of transport in Africa, Bangladesh, and parts of India acts as a serious drag on economic development and a burden on women who carry water, fuel, and crops over hundreds of kilometers. In other parts of the world, too much transport is the cause of serious gridlock, congestion, pollution, and health damage to children. Seriously congested traffic conditions in California, southeast England, and around Frankfurt in Germany result in the loss of billions of U.S. dollars to the economy and serious disruption to everyday life. Road traffic accidents—currently the cause of over 3,000 deaths per day—are a significant public health problem and a serious blight on the lives of pedestrians, cyclists, and bus users in developing countries.

All these problems are getting worse over time as the world's population moves higher and higher up the "mobility ladder" and demands more and more investment in roads, railways, and airports. These investments gobble up a significant slice of the tax-dollar pie. At the same time, all over the world, health care, pension, and social infrastructure costs are rising fast, and the pressure on every tax dollar has never been greater.

The growth in the demand for transport is a major factor in the growth of greenhouse gas emissions and the problems associated with climate change. Transport growth has shown itself to be immune from most of the discussions surrounding sustainable development or the promotion of environmentally friendly modes of transport, especially walking and cycling. Every year, most of us expect to travel farther to carry out even the most mundane activities, such as traveling to school or work. Even if we use the latest low-emission vehicles, our increased kilometers of travel cancel out those gains and push us further up the greenhouse gas emission curve. The more we want to travel, and the more we abandon the environmentally friendly modes, the more we damage the global environment.

Environmental problems are also created by the enormous growth of new road and motorway building. New motorways in the ex-Communist countries of Eastern Europe are already encroaching on sensitive environmental sites and polluting the atmosphere and watercourses. New roads in the United Kingdom, such as the Birmingham Northern relief road, are cutting a huge swathe of tarmac and concrete through open countryside, which has traditionally been protected. These new roads, in their turn, encourage yet more traffic and switch more trips from public transport to private cars. New roads also change geography and contribute to the "spreading out" of everything we do. The spread of business parks, out-of-town shopping centers, and low-density suburban housing in the United States, the United Kingdom, and even in Kolkata, India, is a consequence of new highway investment.

The British government has shown that most transport investment benefits those who are already wealthy. Transport spending is inequitable and socially divisive. Rich people fly more and drive more often than poor people, and most public investment supports those who go farther and faster than anyone else. In the United Kingdom, the taxpayer supports those who fly to the tune of about £10 billion a year (about 1.60 billion dollars). The environmental impact of flying—air pollution and noise, for instance—affects poorer people more than richer people.

The policy response to serious traffic and transport problems has traditionally been to invest money in more infrastructure. More roads, more airports, and more high-speed trains are still the norm. Even in developing countries, where the pressure on tax dollars is particularly acute, the bias in investment is toward these "wealthy" modes. This, consequently, takes investment dollars away from the needs of the rural poor and away from the needs of the poorest in Mumbai or any other Asian or Indian city.

Recent events in Bogotá, Colombia, show that a completely different path is now possible. The former mayor of Bogotá has carried through a series of radical transport policies and funding reallocations to serve the needs of the poor and those who travel short distances to work and school in their own neighborhoods. Bogotá has developed the tradition of closing its main arteries to motor vehicles for seven hours every Sunday so that people can use the roads for cycling, jogging, and meeting up. The total amount of road space closed to traffic has doubled to and now represents 120 kilometers (about seventy miles) of main city roads. Investment in cycling has resulted in the building of more than 300 kilometers of protected bicycle paths, with a steady increase in cycling. Public investment of $5 million per kilometer in the TransMilenio passenger buses now means that city residents now take more than 540,000 daily trips by bus.

16 17 18 19 20 21 22 23 24 25

The Bogotá approach has set a new standard for environmentally friendly, sustainable, and socially just policies. London's congestion charge, introduced in February 2003, sets a similar standard. The charge of £5 (or about $8.00) currently raises about £500,000 (or $800,000) per day, the majority of which will be spent on buses, walking, and cycling in London. This is socially just. Only 20 percent of the trips into London everyday are by car. This minority of trips imposes a serious burden on poorer Londoners, who cannot afford the high price of a house in the country where they, too, could escape the pollution. To everyone's benefit, the congestion charge has reduced traffic by about 20 percent, and conditions for cyclists and bus users are the best they have been in the last seventy years.

It is possible to improve living and working conditions in all the world's cities as well as in the poorest rural areas. There is nothing to lose and everything to gain. The only serious barrier is political conservatism, that is, the distorted perceptions of politicians who think that moving in an environmentally and socially just direction will lose them votes. It will not. The time has arrived for a new transport paradigm, one that rewards the cyclist, the pedestrian, and the bus user while making motorists and trucking companies pay the full cost of the environmental damage they impose on society. There are signs that this is happening, but it will require individuals as determined and clear-sighted as Enrique Peñalosa (former mayor of Bogotá) and Ken Livingstone (current mayor of London) to make it happen. One thing is clear. We have all the relevant transport expertise we need, and we understand the problems and the solutions that will deliver a sustainable, vibrant, socially just, and child-friendly future. We lack only the political vision and the political leaders to make it happen.

JOHN WHITELEGG and GARY HAQ
Stockholm Environment Institute, University of York, England

26	27	28	29	30

SEPTEMBER

01

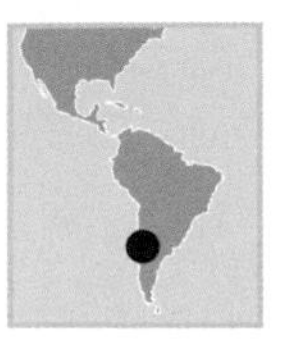

Loading sawdust north of Calbuco, Chile (41°45' S, 73°10' W).

The loading machinery fills the boat with sawdust, which will be used to make cellulose, paper, and laminated furniture. Mainly intended for export, it comes from the region's rainforests but also from plantations. Since the 1970s, economic and fiscal incentives have encouraged the replanting of former forest areas that had been felled. But although these plantations provide 85 percent of the wood that Chile's wood industry uses, they are not a satisfactory replacement for the old forests. For one thing, they consist mostly of exotic species such as eucalyptus (12 percent) or pine (80 percent), and therefore do not regenerate the original ecosystems that local fauna need to survive. Nor do they make up for the 49,500 acres (20,000 ha) of forest that Chile loses every year. Worldwide, 23.2 million acres (9.4 million ha) of forest per year go up in smoke, are turned into sawdust, or become furniture. If every inhabitant of the planet owned 64,578 square feet of forest (about an acre and a half, or 6,000 m^2), we would each lose about 130 square feet (12 m^2) every year.

AIO PAPER

SEPTEMBER

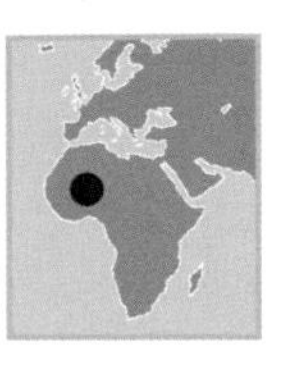

Market gardening near Timbuktu, Mali (16°46' N, 3°00' W).

In the arid Timbuktu region, in the heart of Mali, market gardening is rendered difficult by an infertile sandy soil and extreme climatic conditions: daytime temperatures can reach 122 °F (50 °C), and rainfall barely exceeds 6 inches (150 mm) per year. Water comes from underground, and it is gathered by means of traditional wells—simple holes dug in the ground, whose sides are faced in stone or slathered with concrete to keep them from collapsing. A well's lifespan rarely exceeds 20 years. These "sand gardens," consisting of adjoining patches of about 3 feet (1 m) square, produce vegetables (peas, broad beans, lentils, beans, cabbage, salad, and groundnuts), essentially for local consumption. The growth of market gardening in Mali is a result of the great droughts of 1973–1975 and 1983–1985. These killed large numbers of livestock belonging to the herders in the north of the country, forcing some of them to settle down and turn to agriculture.

02

SEPTEMBER

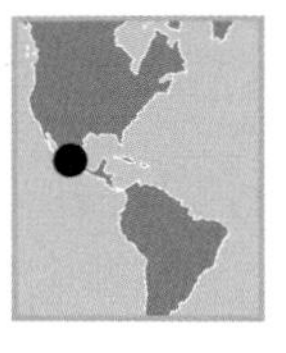

Mexcaltitán, Nayarit state, Mexico (21°54' N, 105°28' W).

Rising from the marshy meanderings of a vast coastal lagoon, the village of Mexcaltitán rests on an isolated spit of sand 1,300 feet (400 m) long on the northwest Pacific coast of Mexico in the state of Nayarit. In September, toward the end of the rainy season, the waters of the lagoon flood the village streets, forcing the inhabitants to travel by canoe and giving the place the look of a "Mexican Venice." Some historians see the village as the mythical island of Aztlán, where the Aztecs reputedly originated. Half land, half water, Mexcaltitán reflects the rich natural heritage that surrounds it—a network of canals threads through the mangroves, home to more 300 species of birds. Mexico's biodiversity is among the richest on the planet. With just 1.4 percent of the world's land surface, it has more mammal species than any other country (450). Mexico is also home to 10 percent of the known species of each animal and vegetable genus.

03

SEPTEMBER

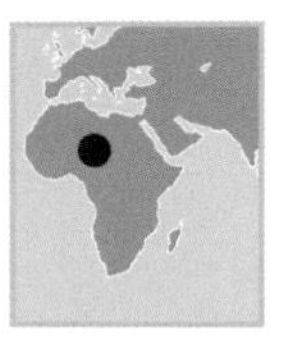

Waste from the Arlit uranium mine, Aïr Massif, Niger (19°00' N, 7°38' E).

The Aïr Massif is the backbone of Niger and the native land of the Kel-Aïr, a tribe of nomadic Tuaregs who have come to dominate the whole of the western Sahara. In 1965, the discovery of uranium deposits in this arid region lured people from all over West Africa with the hope of work. Since then, the Arlit mines have produced almost 3,000 tons a year of this precious mineral, or 8 percent of world output, putting Niger third, after Canada and Australia, among world producers. Niger's export revenue, which is heavily dependent on uranium, fell by 16 percent between 2000 and 2001. This was chiefly the result of the fall in the price of uranium, which supplies Europe's nuclear industry. Despite its underground riches, which are essential to some Western countries' energy supplies, this country in the Sahel is the world's second poorest.

04

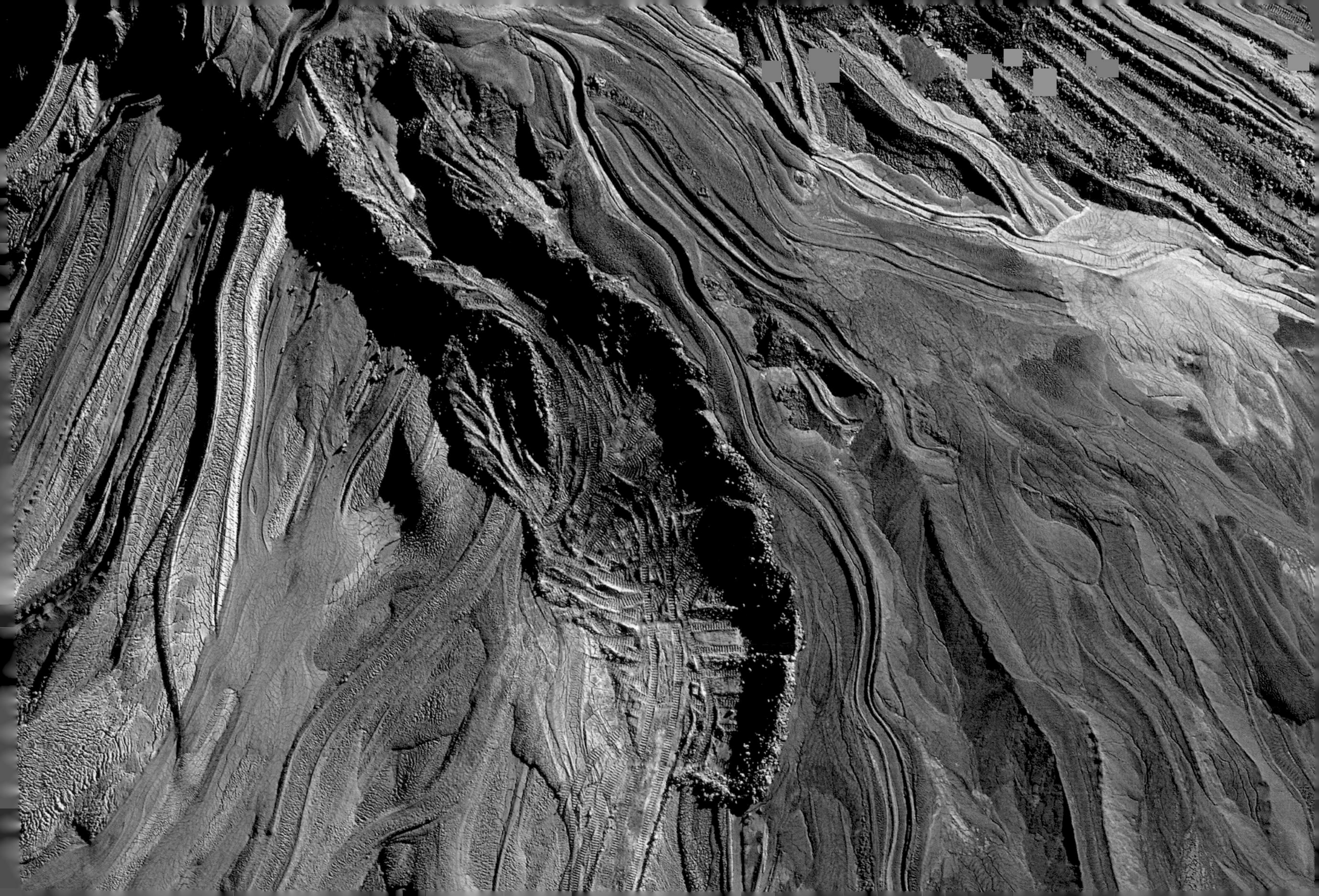

SEPTEMBER

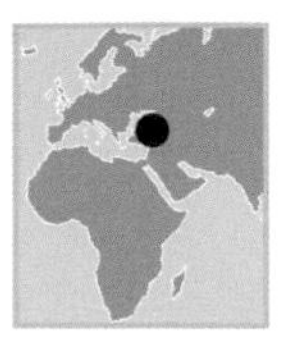

Snowdrifts on the Mount Lebanon range, Lebanon (34°09' N, 35°59' E).

On the western slope of the Mount Lebanon range, whose highest point is 10,109 feet (3,083 m) above sea level, the spring sunshine slowly erases the zebralike stripes formed by these snowdrifts. The grooves in which the snow piles up are worn by erosion that furrows these sandy slopes, denuded by overgrazing and deforestation (at a rate of 3,212 acres, or 1,300 ha, per year). Of the country's surface area, 65 percent is affected by erosion. In this area, winter sports are an attractive development, and are contributing to the recovery of the tourist industry, which is growing at a rate 10 percent per year—in 2000, there were 742,000 visitors. Before the war, tourism accounted for 20 percent of gross domestic product—in 1974, there were 1.5 million tourists. For the 2002 summer season, a majority of visitors to the country was expected to come from the Arab world. Indeed, these tourists, less at ease in Europe and the West since the attacks of September 11, 2001, are favoring Lebanon, whose blend of Arab and Western culture is a considerable asset for tourism.

05

SEPTEMBER

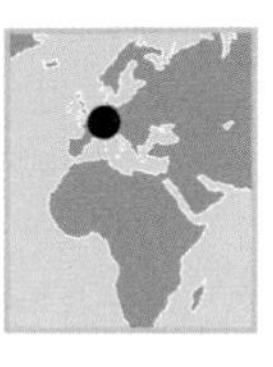

The Reichstag and its dome, Berlin, Germany (52°31' N, 13°25' E).

The new Berlin is being unveiled as building after building is inaugurated. Since the country was reunified, 12 years of reconstruction have transformed the German capital. What was once the symbol of the division of Europe is now a meeting place between the Continent's east and west. In this historic yet modern city, glass and steel mingle with nineteenth- and twentieth-century architecture. The renovation of the Reichstag, the seat of the German Bundestag built between 1884 and 1894 to a plan by Paul Wallot, was entrusted to the British architect Sir Norman Foster between 1995 and 1999. The original outline has been restored to the building, which lost its imposing dome when the Nazis set fire to it in 1933. A staircase spirals its way up inside the translucent glass dome, leading to its 154-foot (47-m) apex. The day it opened, 24,000 people flocked to see the view over the city.

06

DEM DEUTSCHEN VOLKE

SEPTEMBER

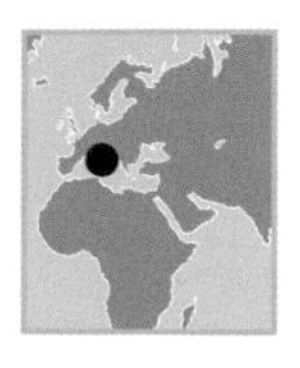

Countryside around Siena, Tuscany, Italy (43°19' N, 11°19' E).

Washed by the Tyrrhenian Sea, Tuscany in central Italy is one of the loveliest regions of the entire peninsula. Tuscany owes part of its fame to its hills, covered with vineyards, olive trees, corn and barley fields, and punctuated with medieval villages. Tuscans developed "ecotourism" at an early date. Local enterprises, especially artisans, now adhere to Social Accountability 8000 (SA 8000), an international standard based on the principles of human rights delineated in International Labour Organization Conventions, the U.N. Convention on the Rights of the Child, and the Universal Declaration of Human Rights. SA 8000 is a voluntary standard that ensures that economic development occurs through the ethical sourcing of goods and services.

07

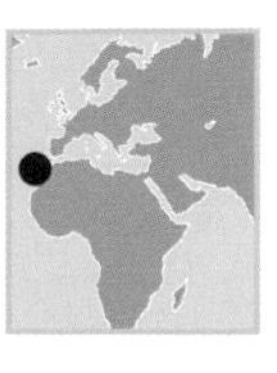

Montaña de Taco, near Buenavista, Tenerife, Canary Islands, Spain (28°21' N, 16°48' W).

This 1,055-foot (322-m) volcanic dome appears to stand guard over the farming region of Buenavista. Its ancient crater has been converted into a reservoir to supply the banana plantations that cover the coastal plain of northwestern Tenerife. Banana plantations, which thrive on the fertile volcanic soil and subtropical climate, account for a third of the Canaries' agricultural produce. However, the islands suffer a serious shortage of fresh water, chiefly because rainfall is low and there are no rivers. Two of the seven islands depend almost entirely on desalinated seawater. The shortage affects more than 20 percent of the islands' population, and the influx of tourists, which is already considerable, is relentlessly increasing demand.

SEPTEMBER

09

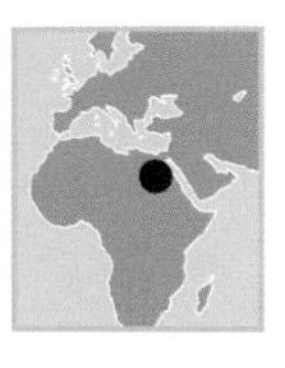

Modern graves in a cemetery at Asyut, Nile valley, Egypt (27°11' N, 31°11' E).

The idea of eternal life, so dear to the ancient Egyptians, is conveyed through a style of funerary architecture that stands the test of time. These tombs are divided into two sections, one representing the life of the deceased and the other containing the person's remains and the objects customarily regarded as making life in the hereafter more pleasant. The world of the living coexists with that of the dead, and cemeteries are close to towns. An Egyptian city of the dead can stretch over several miles (or kilometers) and is laid out like a town, with a rich variety of open spaces and architecture. With the passage of time, the commingling of the worlds of the living and the dead becomes more obvious. A combination of deregulation of the rental market, a serious shortage of social housing, and evictions, with no system for compensation or finding alternative accommodation, have driven some people to live in cemeteries. In Cairo, a megalopolis of 16 million people, the famous City of the Dead is thought to accommodate between 500,000 and 1 million underprivileged people—at least 20,000 of whom are living in the tombs themselves.

SEPTEMBER

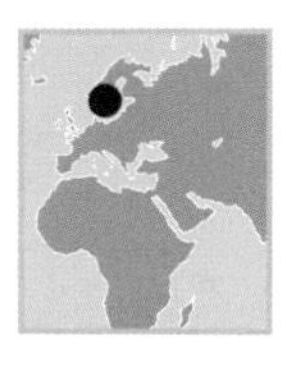

Gamla Stan, Stockholm, Sweden (59°20' N, 18°03' E).

Where the graceful city of Stockholm stands, the fresh waters of Lake Mälaren run into the salty undertow of the Baltic, and the land looks as if it has been torn into shreds. For the Swedish novelist Selma Lagerlöf, Stockholm is "the city that floats on the water." Indeed, the capital is built on fourteen islands, linked by forty bridges and countless ferries. These islands belong to the vast Skärgården archipelago, whose 24,000 islands are scattered over the Baltic. This photograph immortalizes Stockholm's 750th anniversary, which thousands of its 750,000 inhabitants flooded the cobbled streets of Gamla Stan to celebrate in the spring of 2002. The "old town," the city's medieval heart, contains the royal palace (Kungliga Slottet), whose imposing square shape is visible. In 1972, Stockholm hosted the first United Nations conference on the environment. Since then, humans have still not given up certain practices and policies that are unviable in the long term and which, in attacking the environment, "affect the well-being of populations and economic development throughout the world."

10

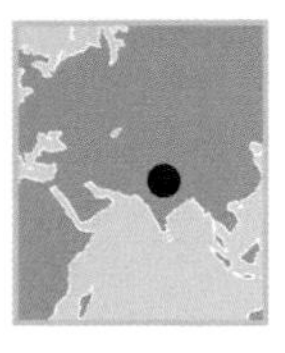

Rice paddies west of Katmandu, Nepal (27°45' N, 88°15' E).

Nepal's subtropical climate, heavy monsoon rains, and fertile alluvial soil are ideally suited to growing rice, the country's main agricultural product. However, investment is lacking, and the division of land into small plots imposes a system of subsistence farming in an economy based on barter. With 42 percent of its population living below the poverty threshold, Nepal is one of the poorest and least developed countries on the planet. In 2000, the average life expectancy at birth was only fifty-nine, compared to seventy-nine in Switzerland. Children are especially affected by poverty and poor living conditions: out of 1,000 births, eighty-two children will die before the age of a year; in France, for example, the rate is 4.5 per thousand. Although Nepal's constitution prohibits children from working in industry before the age of sixteen or in farming before fourteen, of the country's 6.2 million children aged five to fourteen, 40 percent work. The remaining 60 percent combine work with schooling.

11

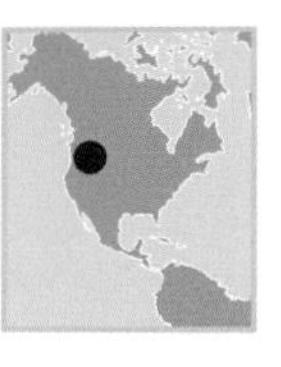

Power lines in a field near Idaho Falls, Idaho, United States (43°28' N, 112°02' W).

The population of the United States is concentrated on its Atlantic and Pacific coasts, leaving the vast lands in between almost empty by comparison. The Great Plains and the Rocky Mountains are exploited to produce energy and food for the coastal inhabitants. Thus, in Idaho, north of the Rocky Mountains, the Snake River has been divided up by dams, which produce electricity and retain water to irrigate the rich surrounding farmland. The dams are built, and the land farmed, on a large, highly mechanized scale by big companies or cooperatives. Paradoxically, these projects recall the unlamented *kolkhoz* (collective farms) of the former Soviet Union. High yields and small workforces ensure that America is competitive and make it the world's biggest cereal exporter, dominating the market—a fact the country does not hesitate to use as a political weapon.

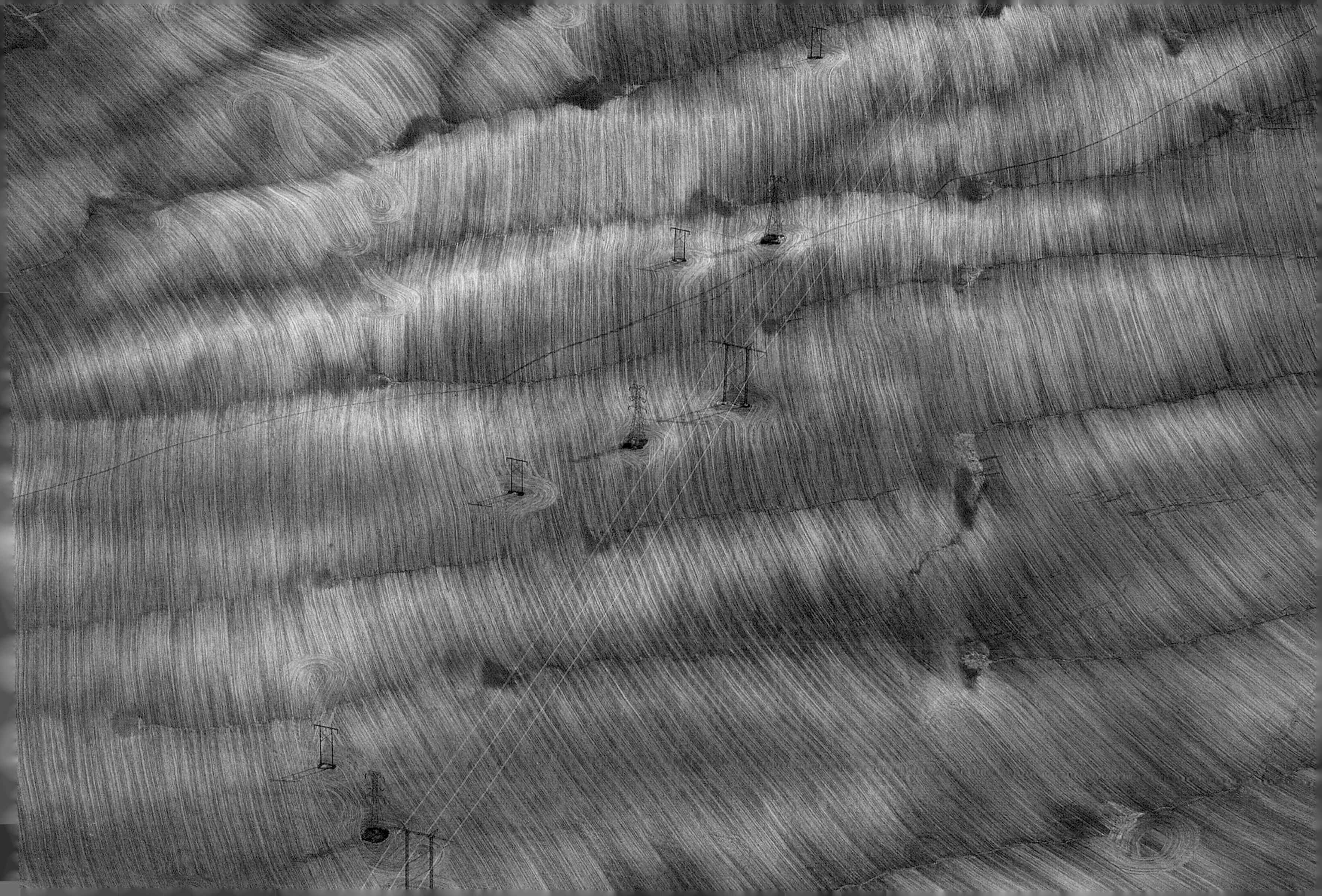

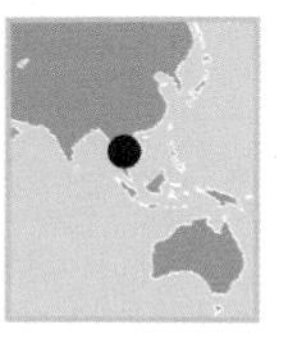

Phumi Kantrab, village on stilts in the Tonle Sap lake, Cambodia (13°11' N, 103°57' E).

A veritable inland sea, classed as a biosphere reserve by UNESCO in 1997, the lake of Tonle Sap undergoes wide variations in water level. Its surface area ranges from 1,158 square miles (3,000 km^2) in the dry season to 3,860 square miles (10,000 km^2) during the monsoon; as a result, it plays a part in regulating the levels of the River Mekong. This natural cycle, combined with the abundant forest cover, makes it one of the most prolific freshwater ecosystems in the world. The lake is believed to be home to 850 species of fish, including the rare Mekong giant catfish, which can grow to 10 feet (3 m) in length, while the flooded forests contain Southeast Asia's biggest water-bird colony. With its abundant food and fertile land, the Tonle Sap accommodates about 170 floating villages or villages built on stilts. However, the region is threatened by heavy sedimentation (a result of soil erosion exacerbated by deforestation), while numerous dam projects along the Laotian and Chinese tracts of the Mekong threaten to disrupt the river's flood cycle, which is essential to the lake's well-being.

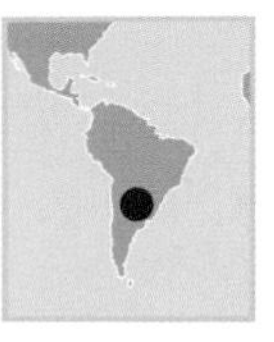

Gaucho horsemen, Neuquén province, Argentina (39°00' S, 70°00' W).

Gauchos are inseparably associated with the vast expanses of the pampas, often sporting their traditional dress of black hat and two-colored Indian poncho. These experienced horsemen are descended from mixed marriages between Indians and Spanish colonizers who were often fugitives or adventurers. Originally they were nomads who loved their freedom, but since the end of the eighteenth century, the increase in private land ownership has obliged them gradually to become sedentary, and they have turned to herding animals. Gaucho culture, which survives in remote areas of Argentina, is still tied to the *estancias*, vast farms that helped make the country the world's biggest beef exporter for many years. Although Argentina is still the world's fourth-biggest food exporter, 20 percent of children in the country were malnourished in 2002. Ever since the economic and political crisis that came to a head in December 2001, social conditions have progressively worsened. Today, more than half of all Argentines live below the poverty line—a 40-percent increase between 2001 and 2002.

SEPTEMBER

15

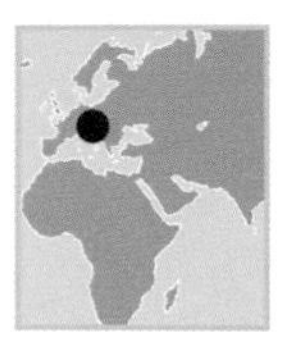

Sunset over the Grossglockner Massif, Hohe Tauern range, Alps, Austria (47°04' N, 12°42' E).

The year 2002 was the International Year of Mountains, a United Nations initiative intended to draw attention to the problems facing fragile highland areas on which whole communities depend, as well as to the need to manage the planet's mountains sustainably. The Alps are Europe's biggest mountain range, stretching for more than 620 miles (1,000 km) from the Mediterranean to Vienna. In Austria—where more than two-thirds of the country is taken up by the eastern Alps and their 680 peaks that rise above 9800 feet (3,000 m)—mountains are a culture in their own right. The country has more than 13,660 miles (22,000 km) of downhill ski *pistes*, close to 10,000 miles (16,000 km) of cross-country ski trails, and more than 500 high-mountain huts. The Hohe Tauern National Park, set up in 1981, contains 695 square miles (1,800 km^2) of mountain slopes on the eastern side of the Grossglockner, at 12,450 feet (3,797 m) the highest peak of the Austrian Alps. Here, tourism coexists harmoniously with the preservation of cultural and natural heritage.

SEPTEMBER

16

Slash-and-burn farming near the artificial reservoir of Guri, Bolivar state, Venezuela (7°30' N, 62°50' W).

Around the vast lake formed by the world's biggest hydroelectric dam—Guri, in Bolivar state, southeastern Venezuela—small farmers grow cassava and maize on patches of land cleared by burning. Slash-and-burn is an itinerant form of agriculture, in which fire is used to clear land to grow crops for sale or for subsistence. After two or three years, the land is abandoned because it has become infertile. Slash-and-burn is seen as the prime cause of deforestation in Latin America. Although almost half the region is still covered by natural forest, deforestation is proceeding rapidly. About 18.5 million acres, or 7.5 million hectacres, (0.77 percent) of the forest disappears each year, and over the last thirty years almost 470 million acres, or 190 million hectares, (the equivalent of the whole of Mexico) have been destroyed. Central America and Mexico have the highest deforestation rate in the world: 1.6 percent per year.

SEPTEMBER

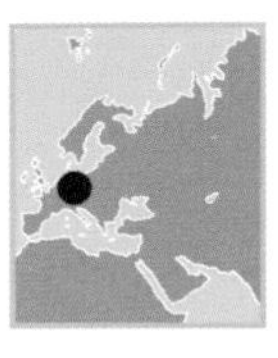

Heuersdorf lignite mine, Saxony, Germany (51°10' N, 12°22' E).

The landscapes produced by mining, which cover large areas south of Leipzig in eastern Germany, are a veritable cultural heritage. Lignite is formed by the decomposition of vegetation and is used to fuel power stations. Since it does not lie deep underground, it can be extracted by vast opencast mines, as here at Heuersdorf. In autumn of 1990, Germany took on the challenge of transforming the centralized, obsolete Communist system into a competitive market economy. The country spent huge sums on cleaning up industrial areas, improving energy efficiency, applying environmental standards, and harmonizing wages. The gulf that separated East and West Germany for more than 40 years is still not completely bridged: for example, unemployment in the former East Germany stands at 18 percent, twice the rate in the western *Länder* (states). The German federal government has invested 600 billions euros (about $690 billion) over 10 years on reunifying the country. The European Union will spend only a fifteenth of that sum (no more than 0.08 percent of its annual gross domestic product) during the first three years after it welcomes ten new member states in 2004.

17

SEPTEMBER

18

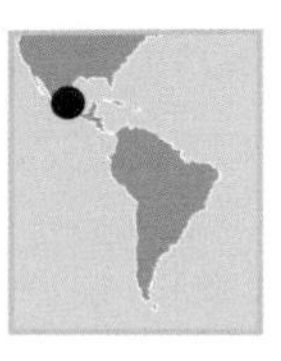

Mural, Mexico City, Mexico (19°20' N, 99°45' W).

The muralist movement sprang from the Mexican revolution in 1910, with roots in the country's history and its workers' movement. An expressionist and nationalist school, muralism promotes a typically Mexican art form that has astonished, even scandalized, the world. Muralists initially used walls to evade censorship, but they became almost official artists when, in 1921, the new government asked them to help launch a popular development campaign. Gigantic frescos appeared on the walls of government buildings. Bearing a message for the masses, muralism allowed Mexican art to assert itself. Today it is seen as a breeding ground for ideas and modes of expression. Its influence has been enormous throughout Latin America and in some U.S. cities, notably San Francisco.

SEPTEMBER

19

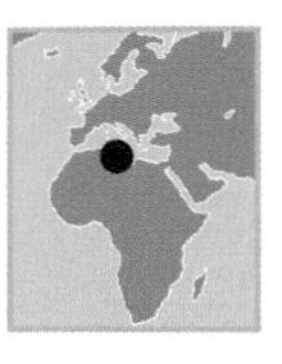

Field on hillsides of the Ksar valley, Tataouine governorate, Tunisia (33°08' N, 11°25' E).

This animal shape traced on the hillsides is a plot of land used for cultivating grain. The furrows and outline of the field, which is terraced, follow the contours of the ground and retain water from the rare but torrential rains. In this way runoff is limited, water is made available for farming, and the precious soil is preserved. Limiting soil erosion and gathering rainwater are top priorities for the Tunisian government. The country loses more than 37,000 acres (15,000 ha) of soil per year, and it is estimated that close to 654 million cubic yards (500 million m^3) of rainwater enter the sea through runoff. All over Africa there is a significant risk of soil degradation. Population growth, and poverty in certain regions, are raising fears that large areas of forest will be felled to meet demand for food, fuel, medicines, and building land. The soil thus exposed would be in greater danger of being eroded.

SEPTEMBER

20

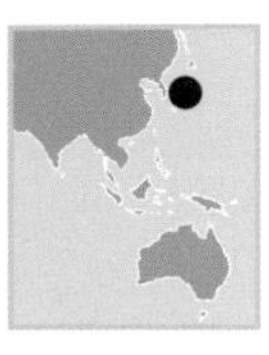

The Golden Pavilion, Kyoto, Japan (35°00' N, 135°45' E).

Kyoto, formerly Heian, was the seat of the imperial court for eleven centuries, from 794 to 1868. During the Heian period, around AD 1000, Murasaki Shikibu, a lady at the court, wrote *The Tale of Genji*, one of the great works of Japanese and world literature. This chronicle of court life is famous for the psychological subtlety of its characterization, which made its author the first novelist in history. For centuries, Kyoto was a place of refinement and a thriving center for the arts, even during Japan's periods of internal upheaval. The Golden Pavilion bears witness to this. The shogun (first minister) Ashikaga Yoshimitsu built the pavilion in 1397 in the *shoïn* style, with tatamis and movable walls and partitions that were imitated in many later buildings. The refinement and subtlety of Japanese art greatly influenced the West, especially at the end of the nineteenth century with the development of the Art Nouveau and later the Art Déco styles, whose aesthetic was directly inspired by that of the Land of the Rising Sun.

SEPTEMBER

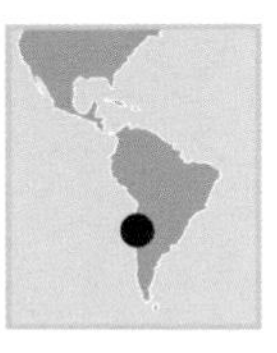

Chuquicamata copper mine, Chile (22°19' S, 68°56' W).

Chuquicamata is the world's biggest opencast copper mine. A mile and a quarter (2 km) wide by nearly 2 miles (3 km) long, it delves 2,300 feet (700 m) deep into the earth. Its highly sought-after ore is the world's richest in copper but, unfortunately, it also contains a high level of sulfate. This toxic substance is present in the clouds of dust that the trucks release with their constant coming and going, as well as by the explosions used to detach the ore from the pit's walls. The mine workers breathe these sulphurate particles all day long and, although they are not allowed to work in the mine for more than three years, they still risk dying from lung cancer. The region's other inhabitants are not spared this pollution either. There is a high rate of respiratory illness in the town near Chuquicamata. Every year, 500,000 people in developing countries die from the high level of sulfur dioxide (SO_2) in the atmosphere.

21

SEPTEMBER

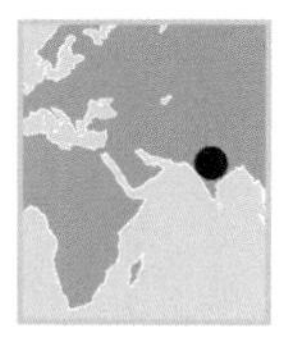

Well at Fatehpur Sikri, Uttar Pradesh, India (27°06' N, 77°40' E).

The Great Mogul Akbar, emperor of India, built the city of Fatehpur Sikri in 1573 to celebrate his victory over the Afghans. He installed his court magnificently in red sandstone palaces built on a rocky plateau, 897 miles (38 km) from the imperial city of Agra. Fatehpur Sikri has often been described as a Versailles to Agra's Paris. The similarity ends there, however, for Fatehpur Sikri was abandoned 15 years after it was completed. This was probably due to the exhaustion of ground water supplies, which were used to maintain parks and fill pools—as testified by the depth of the wells now used by farmers. Water consumption in India is greater than the natural replenishment of ground water; consequently, the water table has fallen by between 3.25 and 9.83 feet (1 and 3 m) over 75 percent of the country. This deficit is endangering Indians' food supplies, and is a threat to lakes, rivers, and other wetland ecosystems.

22

The medieval city, Dubrovnik, Croatia (42°39' N, 18°04' E).

"Freedom is not for sale—not for all the gold in the world," reads the motto at the gate of Dubrovnik, a city on the Dalmatian coast founded in the seventh century. The "Pearl of the Adriatic," Dubrovnik was Venice's formidable rival until the eighteenth century; the city succeeded in retaining its autonomy until 1808, when Napoleon invaded. Described as "heaven on Earth" by the Irish dramatist George Bernard Shaw, this city's imposing walls enclosed churches, monasteries, and Gothic, Renaissance, and Baroque palaces. In the autumn of 1991, more than 2,000 shells smashed into these architectural treasures, hitting 563 of the 824 buildings. This bombardment, carried out by the Serbian army of Slobodan Milosevic in response to Croatia's declaration of independence, led UNESCO to put the city on its list of endangered World Heritage sites. By 1998, after tireless work by architects, sculptors, and restorers, this unique architectural legacy had been saved. But the six months of fighting in Croatia killed 13,000 people, wounded 40,000, and displaced many thousands more.

SEPTEMBER

Surtsey island, Vestmann islands, Iceland (63°16' N, 20°32' W).

The island of Surtsey is the world's youngest. On November 14, 1963, about 12.4 miles (20 km) from the Vestmann archipelago, the waters boiled and, amid cascading volcanic eruptions and clouds of ash, a gap appeared. The eruption lasted four years. Four volcanic islands rose from this fissure but only one, buttressed by the large volume of lava that formed it, survived the combined action of wind and waves. The island is named Surtsey after the Norse god Sutr, who spreads fire round the world. Forty years on, erosion has obliterated almost half of this little scrap of land, which now amounts to less than three-quarters of a square mile (2 km^2). Surtsey is a protected area, affording scientists the opportunity to study the establishment of an ecosystem in a virgin setting untouched by any form of life. In this natural laboratory, fifty types of plant have taken root. Seeds have been brought by the wind, the sea, and by birds, of which seven species have nested and thirty species have frequented the island.

24

SEPTEMBER

25

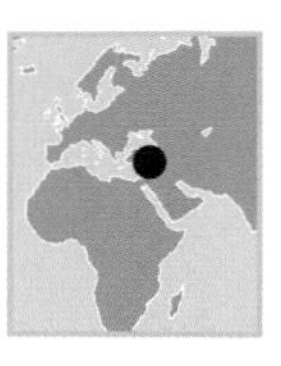

Remains of an ancient trap or "desert-kite" between As Safawi and Qasr Burqu, Mafraq, Jordan (32°28' N, 37°34' E).

The 700 to 800 "desert kites" dotted throughout the Middle East owe their name to British postal service pilots of the 1920s. They were built by hunters in the Neolithic period, probably nomadic, who would drive the herds of gazelles living in the valleys between their walls, 0.6 to 1.2 miles (1 or 2 km) long. Their funnel shape led the game into an enclosure about 1,000 feet (several hundred meters) in circumference, often hidden behind the crest of a hill. The panic-stricken animals would scatter in this circle, round which groups of hunters armed with spears hid, ready to ambush them. Petroglyphs depicting such scenes can be found from the Caucasus to the Sinai. The engravers used the rock surface itself as a relief of the landscape, thus producing a model of the "desert kite."

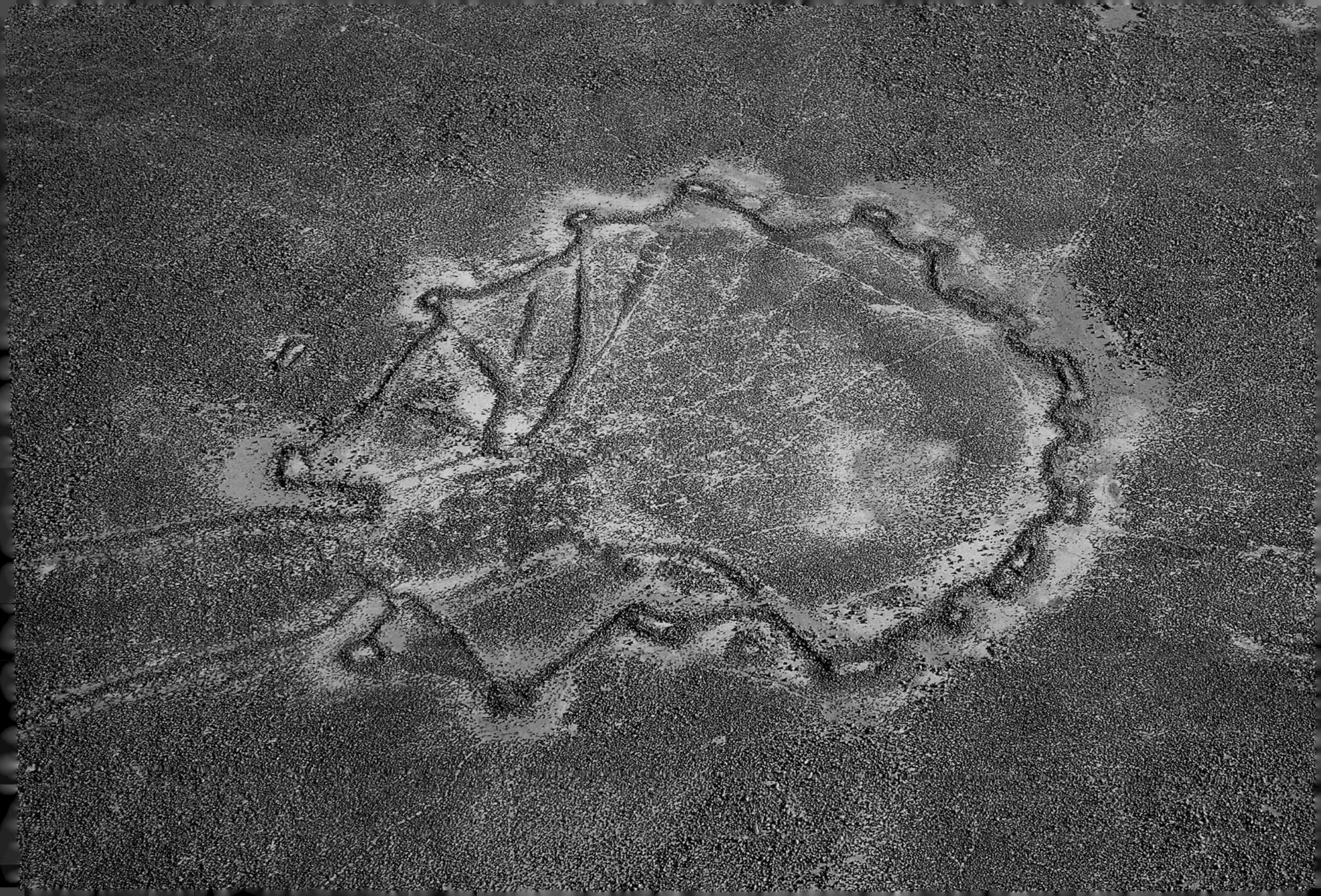

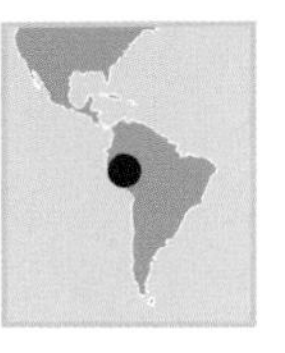

The Inca city of Machu Picchu, Cuzco region, Peru (13°05' S, 72°35' W).
Machu Picchu was built in about 1450 in the Urubamba valley, about 62 miles (100 km) northeast of Cuzco. Lying at about 8,200 feet (2,500 m) above sea level, this sanctuary built of stone blocks straddles the crest of a spur of the Andes. The land was terraced to allow building and agriculture, and the whole is wonderfully integrated with its surroundings. Since 1982, the site has been on UNESCO's World Heritage list, on the grounds of both its cultural and natural history value. However, it is fragile and constantly threatened with degradation. Rain perpetually causes erosion in this precipitous region but, more worrying, there is pressure from tourism. The authorities must choose whether to protect one of the most outstanding examples of humanity's heritage, or to enjoy the revenue from the country's prime tourist site.

26

SEPTEMBER

27

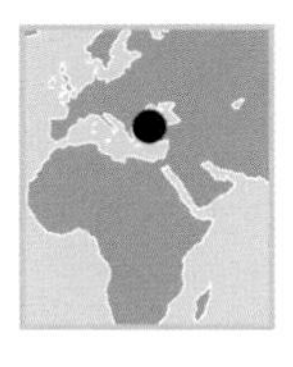

Süleymaniye Camii Mosque, Istanbul, Turkey (41°00' N, 28°57' E).

The imperial mosque, the most impressive building in the Ottoman Empire, was built by Sinan on the orders of Suleiman the Magnificent between 1550 and 1557. Towering over the Golden Horn, the mosque is at the center of a vast complex containing five primary and secondary schools, a charitable public kitchen, a free hostel for travelers, a hospital, and baths. This brilliant intellectual and artistic center recalls the time shortly before its decline when the Ottoman Empire was a shining example to the Arab world and Europe. Now, Turkey is once again at a turning point in its history. Its candidacy for entry to the European Union was accepted on principle in 1999, but the Turkish government has not always taken the steps required. Also, there are thorny problems—such as the Kurdish question, Turkey's relations with Greece, and its human rights record—which may well jeopardize this ambition. Although the use of torture and arbitrary detention decreased during the first half of 2000, there still remained 10,000 political prisoners in Turkey.

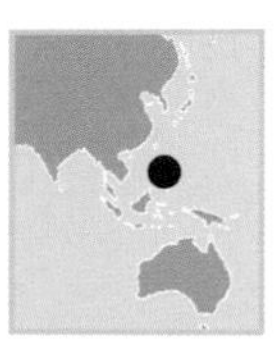

Mount Pinatubo, a volcano north of Manila, Luzon, Philippines (15°08' N, 120°21' E).

Mount Pinatubo's eruption in 1991, the biggest of the twentieth century, blew 30 million tons of sulphates into the atmosphere to altitudes of up to 15.5 miles (25 km). These formed a veil of aerosols that temporarily lowered the amount of solar radiation received by the planet, from 200 watts to 196 watts per square meter. The result during 1992 and 1993 was a global fall in ground temperatures of several tenths of a degree. The consequences of two other violent eruptions—Mount Agung, Indonesia, in 1963 and El Chichón, Mexico, in 1982—has already been noted. However, the atmospheric and climatic effects of such events are temporary and should not divert attention from the global warming induced by human activity, especially deforestation and growing consumption of fossil fuels. The latter has increased fourfold over the last 50 years, while the world's population has only doubled.

SEPTEMBER

29

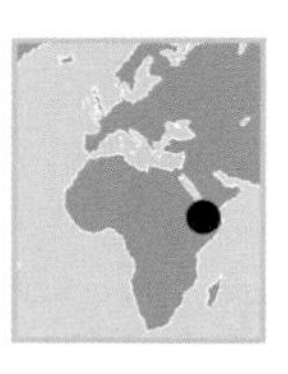

Herd of goats among the chimneys of Lake Abbé, Djibouti (11°06' N, 41°50' E).

Geographic conditions make life difficult in the republic of Djibouti. The desert climate, with its extremely erratic rainfall, has caused many droughts, including the severe one of 1980, which killed almost all the country's livestock. The droughts are the reason for the sparse vegetation, consisting of shrubs and prickly bushes, which can barely feed the herds of sheep, camels, and goats. Droughts are on the increase: over the last 30 years, rainfall has dropped by an average of 6 to 15 percent. This has led to a progressive decline in the nomadic lifestyle of the almost 80,000 shepherds belonging to Djibouti's two main ethnic groups: the Afars (37 percent of the population) and the Issa-Somali (50 percent). The end of colonialism and the decline of traditional trade routes with the Orient have hit Djibouti hard. They have been even more damaging to Aden, on the opposite shore of the Red Sea, which was an essential port of call for British ships en route to the Indies.

SEPTEMBER

30

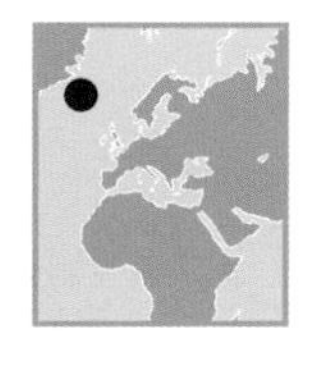

Pingvellir Fault east of Reykjavik, Iceland (64°18' N, 21°08' W).

Iceland is situated at the emergence of the mid-Atlantic submarine dorsal and thus stands at the junction of two tectonic plates. The island is distended by the force of the volcanic action of this rift, which, through the production of magma, is driving Europe and North America apart at an average rate of nearly one inch (2 cm) per year. The narrow road bordering the cracks of the rock, shattered by enormous tectonic stresses, is like a house built at the water's edge. It reflects the customary hardiness of human societies, and a surprising faith in nature. The motion of the Earth's plates will continue far longer than the life span of ten generations. The rhythm of human life differs fundamentally from that of the movements of our Earth.

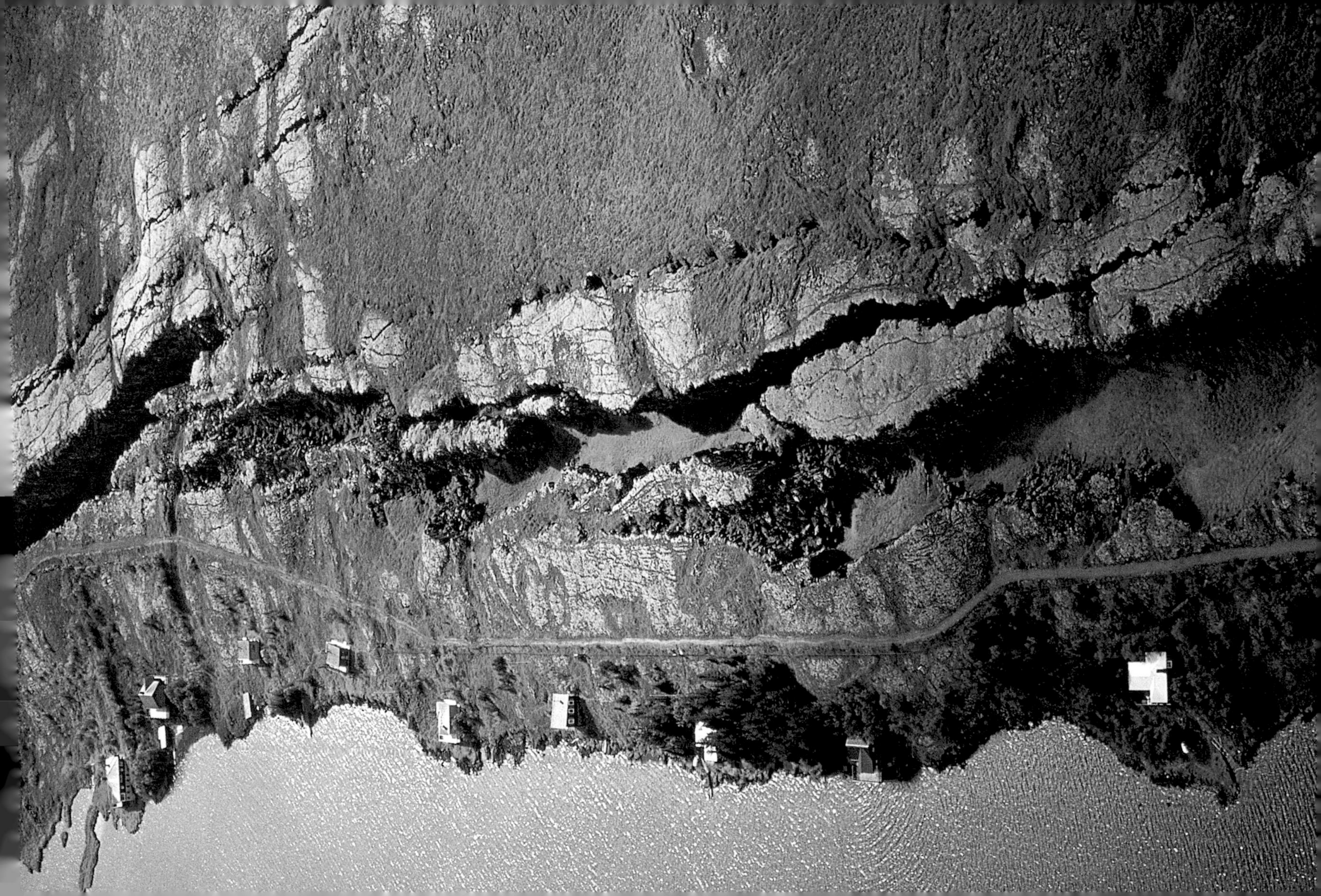

ABOLISHING POVERTY

OCTOBER

01	02	03	04	05

We live in a world that abounds in economic and natural resources and in an age of unprecedented wealth and technical prowess. And yet, today, 1.2 billion people—men, women, and children—live in abject, dehumanizing environments. Over 400 million people over the age of sixty-five dwell in impoverished conditions. Tens of millions of the world's urban children and adults are actually homeless, sleeping in unsafe public places, on pavements, in construction sites, or graveyards. Many low-income countries have child mortality rates as high as 100 to 200 per 1,000 live births. Worldwide, over 8 million children die every year from diseases caused by dirty water or poisoned air. And it is a known fact that these figures underestimate the situation.

In response to this sobering, unacceptable picture, the world community in the year 2000 has resolved to halve, by the year 2015, the proportion of the world's people whose income is less than one dollar a day. This laudable target has been chosen on the basis of definitions made by international financial institutions, who reckon that a person is poor who lives on less than one U.S. dollar per day. From the outset, this definition is conceptually flawed. To consider that someone living on $1.50 or $2.00 per day is not poor is to ignore, unacceptably, the impoverished existence of millions of inhabitants of this planet. Furthermore, the reality of poverty is much broader than what can be defined through income. Poverty is not just a question of income; it is an issue of the poor being denied their basic rights. It encompasses powerlessness, a sense of humiliation, violation of dignity, social isolation, and deprivation. Poverty is both the cause and consequence of a denial of human rights.

The condition of poverty is neither inevitable nor immutable. It is a constructed social and economic reality. The poor are not poor because they are physiologically or mentally inferior to others living in better conditions. Nor are they poor because they have a different set of values. On the contrary, poor people care about many of the things that other people care about: happiness, family, children, livelihood, peace, and security. Their poverty is often a direct or indirect consequence of society's failure to establish its social and economic relations on the basis of equity and fairness.

06 07 08 09 10 11 12 13 14 15

Numerous factors contribute to poverty, and they exist at every level from the most local to the international. At the national level, two factors are among the most complex and politically controversial. The first is the distribution of power, authority, and resources among different levels of government; second is the quality of "governance" in terms of its responsiveness, accountability, transparency, and the quality of its engagement with civil society. But what can be done to make governments in both developed and developing countries more accountable toward the poor?

At the 1993 World Conference on Human Rights, held in Vienna, poverty was deemed to be a human rights violation. Now that it is acknowledged that this is the case, the international community has a responsibility to strive to abolish poverty in the same way it is responsible for ridding the planet of slavery, torture, and, more recently, apartheid. All human rights stem from the precept that fundamental principles should be respected in the treatment of all human beings. States owe a duty to protect the life, liberty, and security of their citizens. They must create the conditions to meet the basic human need to live in peace with dignity.

Poverty dehumanizes half of the of the world's inhabitants, leaving the other half totally indifferent. That such massive and systematic daily violations do not shake the conscience of millions is certainly one of the most important moral issues in this century. On the one hand, equality in the attribution of rights is proclaimed, while on the other, inequality in the distribution of goods persists, maintained by unjust political and social policies at national and global levels. Treating poverty as a human rights violation implies realizing the concept of international justice. This subjects relations between states and nations and their citizens to one stricture of global justice, governing relations among human beings. All people would be equal, regardless of race, religion, creed, gender, or age. All people would live in a global society and benefit from absolute and indivisible rights—like the right to life, guaranteed by the international community. These rights are entitlements, and individual nations are the guarantors or duty bearers. But a moral obligation weighs on us all—particularly in the case of failed or impoverished nations—to help eradicate poverty. The principle of global justice, therefore, means establishing the conditions for a more equitable distribution of resources among the world's inhabitants so as to satisfy certain absolute rights.

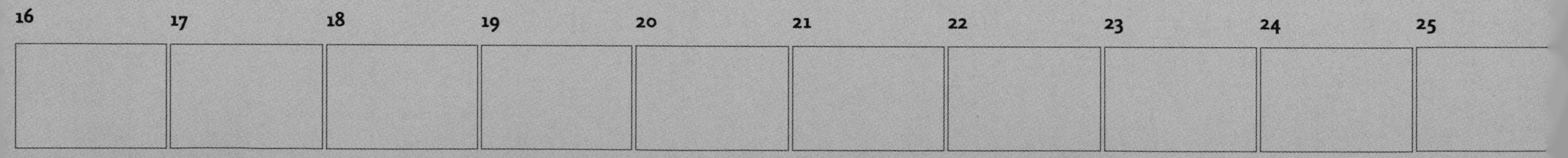

If we were to abolish poverty on the grounds that it is a systematic and continuous violation of human rights, the condition of poverty would have a new status. Instead of being a deplorable consequence or accepted status quo, it would become an injustice. With this shift in perception, the poor would become entitled to claim compensation on the grounds that their impoverished living conditions are a denial of all or part of their civil, cultural, economic, social, and political rights.

This shift, however, requires creating public awareness of the concept of universal justice, a goal that is within our reach, thanks in part to the Universal Declaration of Human Rights and the Rome Conference that established the International Criminal Court. Galvanizing public opinion, to make all citizens concerned with the prevalence of poverty as an affront to universal justice, is one step.

There is also a need to harness political will at the international and national levels. International justice must include a foundation of legislative measures that will hold perpetrators responsible. Such a legal framework would provide the poor with the right to gain compensation for their status.

It is true; we are all different. True, cultural diversity is a fact like biodiversity. True, we are all unique—the world is peopled by 6 billion unique individuals. And this is what makes our humanity. It is a moral, ethical, and legal imperative to abolish poverty worldwide so as to secure long-lasting peace and prosperity on this planet. And this goal must be embedded in the conscience of all.

PIERRE SANÉ
Assistant Director-General for Social and Human Sciences, UNESCO, France

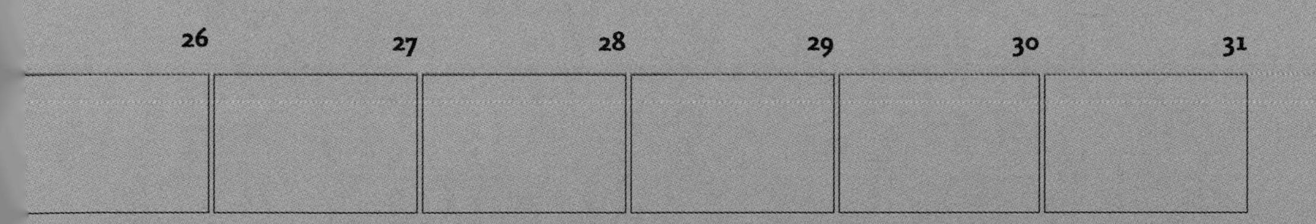

OCTOBER

Boat on the Pink Lake (Lake Retba), near Dakar, Senegal (14°45' N, 17°25' W).

These men are not paddling. Instead, they move by using a stick to break the salt crust that has formed under the surface of the lake, whose water owes its color to microorganisms living in it. Twenty years ago, the Pink Lake, once known as Lake Retba, was popular with fishermen because of its teeming waters, which were fed by the winter rains that flowed gradually off the surrounding dunes. But persistent drought interrupted this supply of fresh water and considerably reduced the lake's area. As a result of heavy evaporation, its salinity rose to a level comparable with the Dead Sea's—320 grams of salt per liter (the Atlantic contains 30 grams per liter). Fishing naturally gave way to salt extraction, at the rate of some 30 tons per day. Like other countries in the Sahel belt, Senegal is threatened by desertification. Nevertheless, it is one of the countries in Africa that is best supplied with drinking water—78 percent of the population has access to this vital resource.

01

OCTOBER

02

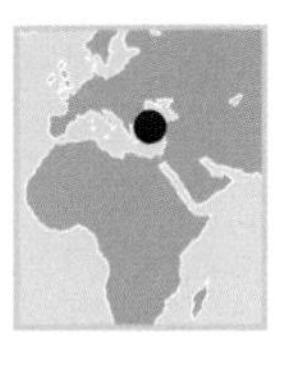

Earthquake at Gölcük, on the coast of the Sea of Marmara, Turkey (40°43' N, 29°48' E).

The earthquake that struck the region of Izmit on August 17, 1999, at 3:02 a.m., registered 7.4 on the Richter scale (9 is the maximum). Its epicenter was at Gölcük, an industrial city with a population of 65,000. The quake had an official death toll of at least 15,500 people, many buried in rubble while they slept. The partial or total collapse of 50,000 buildings led to outrage against building contractors, who were accused of disregarding earthquake-proof construction codes. Southern and northern Turkey are sliding along the North Anatolian fault at an average relative speed of 1 inch (2.5 cm) per year, but the motion actually occurs quite abruptly, in the form of earthquakes—the Earth moved nearly 10 feet (3 m) in less than a minute during the Izmit earthquake. Regions bordering tectonic plates, such as the trans-Asian zone running from the Azores to Indonesia by way of Turkey, Armenia, and Iran, are particularly exposed to seismic risk. Although they are rarer than storms and floods, earthquakes claimed 169,000 victims throughout the world between 1985 and 2000.

OCTOBER

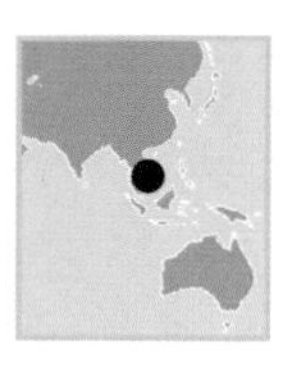

Phi Phi Le, near Phuket island, Thailand (8°00' N, 98°22' E).

The Phi Phi archipelago, 24.8 miles (40 km) off the coast of Thailand, consists of the islands of Phi Phi Don and Phi Phi Le. The latter is the more unspoiled of the two and is uninhabited as a result of the Thai government's efforts to combat the illegal traffic in swallows' nests. Consisting of threads of hardened saliva, the nests are highly prized for their tonic properties. These birds—in fact a type of swift—nest in karstic caves in cliffs that rise to a height of 1,226 feet (374 m). Fishermen come to gather this rare foodstuff by erecting fragile bamboo scaffolding. This "white gold" fetches up to $9,243 a pound (3,000 euros a kilogram); in the early 1990s, the trade was thought to have been worth $74.75 million (65 million euros). Between 1995 and 1999, some 1.5 million wild birds, 150,000 animal furs, and 1 million snakeskins were sold every year. The Convention on International Trade in Endangered Species of Wild Flora and Fauna (CITES), which came into force in January 1975, has been ratified by 150 countries, which have committed themselves to controlling trade in the 30,000 species in danger of extinction.

03

OCTOBER

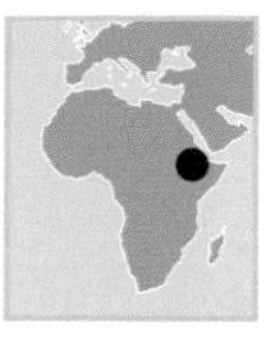

Bieta Ghiorghis monolithic church, Lalibela, Ethiopia (12°02' N, 39°02' E). In the heart of the high plateaus of western Ethiopia is the cruciform church of Bieta Ghiorghis. This 36-foot-high (11-m) building, carved directly out of a sandstone rock, is the masterpiece of the rock architecture of the Christian sanctuary of Lalibela. The sanctuary includes eleven monolithic churches built during the thirteenth century during the reign of King Gadla Lalibela. At the time, Jerusalem had been conquered by the Muslims, and the king wanted to offer Christians a new place of pilgrimage as a sort of replacement. Today, tens of thousands of the faithful, mostly Coptic Orthodox, still flock to this complex of rock buildings, which was added to UNESCO's World Heritage list in 1978. Orthodox Christianity is the majority religion in Ethiopia, which has more than 30 million Christians and 25 million Muslims. In this ecumenical country, almost all women, regardless of their religion, are circumcised.

04

OCTOBER

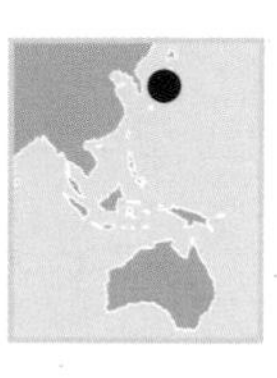

Pedestrians in the streets of Tokyo, Honshu, Japan (35°42' N, 139°46' E). Formerly known as Edo, the city was renamed Tokyo, or "eastern capital," by the emperor Meiji in 1868. With 28 million inhabitants, Tokyo is now the world's biggest megalopolis, stretching over a radius of 87 miles (140 km) along the coast. Repeatedly destroyed by fires, earthquakes, and especially by bombing in World War II, Tokyo is constantly undergoing change and boasts some bold architectural projects. But beyond the great arteries, beyond the sky laced with freeways is hidden a Tokyo of villages, with individual houses and small buildings, where the pedestrian and the bicycle reign. In this constant transition from anonymous megalopolis to convivial neighborhood living, Tokyo continues to surprise, with its houses lacking an address, its safety (the crime rate is among the world's lowest), and the civic sense of its inhabitants, who will restore to its owner property lost in a shop, train, or metro carriage.

05

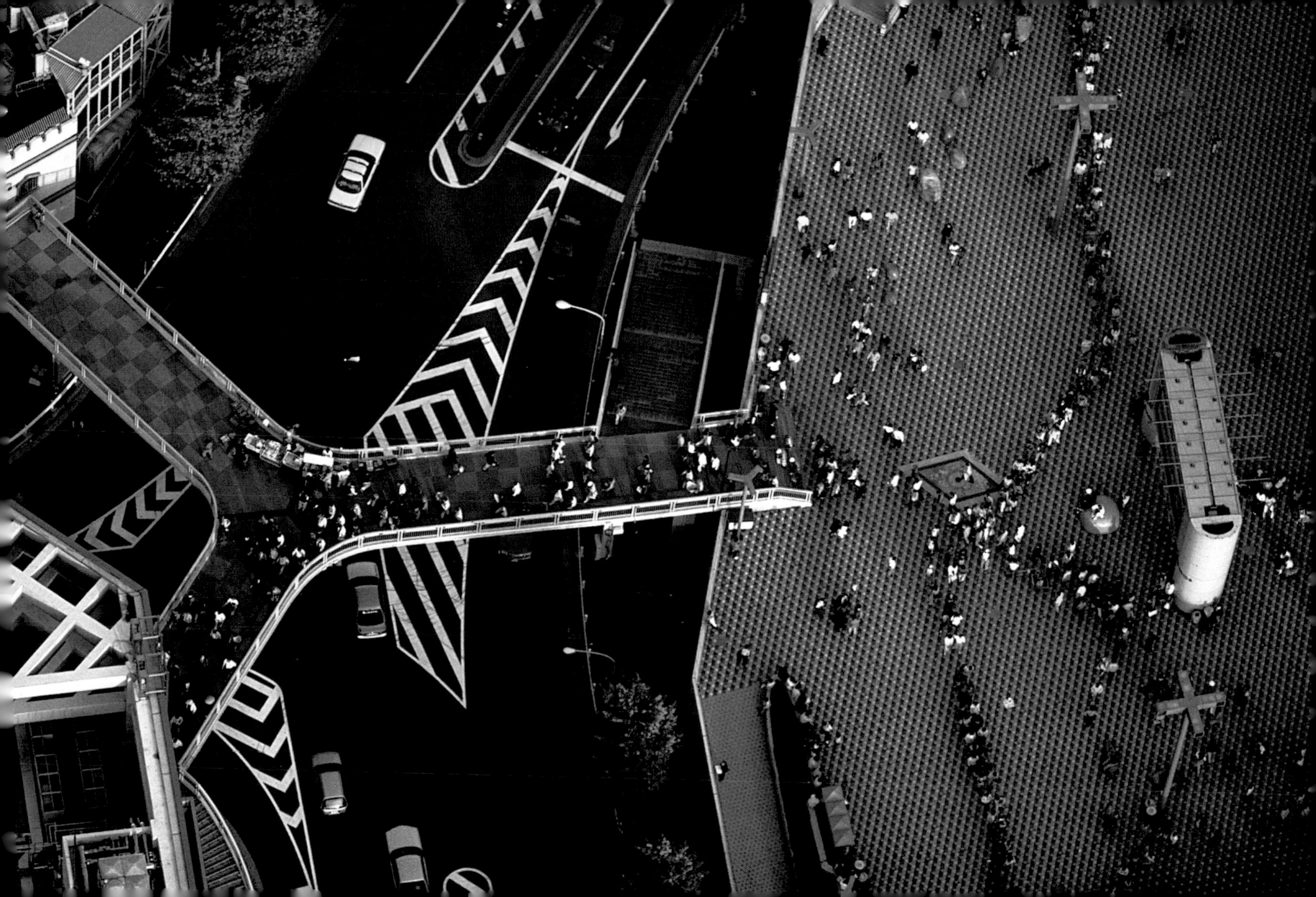

OCTOBER

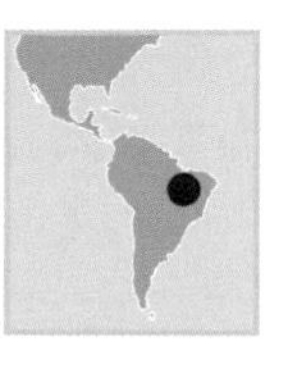

Gold mines near Pocone, Mato Grosso do Norte, Brazil (16°15' S, 56°37' W). The astonishing luster of this mine is due to the use of mercury to enrich gold-bearing rock—a technique that is widely used because it is cheap, simple, and effective. However, exposure to mercury vapor is extremely harmful for those who work there, from the gold-washers to the refiners, who risk poisoning—all the more so since hygiene is not given a high priority in mining. Mercury's toxicity is often unrecognized, and protective equipment is too expensive for small groups of gold-washers. Moreover, it has been estimated that 5,000 tons of mercury have been dumped in Latin America's forests and urban areas since the end of the 1970s—an amount almost equal to the quantity of gold produced.

06

OCTOBER

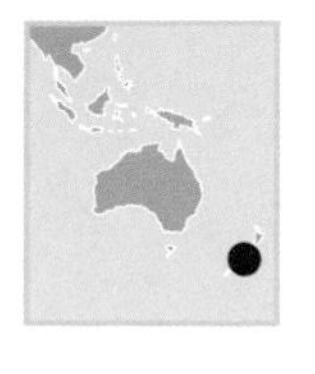

Rakaia river near Canterbury, South Island, New Zealand (43°20' S, 171°26' E).

East of the Southern Alps, the country's main mountain range, the Rakaia River is a network of channels, meanders, and great gravel banks. Several watercourses in the plains of Canterbury retain this kind of winding course, due to a drainage pattern that remains highly variable because humans have not tampered with it. They form a great system of "free" rivers, a rare ecosystem that is only found on such a scale in the Himalayas and in North America. The myriad alluvial islands are home to twenty-six species of waterbirds, which come to lay their eggs on the pebble beach or to feed on the abundant insects and fish. Several of these are waders found nowhere else in the world and, sadly, in danger of extinction. Only forty-eight specimens of the elegant black stilt (*Himantopus novazelandiae*) remained in 2000. The building of embankments and dams is chiefly responsible for the degradation of this natural habitat.

07

OCTOBER

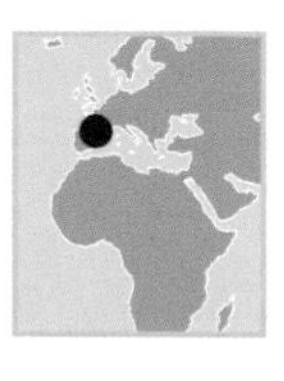

Guggenheim Museum Bilbao, Bilbao, Basque country, Spain (43°15' N, 2°58' W).

The Guggenheim Museum Bilbao, inaugurated in 1997, is part of a program of urban renewal in this industrial city. Built at a cost of $100 million, the structure was designed by the California architect Frank O. Gehry, with the help of a computer program used in aeronautics. Its glass, steel, and limestone construction, partially covered with titanium, echoes the city's shipbuilding tradition. Encompassing a total area of 250,000 square feet (24,000 m^2), the museum has 118,000 square feet (11,000 m^2) of exhibition space divided among 19 halls, including one of the world's largest galleries (310 by 100 feet, or 130 by 30 m). This cultural attraction has raised the number of visitors to Bilbao from 260,000 to more than 1 million each year. By energizing the local economy (the gross industrial product of the Basque region grew fivefold), the museum has also brought new life to the city.

08

OCTOBER

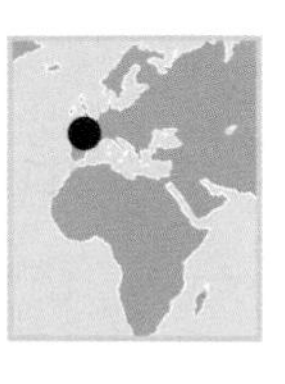

Low tide in the Gulf of Morbihan, Morbihan, France (47°34' N, 2°49' W).

These exposed mudflats hold in their grip the sailing boats left abandoned by the tide. The tides are the main agent in the Gulf of Morbihan's geomorphology, constantly reshaping this inland sea, which is 12.4 miles long (20 km) and almost as wide. The cyclical variation in the level of the oceans is due to the gravitational pull of heavenly bodies. Since the moon is the nearest, it is chiefly responsible for moving vast masses of water. But when the moon and sun are in alignment, the phenomenon is greatly accentuated, producing extreme tidal movements. In the Gulf of Morbihan, tidal flows are slowed down by the second-biggest bed of sea wrack in France. This aquatic plant helps to feed the 130,000 or so birds that spend the winter here. This richness has led to the site's listing under the Ramsar Convention, which covers wetlands of international importance.

09

OCTOBER

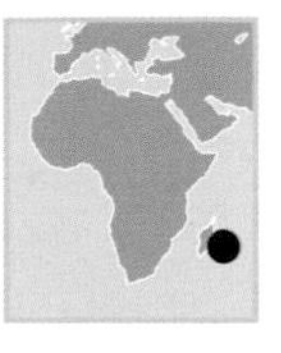

Helicopter in the Cirque de Salazie, Île de la Réunion, France (21°01' S, 55°32' E).

In the mountainous center of Réunion island, waterfalls streak the vegetation that clings to the precipitous walls of the Cirque de Salazie. Dominated by the Piton des Neiges (at 10,063 feet, or 3,069 m), this vast, 38.6-square-mile (100-km^2) crater is the most accessible of the three cirques on Réunion. Constant volcanic activity has given this island the most rugged relief in the Mascarene Islands, the group comprising Réunion, Mauritius, and Rodrigues. This region has been notorious for the serial extinctions of animal and plant species that have happened here ever since the first Europeans arrived in the sixteenth century. Although the dodo, devoured by the greed of sailors on Mauritius, is the most deeply symbolic of these extinctions, at least six other bird species and 100 plants, all endemic, have disappeared from the region, and therefore from the world, forever. A small bird of prey, the Mauritius kestrel, was brought back from the brink: its population rose from just four individuals in the wild in 1974 to 400 pairs in 1996, thanks to habitat conservation and the release into the wild of birds bred in captivity.

10

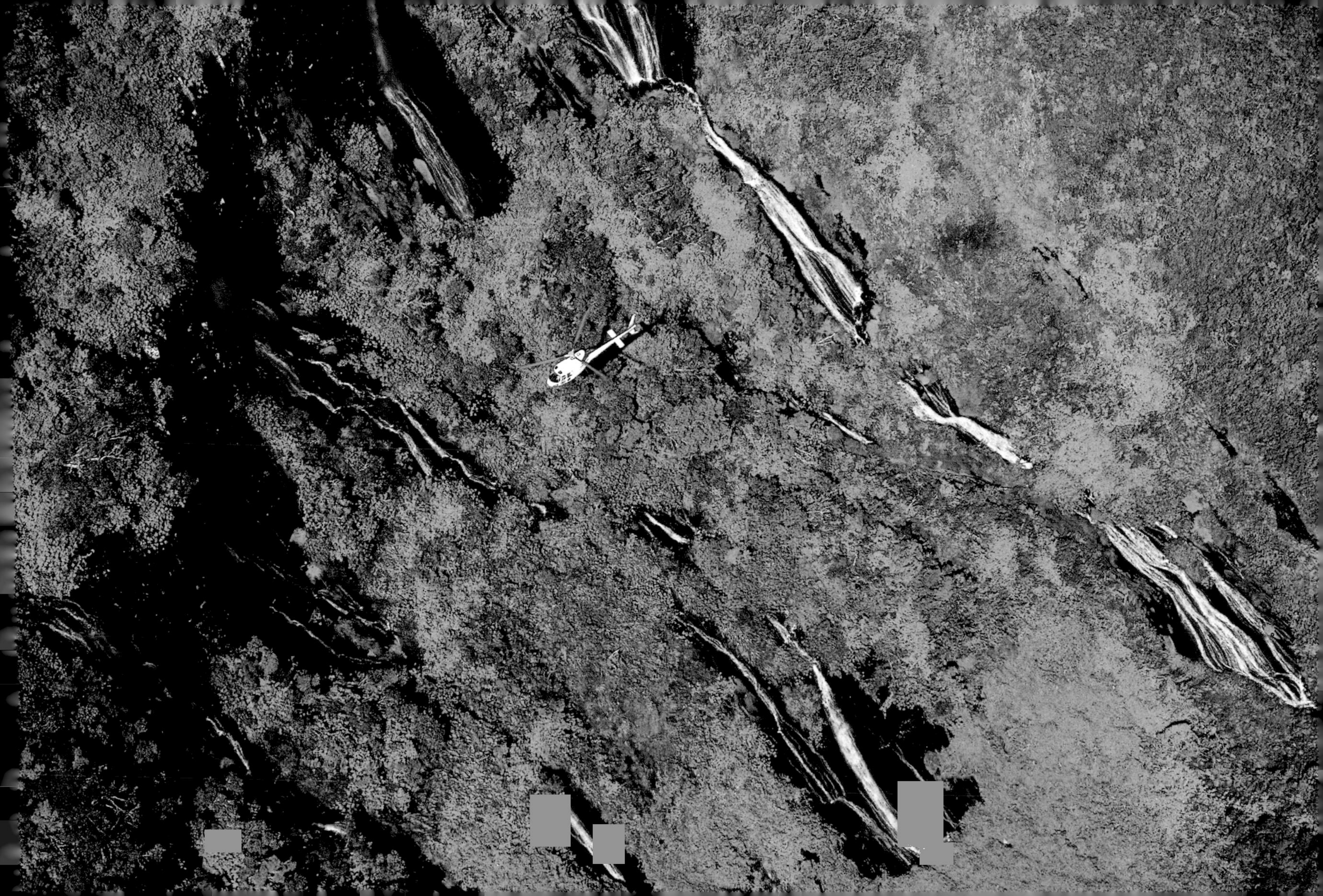

OCTOBER

Lechwe in the Okavango Delta, Botswana (18°45' S, 22°45' E).

Two million years ago the Okavango river flowed into the Limpopo and emptied into the Indian Ocean, but the faults created by tectonic movement diverted the river from its original course. The “river that never finds the sea” now ends in Botswana, in a vast interior delta of 9,300 square miles (15,000 km^2) at the entrance of the Kalahari Desert. This labyrinth of swamps is home to 400 species of birds, 95 reptiles and amphibians, 70 fish, and 40 large mammals. Hidden in the islets of vegetation where they find food and protection from predators, lechwe (*Kobus leche*)—an antelope typical of swampy environments—exist in abundance in the waters of the Okavango delta. Since 1996 the Okavango Delta has been protected by the Ramsar Agreement, which concerns 1,075 wet zones of international importance throughout the world.

11

OCTOBER

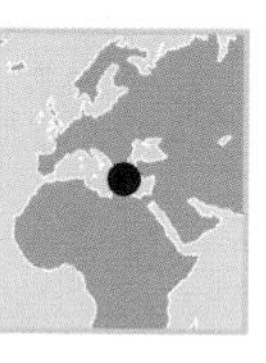

Peasant working his field, Lassithi Prefecture, Crete, Greece (35°09' N, 25°35' E).

Crete's hilly, rocky terrain creates problems for agriculture and also makes access to the fields difficult. The donkey, a traditional mode of travel, transport, and towing, is the animal best suited to the topography of the island. It remains widely used today, as in this fertile plain of the Lassithi plateau. The local climate, considered one of the healthiest and gentlest in Europe, no doubt favors the exceptional longevity of the inhabitants of Crete. Yet the virtues of the Cretan diet, in which olives and olive oil reign supreme, surely also play a role. Cretans, however, are not the only group of people who commonly live a full century; the Vilcabamba Valley in Ecuador also has many centenarians. Medical progress and improved health conditions around the world are gradually extending the average human life expectancy, which now stands at 66 years. But the duration of life on Earth remains very uneven; in Japan and Canada, people on average live to the age of 80, whereas in the least advanced countries three out of four persons die before 50.

12

OCTOBER

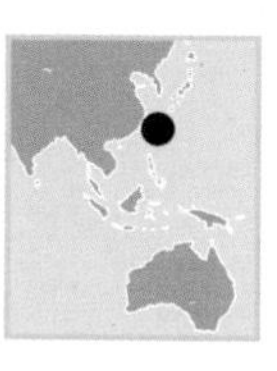

Wreck aground in the Pescadores archipelago, Taiwan (23°37' N, 119°33' E).

The name of the Pescadores (or "fishermen") Islands, wedged between Taiwan and China, recalls the Spanish colonizers of the sixteenth century. This wreck, suspended between the sea and basalt, bears witness to the density of sea traffic in Taiwan, which has the fifth biggest port facilities in the world. On January 14, 2001, a Greek oil tanker sank close to the Lungkeng marine nature reserve, in the south of the island, wiping out coral and fish for several years. Taiwan's coral reefs, one of the ten places in the world where coral is found, are among the richest, with 300 species of coral and 1,200 of fish. But they are less threatened by oil spills than by destructive fishing methods that make extensive use of dynamite. Worldwide, 27 percent of coral-bearing zones have been destroyed, and a further 14 percent are expected to meet the same fate within 10 to 20 years.

13

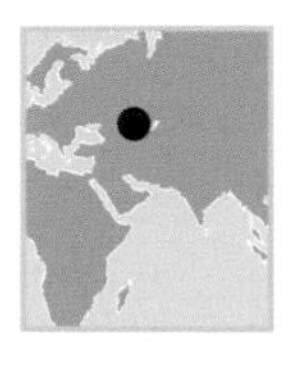

Stubble burning near Kazalinsk, Kazalinsk region, Kazakhstan (45°46' N, 62°07' E).

The steppes of Kazakhstan stretch almost to infinity, covering 993,600 square miles (2,700,000 km^2) from the Caspian Sea to northern China. Here roam about 9 million nomads, descended from Turki and Mongol tribes who were Islamized in the fifteenth century. Today, Kazakhs make up only half the country's population, living alongside Russians, Ukrainians, Uzbeks, and Germans who were sent or deported there by the Soviet Union, which occupied the country until 1991 (an extension of the Russian empire's occupation of the region, which began in the mid-1700s). The Germans introduced cereal farming, which briefly raised the hope that a new breadbasket had been opened up in the East. But the constant plowing up of the land helped to denude it of an already thin and wind-eroded soil. Between now and 2025, Kazakhstan is expected to lose almost half its cultivable land, putting the livelihoods of its arable and livestock farmers in jeopardy—20 percent of the population. Worldwide, at least 1 billion farmers and stockbreeders will suffer the consequences of soil degradation through erosion or other factors.

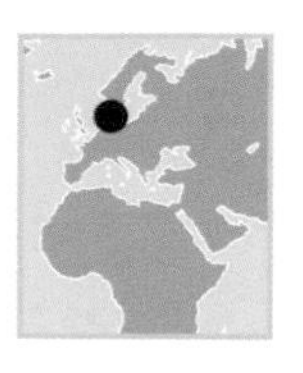

Cereal cultivation east of Kalundborg on the island of Sjaelland, Denmark (51°41' N, 11°06' E).

Danish farmers are among the most productive in the world: each farmer can feed 140 people. Denmark practices high-technology agriculture that is in the forefront of efforts to minimize environmental impact. Strips of grass are planted along waterways, woods, and roads, to limit rainwater runoff carrying pesticides and fertilizers toward rivers and ground water that are tapped for drinking water. Pollution by agricultural, industrial, and domestic fertilizers speeds up growth of algae in rivers and estuaries, choking water life and encouraging the growth of toxic algae. Illegal dumping of rubbish and overconsumption of cleaning materials are also harmful to water supplies. Combating this pollution is one of the challenges facing developed countries.

OCTOBER

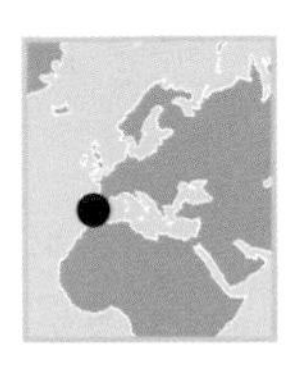

Olive groves by the Vadomojon reservoir, Jaén province, Spain (37°37' N, 4°12' W).

The reservoir of Vadomojon straddles the Andalusian provinces of Jaén and Cordoba, an island dotted with olive trees in a sea of arid lands. Jaén and Cordoba alone provide 40 percent of the world's olive oil, putting Spain far ahead of other Mediterranean countries. The olive groves cover a third of the province of Jaén and are an essential source of work: unemployment stands at 45 percent during summer but falls 10 percent during the harvest. To take maximum advantage of this green gold, the region doubled the area under irrigation between 1990 and 2000, largely due to grants from the European Union. However, Andalusia already faces severe water shortages, and many of its rivers, from which too much water is being pumped, are at a low level. The national hydrological plan, published in 2000, envisaged the pumping of water from northern Spain to the south as well as the construction of seventy new dams. It has caused considerable controversy.

16

OCTOBER

17

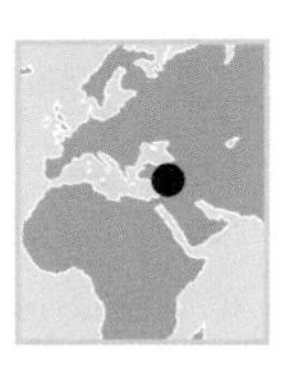

"Snipers' tower," Beirut, Lebanon (33°53' N, 35°29' E).

The interior of this tower, riddled with bullet holes, contains graffiti depicting the dove of peace. Some parts of Beirut still form an open-air museum of the war that raged in Lebanon. Precipitated by a disruption of the balance between the country's Christian and Muslim communities, the civil war lasted 16 years, from 1975 to 1991, cost 150,000 lives, and left the country in ruins. It has risen from these ruins with great energy: the center of Beirut has been almost completely rebuilt, and Lebanon's gross domestic product grew rapidly until 1998. Tensions between communities have eased since power has been more equitably shared, with a Maronite Christian president, as before, and a Muslim prime minister. Despite this, there is still resentment over the 30,000 Syrian soldiers stationed on Lebanese soil, an occupation much better tolerated by Muslims than by Christians. On August 5, 2001, the authorities arrested 200 Christian militants who opposed the policies pursued by Damascus, precipitating a new political crisis.

OCTOBER

Island of Kornat, Kornati National Park, Dalmatia, Croatia (43°50' N, 15°16' E).

The eastern edge of the Adriatic washes the shores of the 150 islands and islets that make up the Kornati archipelago. The largest island, Kornati, is 12.5 square miles (32.5 km^2) in area and accounts for two-thirds of the archipelago's land surface. The fold in the Earth's crust that formed the Kornati mountain range was produced by the collision of the Adriatic and European tectonic plates. The melting of glaciers and rise in sea levels after the last Ice Age, 20,000 years ago, made the mountains into an archipelago which was then eroded by the sea and the wind, exposing fine striae in the limestone. A century ago, the inhabitants of the nearby islands used these rocks to build dry-stone walls to pen in their sheep and keep them from their olive groves and vineyards. Overgrazing has done considerable damage to wildlife and to the thin vegetation. This desertification contrasts with the richness of the waters around the archipelago, where most of the Mediterranean's fish and mollusk species can be found. Nevertheless, these too are threatened by fishing, which sometimes uses illegal methods. Almost 30 percent of the world's fish species are either extinct or facing extinction.

18

OCTOBER

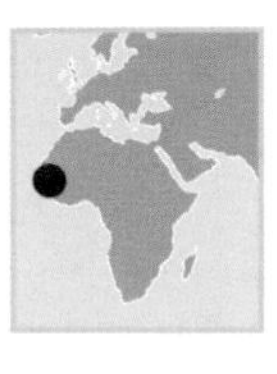

Fishermen returning to Saint-Louis, Senegal (16°02' N, 16°30' W).

Senegal's 435 miles (700 km) of coasts teem with marine life, thanks to the seasonal alternation of mineral-rich cold currents from the Canary Islands with warm currents from the Equator. This local wealth supports coastal fishing, 80 percent of which is done on a small scale, using lines or nets from dugout canoes made out of baobab or kapok logs. But it also attracts European trawlers. Once they have the required permits, these boats can catch fish in far greater quantities and fish intensively, depriving the countries of the region of this resource. Fishing is still Senegal's main industry, producing almost 400,000 tons a year, chiefly destined for the local market. Tuna, sardines, and hake are mostly sold on the beaches where the canoes come in to land. The Senegalese, like 1 billion other people in the developing world, depend on fish as a dietary staple, which provides 40 percent of the population's protein intake.

19

OCTOBER

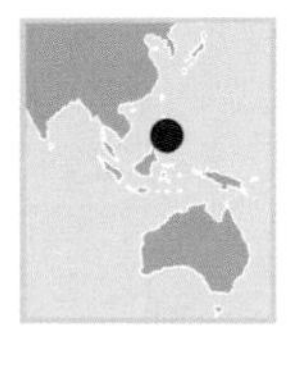

Village of Bacolor under a layer of mud, the island of Luzon, Philippines (14°59' N, 120°39' E).

In 1991 the volcano of Pinatubo, on the island of Luzon in the Philippines, began to erupt after nearly six centuries of dormancy, projecting a 66-million-cubic-foot (18-million-cubic-meter) cloud of sulfurous gas and ash to a height of 115,000 feet (35,000 m) and destroying all life within a radius of 9 miles (14 km). In the days that followed, torrential rains from a hurricane mixed with ashes scattered over several thousand kilometers, causing devastating mudflows, which engulfed whole villages. Before the cataclysmic eruption on June 15, 1991, the evacuation of 60,000 people limited casualties to 875 dead and 1 million injured. Close to 600 million inhabitants of our planet live under the threat of volcanoes, but despite their force, volcanic eruptions are not the deadliest threat to humans. In the past fifteen years, 560,000 persons perished from major natural catastrophes (120,000 in 1998 and 1999 alone); 15 percent of the deaths were due to storms, 30 percent to earthquakes, and half to floods—a natural phenomenon that has become even more devastating as a result of human intervention in the environment.

20

OCTOBER

21

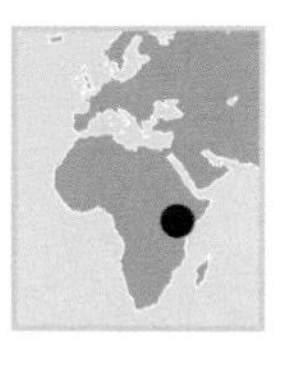

Sunken boat off Lamu, Kenya (02°16' S, 40°55' E).

The bridge is the only part of this wreck that is not rusted. It protrudes from the water because the boat, which probably sank after hitting a coral reef, lies in shallow water. Worldwide, on average, one large ship is wrecked every three days. But ships do not always sink by accident. Sometimes their owners sink old ships in the ocean without bothering to remove polluting materials from them beforehand. Often these vessels are too far gone to interest the scrap metal merchants who strip reusable metal from ships, cars, and domestic appliances for recycling. Compared to the resources required to produce a ton of steel in a regular foundry, recycling the same amount in an electric foundry saves 1.5 tons of iron ore and 0.5 tons of coke. It also uses a third of the energy: 0.2 tons of oil equivalent (TOE) as against 0.6 TOE. In this way recycling reduces waste as well as limiting the environmental impact of steelworks.

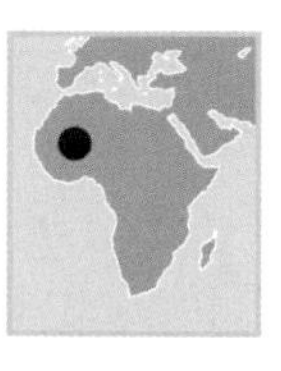

Village near Mopti, Mali (14°40' N, 4°15' W).

The houses and central mosque of this village in the Mopti region are built of *banco* (a mixture of earth and vegetable fibers) and seem to spring straight from the ground. To protect it from erosion by rain, *banco* needs to be repainted every year, which involves covering the mosque in scaffolding that rests on the palm wood pinnacles adorning its facade. After the harvest, the flat roofs are used to dry sorghum. Along with millet, sorghum accounts for 41 percent of Mali's cultivated land. But the country's food self-sufficiency is precarious, for it is caught in a vice between population growth of almost 3 percent per year and desertification. The desert is advancing at the rate of more than 3 miles (5 km) a year across a front 1,242 miles (2,000 km) wide. However, since democracy was established in 1991, the country's general situation has improved, especially in regard to education. Primary schools have been established, as well as secondary schools and even a university in Bamako, which was inaugurated in 1996. But progress is hesitant; the country is strangled by its foreign debt, which eats up 48 percent of its tax revenue.

OCTOBER

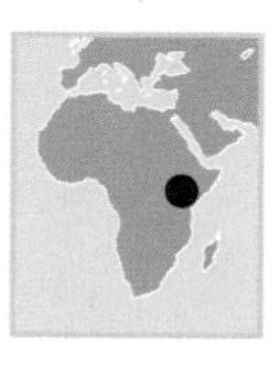

Storm over the Loita Hills, Kenya (1°50' N, 35°80' E).

Kenya's rainfall is extremely irregular, with long periods of wet weather from April to June, and shorter periods between November and mid-December. These rains are often heavy and accompanied by spectacular storms, as here on the Loita Hills, which appear to be attached to the sky by fearsome columns of water. However, Kenya frequently has long droughts, too; the country is one of the eight countries in Africa most affected by lack of rainfall. Its economy depends heavily on agriculture and natural resources and is thus highly vulnerable to the whims of the weather. In 1999 and 2000, stocks of fresh water were low because of drought. Hydroelectric output dropped, requiring both electricity and water to be rationed—with the result that the Kenyan gross domestic product fell sharply. Such disasters, which have already led to the death of 1 million Ethiopians in 1984, will probably become more frequent as a result of climate change.

23

OCTOBER

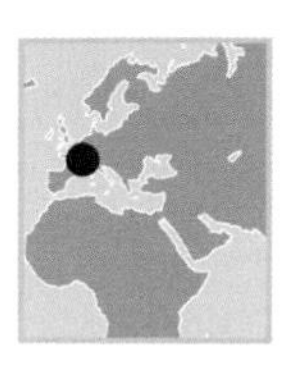

Trees downed by storm in the forest of the Vosges Mountains, France (48°39' N, 7°14' E).

On December 26, 1999, the department of Vosges awoke to find 348 of its 515 communities without electric power, 10 percent of its forests leveled, railway traffic totally at a stop, and 60,000 telephone lines cut. The most severe damage occurred in the Lorraine region, after the storm cut across France causing 79 deaths—an event without precedent in France in recent centuries. Violent winds (up to 105 mph, or 169 kph, in Paris) cut down more than 300 million trees throughout the country, equivalent to three years of harvests for state forests (of which 70 percent is sold). The National Forest Service, which set out to replace these woods, now plans to emphasize forests that are more resistant by nature, without endangering the timber industry, by favoring biological diversity (more adaptive and diverse species) and avoiding systematic alignment. From this viewpoint, not all of the consequences of the catastrophe will prove negative.

24

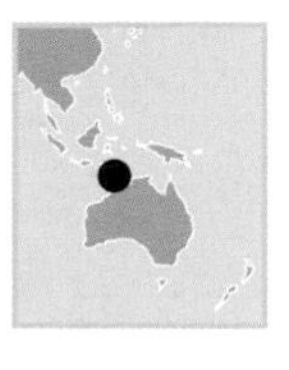

Buccaneer Archipelago West Kimberley, Australia (16°17' S, 123°20' E). Thousands of uncultivated islands, including Buccaneer Archipelago, emerge from the waters off the jagged, eroded coasts of northwestern Australia. Because there is scant agricultural or industrial activity on the shoreline, the waters of the Timor Sea that surround these islands have remained relatively untouched by pollution. This has allowed fragile species such as the *Pinctada maxima* oyster to develop. Harvested in their natural setting, the sea floor, these mollusks are exploited for the production of cultured pearls. Australian pearls, which make up 70 percent of those produced in the South Seas, are twice as large (averaging a half-inch, or 12 mm, in diameter) as those of Japan and, according to experts, finer in appearance. Japan pioneered the pearl industry at the turn of the twentieth century and is the world's leading producer.

OCTOBER

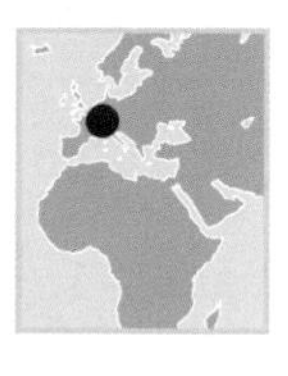

Neuschwanstein castle, Bavaria, Germany (47°35' N, 10°44' E).

The Romantic Road that crosses Bavaria leads up the Lech Valley to Füssen. There, at the foot of the Austrian Alps, the Bavarian king Ludwig II built the mock-medieval fortress of Neuschwanstein. Perched on its rocky spur, this jewel of gray granite bristling with towers and pinnacles draws on the realm of fantasy, and its extravagant architecture inspired Walt Disney. The building was begun in 1869 based on the plans of a theater set designer, Christian Jank; 17 years later, when the king died, it was still unfinished. Ludwig II only spent 172 days in his royal residence; here, on June 10, 1886, he was informed of his removal from office after being declared mentally ill. The castles of Linderhof, Neuschwanstein, and Herrenchiemsee evoke dreams and fairy tales, and they were a financial black hole for the monarch who built them as well as for the state. Now, however, they are an asset to the Bavarian tourist industry. Bavaria is the German *Land* (state) most frequented by tourists, attracting almost a quarter of the country's visitors.

26

OCTOBER

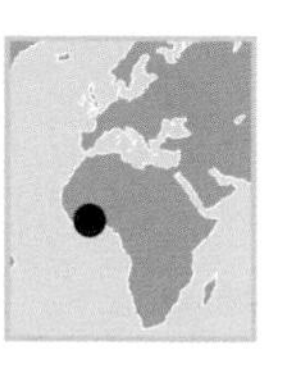

Hydraulic drilling station in a village near Doropo, Republic of Côte d'Ivoire (9°47' N, 3°19' W). Throughout Africa the task of collecting water is assigned to women, as seen here near the regions of Doropo and Bouna, in northern Côte d'Ivoire. Hydraulic drilling stations, equipped with pumps that are usually manual, are gradually replacing the traditional village wells, and containers of plastic, enameled metal, or aluminum are supplanting *canaris* (large terra-cotta jugs) and gourds for transporting the precious resource. The water of these pits is more sanitary than that of traditional wells, 70 percent of which is unfit for drinking. Today 20 percent of the world population is without drinkable water. In Africa this is true for two out of five people, but more than half of the population in rural areas have no access to clean water. Illnesses from unhealthy water are the major cause of infant mortality in developing nations: diarrhea kills 2.2 million children below the age of five. In Africa and Asia improved access to clean drinking water will be one of the major challenges of the coming decades, as their populations grow.

27

OCTOBER

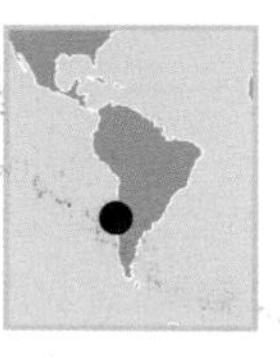

Salar de los Infieles, Aguilar, Andes cordillera, Chile (25°53' S, 68°53' W). The Andes cordillera—a mountain chain formed by the folding of the American tectonic plate under pressure from the Pacific plate—contains a string of salt deserts that have earned it the name "Salt Cordillera." Movements of the Earth's crust have gradually raised areas to high altitudes that were once under the sea. These former seabeds have become enclosed basins that gather water that flows from glaciers. This water, in turn, carries a suspension of salts that originate from volcanoes. The heat of the sun and low humidity in the air mean that the water evaporates, leaving behind the salt deposits, which are rich in lithium and other minerals. This abundance of trace elements has attracted mining companies; however, the development of this new industry would require intensive tapping of the region's ground water, which would threaten its ecosystem.

28

OCTOBER

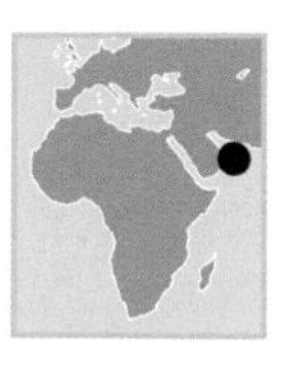

Fishing nets on the beach at Saham, Oman (63°00' N, 24°20' E).

This seine net—a long net that is dragged over a sandy sea bottom—is ready to be used once again. The fishermen have patiently folded it up and laid it alongside their boats, and they now need only pull at the net's two ends to harvest their catch. More than 80 percent of Oman's fish are caught by such traditional methods, but the sultanate would like to modernize this sector. The country is well aware of the limits to its oil reserves (700,000 tons of crude) and would like to diversify its economy. Training fishermen is one of its priorities. With the help of young, qualified staff, it hopes to increase national output while still managing fish stocks sustainably; for while the Gulf of Oman is rich in fish, certain species there are threatened. Overfishing is a worldwide problem, and catches are falling—in the North Atlantic, for example, they have dropped by 25 percent since 1970.

29

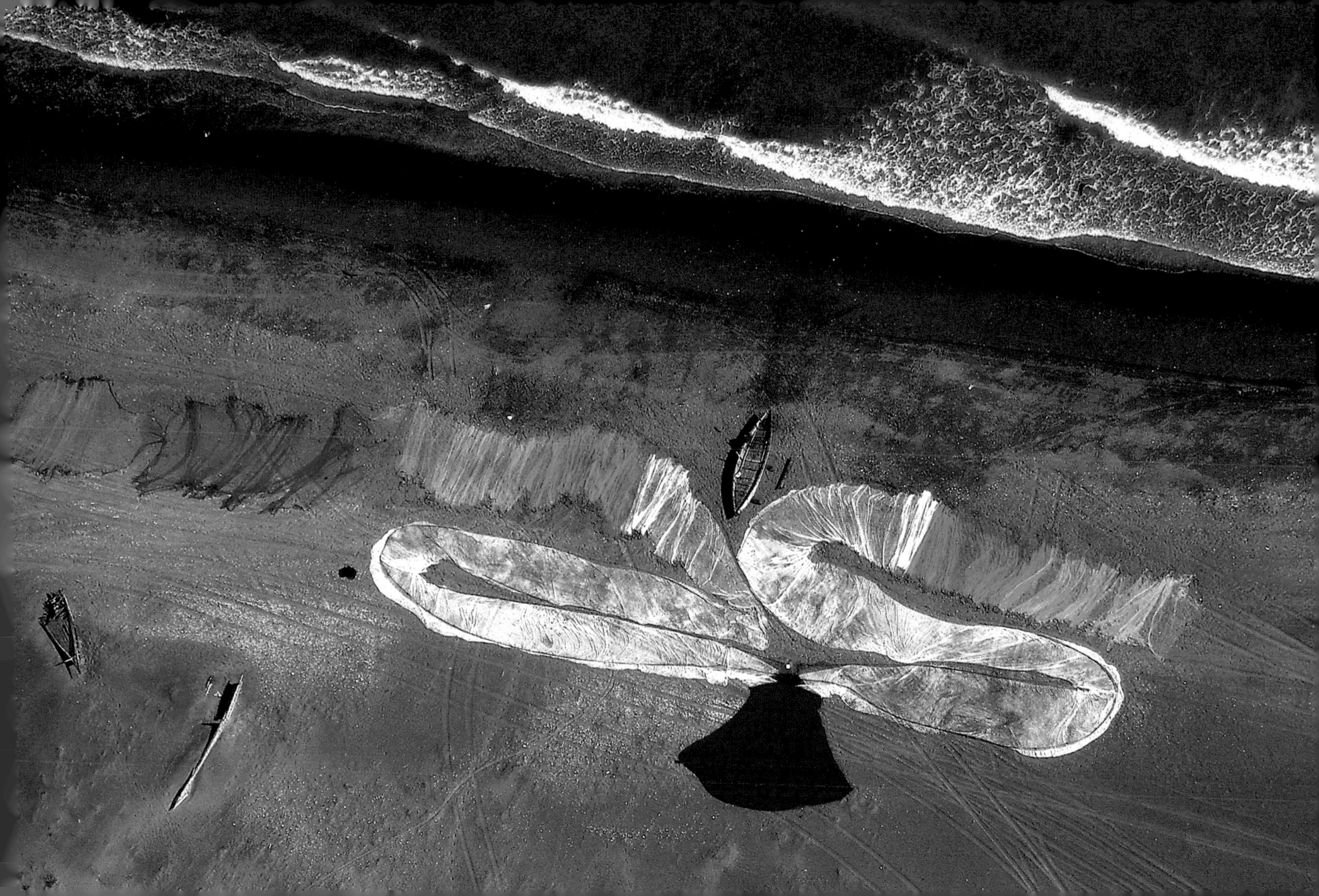

OCTOBER

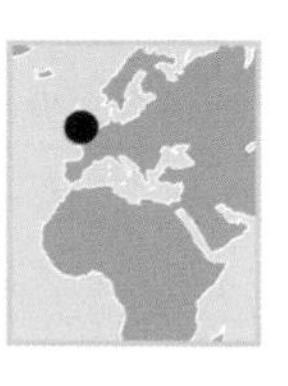

Tower Bridge, London, England, United Kingdom (51°30' N, 0°06' W).

London is one of the world's great cities, a tentacular megalopolis whose heart displays as much of the vigor and political and financial power of the City as leading-edge technology and avant-garde fashion. The city boasts beautiful monuments such as Westminster Abbey, and it can be justifiably proud of one of the most famous bridges in the world—Tower Bridge on the river Thames. After eight years of work, it was completed in 1984; today, 150,000 vehicles cross this bridge every day. More than 900 times a year, this famous structure opens up to let through tall ships, cruise ships, and other large vessels. A visit to its Gothic towers makes for a fascinating historical journey through the original Victorian machine rooms, revealing the hydraulic system of levers driven by enormous pumps that were once powered by steam. More than 11,000 tons of steel were used to build the framework of the towers and walkways, which was then faced in Cornish granite and Portland stone, both for protection and for looks.

30

31

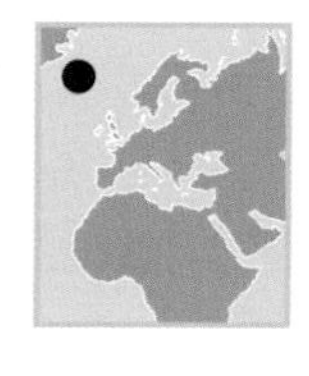

Fold of solidified lava, Mount Maelifell, Myrdalsjökull region, Iceland (63°40' N, 19°05' W).

Mount Maelifell was formed by a subglacial eruption and consists of an accumulation of ash and solidified lava extrusions. This landscape, born of the meeting of fire and ice, illustrates Iceland's unique geology, where the forces that shaped our planet are constantly at work. The fold seen here recalls Iceland's original situation, straddling the North American and European tectonic plates. The country is being stretched continuously, at the rate of some three-quarters of an inch (2 cm) per year, under pressure from the volcanic Mid-Atlantic Fault. This 60-million-year-old island—relatively young in geological terms—is the scene of vigorous volcanic activity with an eruption every five years; of glacial action, with glaciers covering 12 percent of its surface; and of seismic activity, with an major earthquake every 100 years registering about 7 on the Richter scale.

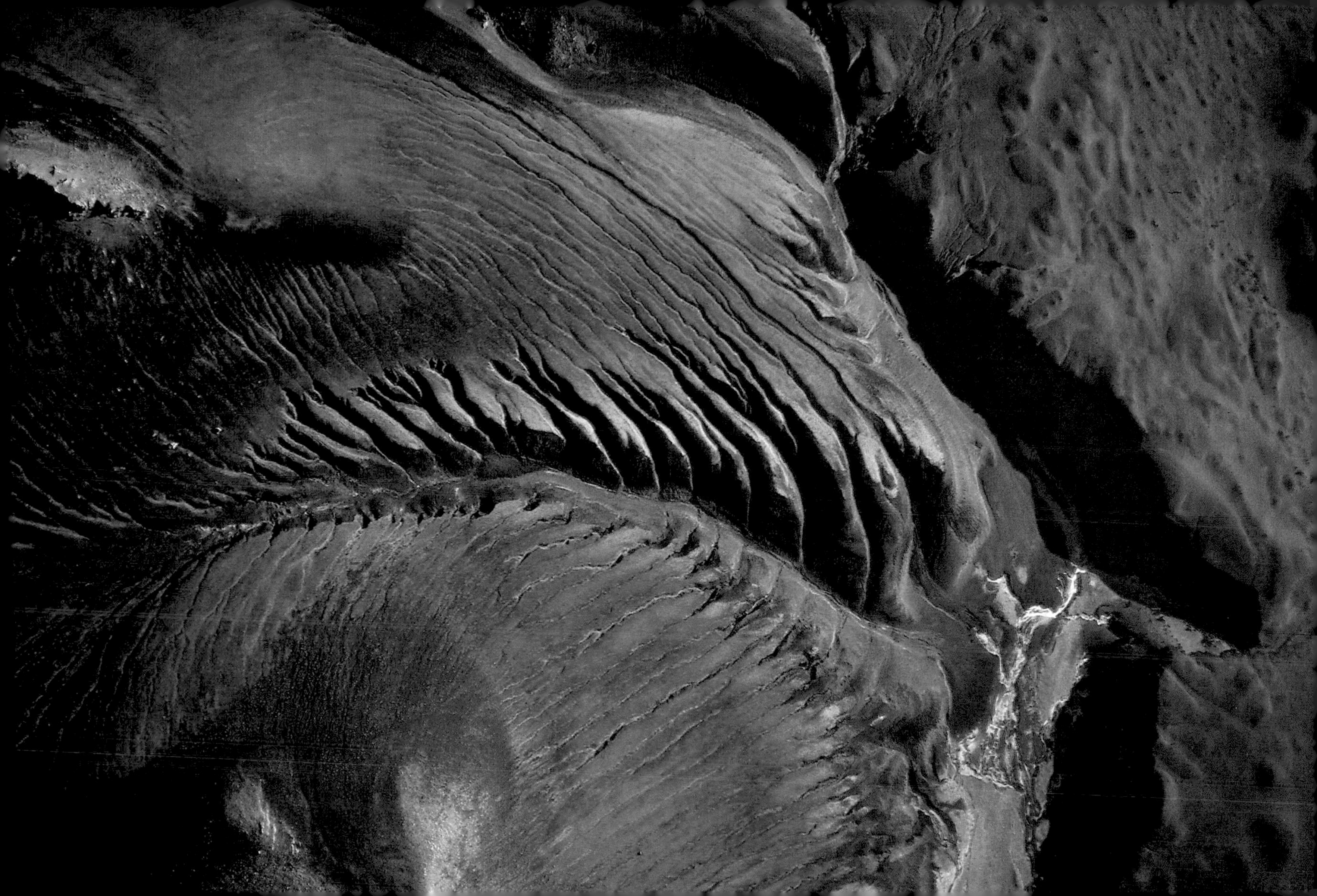

GROWTH, DEVELOPMENT, AND SUSTAINABILITY

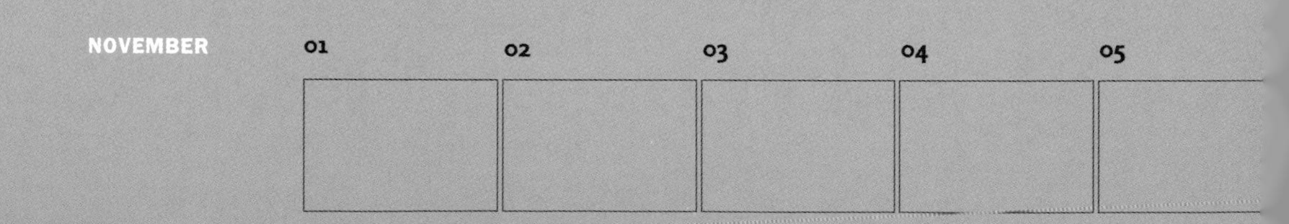

The aim of the economy is to transform nature to better satisfy human needs. Today, because of computers, an unprecedented technological change is shifting the engines of this transformation away from their reliance on energy (such as steam, electricity, and oil) to the new fuel of information and the intangible. In its relationship with nature, therefore, the world is at the crossroads between two eras:

- The era of energy belongs to the past, with its capacity for wreaking havoc and destruction upon the environment.
- The era of information, perhaps the era of the future, offers the prospect of a less wasteful and less damaging model of development.

At the end of the eighteenth century, the German chemist Lichtenberg described a strange dream. In the dream, he boasted that he could identify the nature of any object. An old man appeared out of nowhere and challenged him to analyze a sphere that he pulled out of his pocket. The scholar accepted the challenge. After unracking his brains, pondering, and analyzing, he was soon able to list all the elements that made up the sphere: carbon, hydrogen, oxygen, nitrogen, sulfur, phosphorus, and so forth. "Very good," replied the old man. "But this ball was the Earth." Then the scholar knew that he was face to face with the Creator, and that by considering only the object's physical aspects he had destroyed it.

Earth is far more than a mere object. It is a system of functions and laws that have allowed life to appear and flourish. The economy should therefore not concern itself solely with the material side of things, but also with enduring natural laws, within which all life, and human productive activity, operate.

For a long time, this problem of separating material existence from natural law did not exist. When standards of living were close to the bare minimum, and economic activity did not damage the natural environment, the concept of "more" carried the sense of satisfying basic needs ("more grain"). "More" also meant "better," as it still does for poor people today. At this point, growth—a one-dimensional, quantitative concept referring to the increase of gross national product—was synonymous with "development."

06 07 08 09 10 11 12 13 14 15

Things changed at the beginning of the 1970s, when there was an increase in accidents that caused damage to the environment. In 1972, a report by the Club of Rome[1] asserted that growth at its current rate was destroying nature, from which humanity was extracting its resources and where it dumped its waste. That same year, the Romanian-born American economist Nicolas Georgescu-Roegen[2] emphasized that economic development could not be fully grasped without setting it in the context of the degradation—"entropy"—of the solar energy reaching the Earth. According to him, economic activity could only accelerate this degradation.

Then, in the 1980s, "global" assaults on nature made their appearance: the hole in the ozone layer, the greenhouse effect, the reduction in biodiversity, and so forth. The very laws of nature by which the planet kept itself habitable were under threat. The situation was no longer a case of dysfunction, but of a conflict between two principles: that of economic growth, and that by which the biosphere safeguarded its survival.

Thus, "development" parted company from "growth"—a distinction that François Perroux had already drawn back in the 1960s[3].

For my part, in 1979, I put economic development in the context of a dual process of the *creative destruction* of the sun's rays (damaging, certainly, but also a source of energy that allows life to appear and grow)[4]. Economic activity only destroys the biosphere if it breaches the limits of this process of reconstruction. Looking at in this way, we arrive at the following realization: growth can only be described as development if it respects the mechanisms that ensure the survival of the human and natural spheres within which it takes place.

Finally, in 1987, the "Brundtland Report,"[5] as it is known, vulgarized the concept of *sustainable development*, defining it as "development that meets the needs of the present without compromising the ability of future generations to meet their own needs."

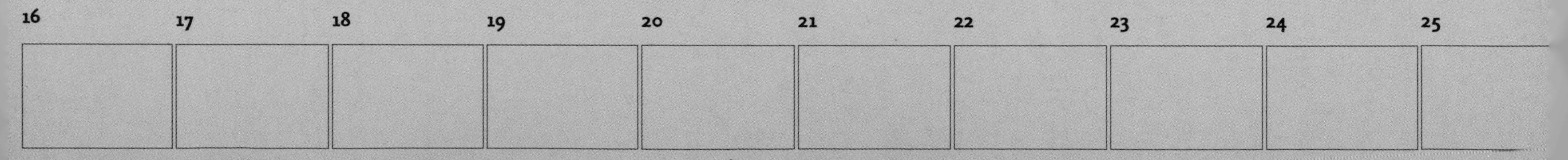

The threats to nature therefore come essentially from productivist ideas that reduce economic—and human—development to mere growth in national product. This confusion, common both to the former planned economies of Eastern Europe as well as to the liberal economies of the West, is worsened today by the "shareholder" system that dominates the planet. From the 1980s on, the priority given to the free movement of capital around the world has led to economic systems being subordinated to an approach whose chief aim is to secure a rapid return on investment. As a result, short-term financial objectives prevail, at the expense of very long-term natural laws. Increasingly, nature, life, and the human race itself have no other value than as a means to this end of making a quick profit. A single conception of economic efficiency, ignorant of the limits of the biosphere that supports it, is leading the world to disaster. It is high time that people and the economy were returned to their rightful roles—now reversed—of end and means.

RENÉ PASSET
Professor Emeritus of Economics at the University of Paris 1-Panthéon-Sorbonne

NOTES

[1] Club of Rome. *Limits to Growth* (Fayard, 1972).
[2] Nicolas Georgescu-Roegen. *The Entropy Law and the Economic Process* (Harvard University Press, 1971).
[3] François Perroux. *L'Économie du XX Siècle* (PUF, 1961).
[4] René Passet. *L'Économique et le Vivant* (1st ed. PAYOT, 1979; 2nd ed. Economica, 1996).
[5] The World Commission on Environment and Development. *Our Common Future* (Oxford University Press, 1987).

Marabouts in Jebel Krefane, Tozeur governorate, Tunisia (33°55' N, 08°08' E).
Ifriqiya (Africa) had been conquered by the Arabs by the end of the seventh century. But Arabization and Islamization were slow to begin with, and those conversions only gathered momentum from the eleventh century on. As in other areas that were late to convert to Islam, Sufism established itself. In Africa, this sect was founded on maraboutism, the cult of holy men that became an essential element of popular devotion: their domed tombs (marabouts) are dotted around towns and the countryside. The word "marabout" comes from *murabit*, which originally described a warrior-monk who lived in a fortified monastery, or *ribat*. Then it came to refer to a person who had distinguished himself for his piety, charity, religious knowledge, or healing powers. The term also denoted his mausoleum, which was a place of pilgrimage where the faithful came to pay their respects with ceremonies of singing and dancing. Today, marabouts have retained a strong spiritual influence, called *baraka*, which can have a bearing not only on the daily lives of the faithful but on the country's political life.

01

NOVEMBER

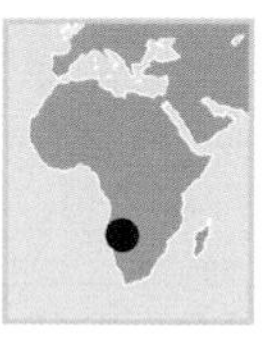

Himba couple, Kaokoland, Namibia (18°15' S, 13°26' E).

The region of Kaokoland in northern Namibia is home to 10,000 to 15,000 Himba, nomadic herders of cows and goats who live along the Kunene River. This people, which has maintained its traditions and lives outside the modern world, suffered twin disasters in the 1980s: a long drought that killed three-fourths of the livestock, and the war between the South African Army and SWAPO (South West Africa People's Organization). Today the Himba face an equal threat—the planned construction of a hydroelectric dam on the Epupa Falls. This project, which would provide power for a water desalination plant in a country that imports nearly 50 percent of its electricity and is seriously deprived of water resources, would also result in the flooding of nearly 160 square miles (400 km^2) of pasture land, destroying a vital resource for the Himba shepherds and endangering their way of life.

02

NOVEMBER

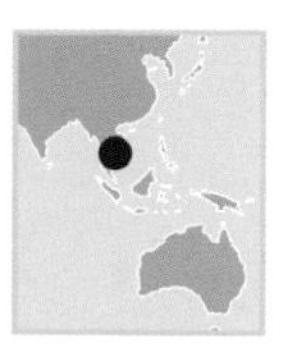

Royal tombs of Wat Phra si Sanphet (Temple of Sanphet), Ayutthaya, Thailand (14°20' N, 100°34' E).

Ayutthaya, an artificial island at the confluence of the Chao Phraya, Prasak, and Lopburi rivers, was the capital of the Kingdom of Siam for four centuries, from 1350 to 1767. Its splendor, dynamic culture, and thriving economy dazzled seventeenth-century Europe. The size of Wat Phra si Sanphet, the city's royal sanctuary since 1491, bears witness to this magnificence. Of the temple complex, only three *chedi*, the equivalent of Indian *stupa*s, survive intact. Symbolizing the stages that must be passed through on the way to Nirvana, their columns contain the ashes of Siamese rulers. Buddhism, the religion of 95 percent of all Thais, binds the country together. The temples, which were formerly centers for education, hospices, and orphanages, are still at the heart of public life. Almost all Thais live as monks temporarily at some point in their lives—sometimes for a few weeks, and often for three months during the monsoon season.

03

NOVEMBER

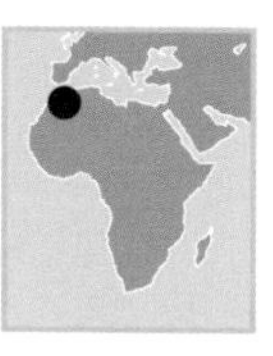

Village in the Ourika valley, Morocco (30°44' N, 6°33' W).

The Ourika valley's lush cultivated terraces contrast with the heights of the surrounding Atlas Mountains, but the Berber houses blend in with the rocky landscape. Built of *pisé*—a mixture of rammed earth, straw, and gravel—they are extremely tough, except in the event of a flood. To protect themselves from these, the Berber villages, also known as *douar*s, cling to the mountainsides. The floods of the Ourika are frequent and feared, and they can be disastrous. When it bursts its banks, as a result of the melting of snows or summer storms, it carries all before it, flowing at up to 1,308 cubic yards (1,000 m^3) per second. After the disasters of 1995, the Moroccan authorities put prevention systems in place that limited the damage during the floods of 1999. These were further refined in 2001 to provide a better warning system.

04

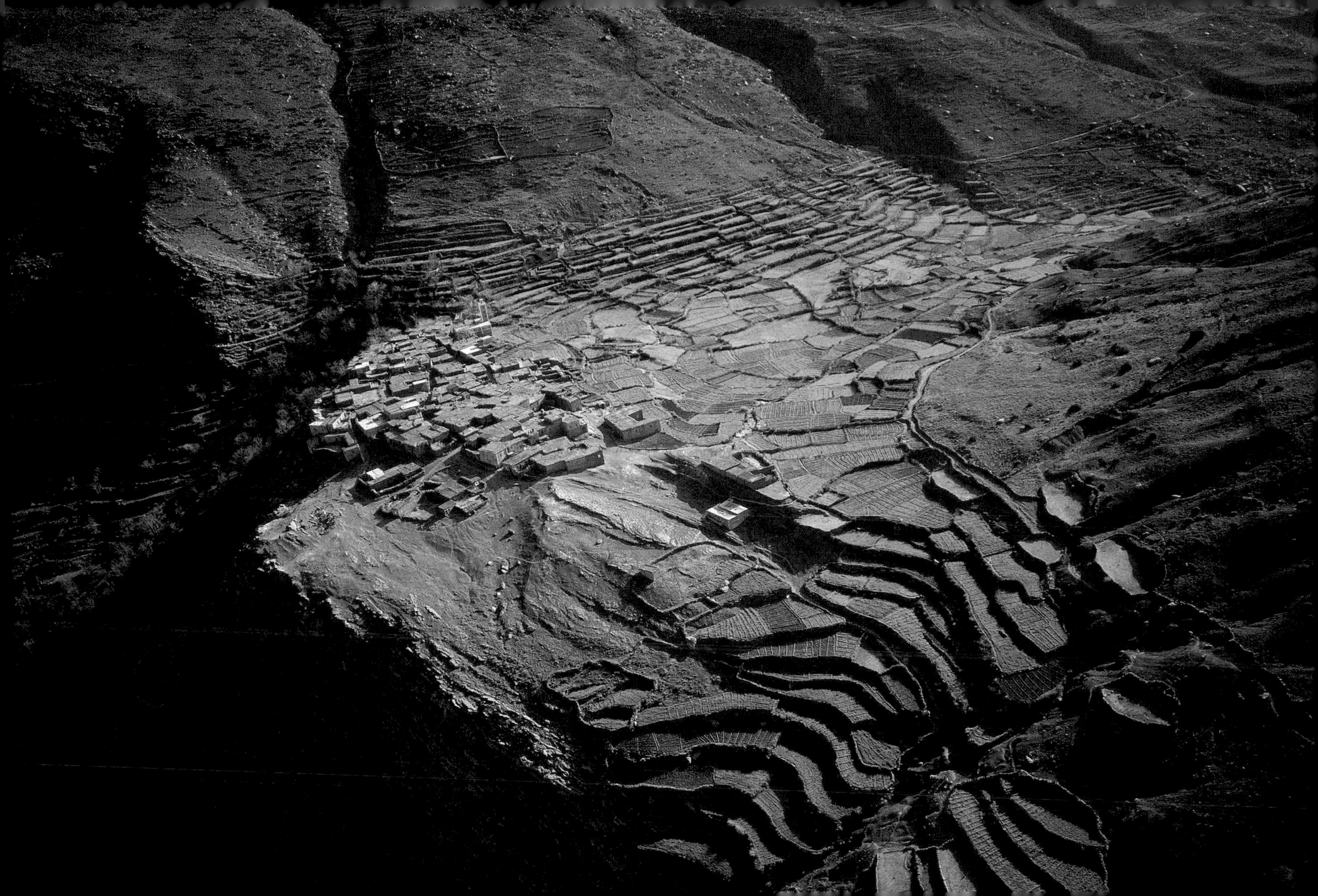

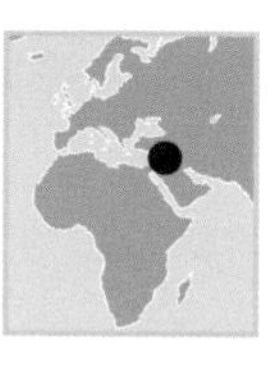

Citadel of Anfeh, El-Koura district, Lebanon (34°20' N, 35°41' E).

Anfeh, a small coastal town south of Tripoli, survives on small-scale fishing and its olive groves, vines, and salt marshes. This ancient Phoenician port, called "Ampi" in the seventh-century BC clay tablets from Al-Amarna, was occupied by the Crusaders in the twelfth century. At that time, its peninsula, which juts out 1,300 feet (400 m) into the Mediterranean, was crowned by the castle of Nephin, a formidable fortress bristling with twelve towers and separated from the rest of the peninsula by a trench cut into the rock. The castle was destroyed by the Mamluks in 1289. Bearing witness to a turbulent history over thousands of years, these protected ruins have escaped the concrete covering endured by the coastline. The coastal region accounts for 16 percent of Lebanon's surface area but accommodates 70 percent of the population—a density of 1,610 inhabitants per square kilometer. The chaotic, uncontrolled development that has been rampant since the 1980s is threatening cultural heritage and the natural environment. From now on, economic renewal and the development of tourism need to take account of the demands of sustainable development.

05

NOVEMBER

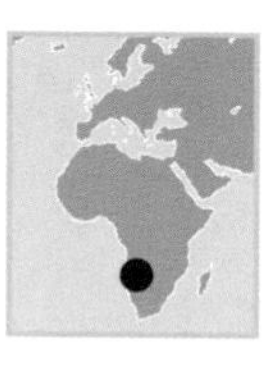

Sandwich Harbor, Swakopmund region, Namibia (23°22' S, 14°03' E).

Sandwich Harbor lies on the Atlantic coast, some 31 miles (50 km) south of the Namibian city of Walvis Bay. It takes its name from a British whaling ship, the *Sandwich*, which operated in the region at the end of the eighteenth century and which periodically stocked up with fresh water from the lagoon. Sandwich Harbor is now protected, and access to it requires special permission. It is frequented by up to 250,000 migrating birds, and especially by 40 percent of southern Africa's pink flamingoes, which makes it one of the most important coastal areas in the south of the continent. Namibia's coast has vast colonies of seabirds, such as cormorants, penguins, and terns; these take advantage of the fish-rich waters of the Benguela Current, a current of cold water that flows off the west coast of southern Africa and that bears abundant plankton, which are manna for birds. Worldwide, however, sea birds are increasingly threatened by certain fishing methods such as *palangres*, fishing lines between 37 and 50 miles (60 and 80 km) long, with thousands of baited hooks. Attracted by the bait, birds are hooked and then drown.

NOVEMBER

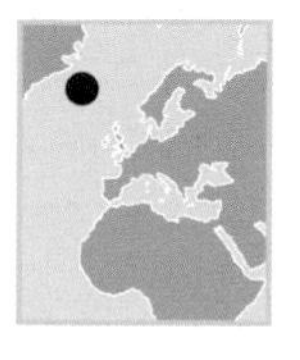

Mouth of the Markarfljót River, Myrdalsjökull region, Iceland (63°32' N, 20°05' W).

Fed by the Myrdalsjökull, a 308-square-mile (800-square-kilometer) dome of ice in the southwest of the island, the Markarfljót River flows round the northern edge of the small Eyjafjallajökull Glacier, tracing a hesitant course over a broad bed of basaltic sediments before ending its journey on a beach of black sand on the Atlantic shore. Like all glacial rivers, the Markarfljót spreads out over a plain from which glaciers have receded, forming a dense, intricate network of channels. Its course changes constantly, and its flow reaches a peak in July and August, when the ice melts. Global warming could disrupt this seasonal cycle. The Vatnajökull ice cap is the country's biggest glacier; with an area of 3,203 square miles (8,300 km^2), it is as big as all continental Europe's glaciers combined. Icelandic glaciologists have warned that the Vatnajökull has been receding at an average of approximately 3.25 feet (1 m) per year for some years, and that snow is replaced by rain at altitudes of less than 3,279 feet (1,000 m). If this trend continues, the glacier could disappear by 2100.

07

NOVEMBER

08

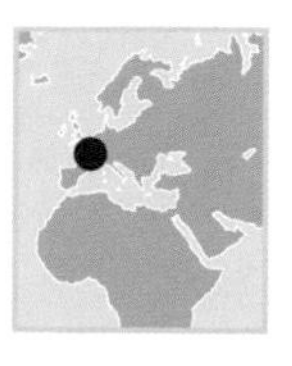

Map of the European Union in the courtyard of the Lycée André Malraux (André Malraux high school), Montereau-Fault-Yonne, Seine-et-Marne, France (48°23' N, 2°57' E).

Only at the Lycée André Malraux in Montereau-Fault-Yonne, southeast of Paris, can students correct their lessons in Greece, munch a sandwich in Portugal, and chat for a while in Denmark. This map of the European Union, painted by the students with their teachers' permission, has adorned the courtyard of the school since April 2002, when twelve of the union's fifteen countries adopted the euro as their common currency. Although clearly visible here, borders between states are gradually disappearing in Europe, while the real, invisible frontiers—between Basques and Spaniards, Walloons and Flemings, Catholics and Protestants in Belfast—endure. The eastern frontier is preparing to open to welcome the next ten member states, who are drawn to this Europe of democracy and prosperity. The union will spend 25 billion euros (about $28.75 billion), or 0.08 percent of its gross domestic product, on these newcomers during the first three years after enlargement. This is less than a tenth of the amount Germany has spent on its reunification since 1990. As for the students, they are ready to take up their brushes again in 2004.

NOVEMBER

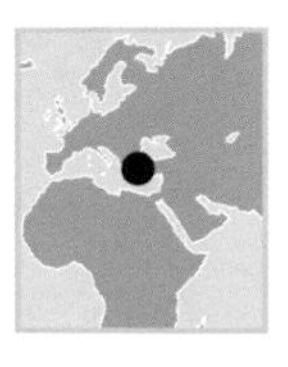

Village at the northern tip of Santorini, Cyclades, Greece (36°27' N, 25°29' E).

The island of Santorini, in the eastern Aegean, is the tip of an ancient volcano that exploded more than 3,500 years ago. The explosion destroyed everything but the rim of a volcanic crater and left behind quantities of ash and debris, on which houses and chapels were later built. In 1967, the Greek archaeologist Marinatos discovered remains on the island indicating an ancient culture similar to that of Minoan Crete. He suggested that before the cataclysmic explosion, Minoan Crete and Santorini had formed a single land mass, Atlantis, where there had lived a dazzling and sophisticated civilization that was engulfed by the waves and by fire, as related by Plato. The volcano's explosion triggered a vast tidal wave and released a cloud of ash that dimmed the light of the sun for several years, ending the Minoans' dominance in that region. But the mystery of the fabulous Atlantis, whose remnants might include Santorini, remains. Although the volcano has been "dormant" since 1950, a future eruption cannot be ruled out. Such an event would threaten the island's 10,000 inhabitants and the many tourists who stay there in summer.

09

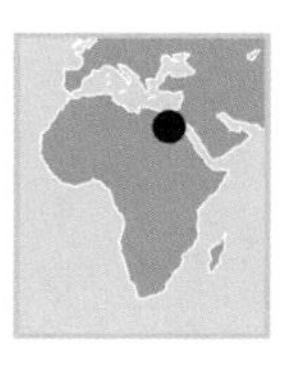

Upturned date baskets, left bank of the Nile, Egypt (25°40' N, 32°35' E).

Almost 800,000 tons of dates are picked and dried each year in Egypt. Although this makes the country the world's second-biggest date producer, it is still not enough. Date palms, like other crops, no longer meet the demand of an Egyptian population that is growing at the rate of 1.69 percent per year. While the population is expected to rise from 69 million people to 100 million by 2025, agriculture will remain confined to whatever land can be irrigated. The country already imports half its food and would like to limit this dependence, but it must also take into account its limited water supplies. Egypt could be facing a shortage of fresh water by 2025, as could two-thirds of the world's people. Population growth, increasing irrigation, and industrial development are likely to increase world consumption by 40 percent, leaving this resource no opportunity to renew itself.

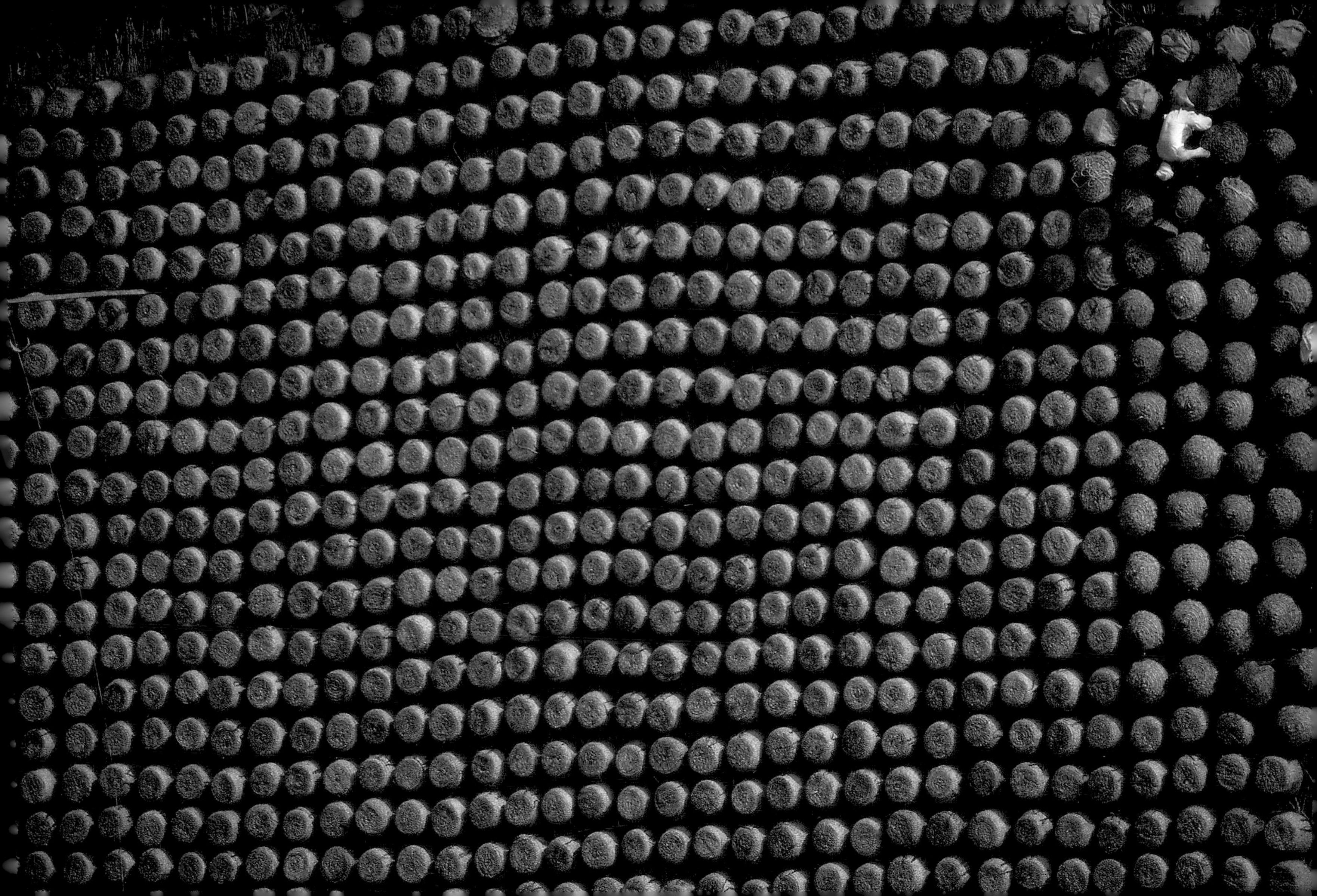

NOVEMBER

11

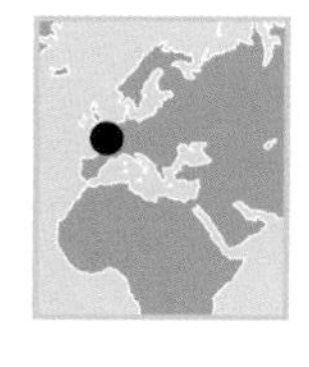

National Military Cemetery of Notre Dame de Lorette, near Ablain-Saint-Nazaire, Pas-de-Calais, France (50°23' N, 02°42' E).

Two major conflicts devastated Europe before the words of Victor Hugo (1802–1885) became a prophecy: "No more armies, no more frontiers, a single Continental currency. . . . The day will come when you will lay down your arms." Before unifying in peace, Europe had to go through two world wars; WWI cost 8 million lives, and WWII cost 45 million. The Battle of Lorette, from October 1914 to October 1915, in which the French and Germans struggled for possession of the strategic plateau of Artois, shed the blood of more than 100,000 victims on the fields of northern France. This military cemetery commemorates the fallen: 20,000 crosses are aligned across 30 acres, and eight ossuaries hold more than 22,000 unknown soldiers. Hugo also said, "The day will come when the only battlefields will be markets open to commerce and minds open to ideas." That day has come, but it is a battle whose outcome remains uncertain.

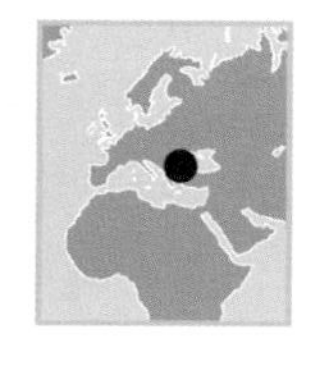

Monasteries of the Meteora in the plain of Thessaly, Greece (39°46' N, 21°36' E).

From the northeastern part of the Thessalian plain rise the Meteora, sandstone peaks sculpted by river erosion during the Tertiary period. Monks established themselves there during the eleventh century, seeking solitude on the summits of these rocky towers. Gradually, there grew up a large community of hermits who, between the fourteenth and sixteenth centuries, built twenty-four monasteries that perch between 655.8 and 1967.4 feet (200 and 600 m) above the Pindus valley. For a long time, access to the monasteries was possible only by means of winches and ropes. Only in 1920 were steps and footbridges built to allow tourists to visit these sites, which have been on UNESCO's World Heritage list since 1988. Most of these *meteorisa monastiria* (suspended monasteries) are in ruins today. Only five, three of which are occupied, are still open to visitors.

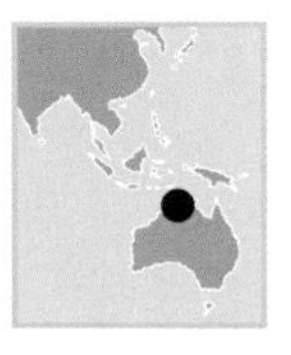

Mount Trafalgar, Prince Regent Nature Reserve, West Kimberley, Australia (15°16' S, 125°03' E).

The wild Kimberley Plateau, between the Timor Sea and the Gibson Desert, is one of the most thinly populated areas on the planet. It the outback par excellence: the remote upcountry of western Australia whose vast area—occupying almost a third of the country, an area five times the size of France—is home to just 1.8 million people. UNESCO has designated the basin of the Prince Regent River a biosphere reserve because of its remarkable, intact habitats: in 2002 there was still not a single road through it. The reserve is surrounded by aboriginal lands—the name "aborigine" means "one who has been there from earliest times." Killed in large numbers by European settlers, the aboriginal population has recovered to stand at 265,000 people, of whom three-quarters are of mixed blood. In aboriginal culture, Mount Trafalgar symbolizes harmony between humans and the Earth, rocks, and other living things, all created by the spirits of ancestors.

NOVEMBER

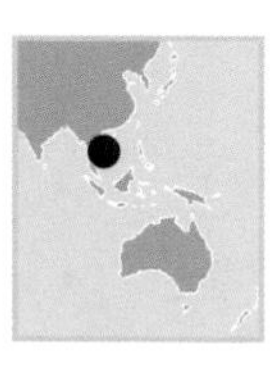

Temple of Angkor Wat, Cambodia (13°26' N, 103°52' E).

Towering above the jungle, the five towers of the temple of Angkor Wat symbolize Mount Meru, center of the world and dwelling place of the gods in Hindu cosmology. This twelfth-century building is the most impressive, and best-preserved, of the 700 remains discovered on the 154-square-mile (400-square-kilometer) site of Angkor, ancient capital of the Khmer empire from the ninth to the fifteenth centuries. Long since overgrown, and at the mercy of looters, this is the largest archeological park on the planet; in 1992, it was put on UNESCO's list of world heritage in danger. It also acts as a driving force in the development of tourism, which is now the most vigorous sector of the Cambodian economy. The number of visitors is expected to rise from 400,000 in 2002 to 1 million in 2010, which would require new infrastructure to be built, to the detriment of the surrounding forest. Although tropical forest still covers 60 percent of the country, its area has fallen from 13 million hectares in 1960 to 11 million today, and illegal clearing is on the increase.

14

NOVEMBER

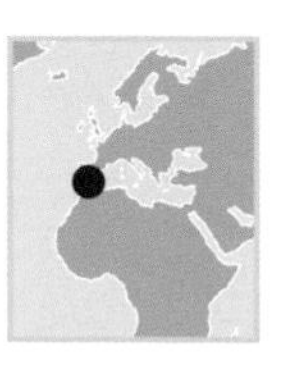

The Rock of Gibraltar, Gibraltar (British overseas territory) (36°08' N, 05°21' W).
At the southernmost tip of the Iberian peninsula, Gibraltar clusters 31,000 inhabitants in the 2.24 square miles (5.8 km^2) around its famous rock. Although claimed by Spain, Gibraltar has been a British territory since it was seized by an Anglo-Dutch fleet in 1704, during the War of the Spanish Succession. Close by, in the eponymous strait that links the Mediterranean with the Atlantic, the coast of Morocco comes within 10 miles (15 km) of Europe, offering a route to Eldorado for would-be illegal immigrants. Every day, hundreds take the risk of crossing, mostly in small, overloaded boats when the sea is calm. Over the last five years, 4,000 bodies of young people who drowned in the attempt, mostly Moroccan, have been washed up on the Spanish coast alone.

15

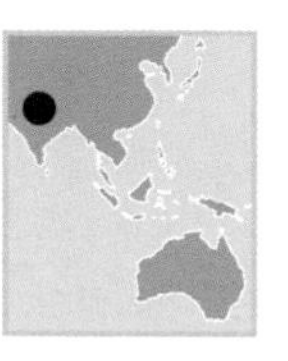

Mount Everest, Himalayas, Nepal (27°59' N, 86°56' E).

In the massif of the Himalayas, which forms the boundary between Nepal and China, stands Mount Everest. Rising to an altitude of 29,028 feet (8,848 m), Everest is the highest point on the planet. In Nepali the mountain is called Sagarmatha, "He whose head touches the sky," and in Tibetan it is called Chomolongma, "Mother Goddess of the world." The name Everest comes from the British colonel George Everest, who in 1852 was assigned the task of drawing up a cartographic outline of India. Since the triumphant expedition by the New Zealander Edmund Hilary and the Nepalese Sherpa Norgay Tensing on May 29, 1953, Everest has inspired more than 300 successful ascents and has claimed some 100 lives. But crowding has caused pollution problems, and the consumption of brushwood for campfires has stripped the slopes and exposed them to erosion. However, in the past ten years, new regulations, clean-up operations, the installation of solar panels, and the introduction of portable fuel for expeditions have helped reverse the degradation of this fragile high-altitude site, declared a national park in 1976, which is vital to the Sherpas.

NOVEMBER

17

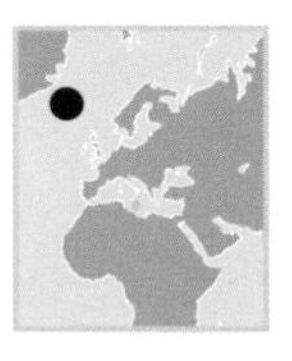

Isolated house near Snaefellsjökull Glacier, Snaefellsness Peninsula, Iceland (64°50' N, 23°00' W).

Because Iceland was cut off from the rest of the world for 1,000 years, its population is highly homogeneous in its genetic makeup. This fact, together with the exceptionally detailed genealogical records that Icelandic churches have kept since the eleventh century, makes it a choice subject for research aimed at deciphering the human genome. These studies should enable genes to be precisely identified, in particular those responsible for illnesses. Under a law passed in 1998 by the Icelandic parliament, the DeCode Genetics company holds a 12-year monopoly on files containing the medical, genealogical, and genetic data on the islanders, as well as exclusive rights to commercialize the results of research on the Icelanders. Such rights, which are common in the scientific fields, take on a different dimension when applied to living things; it is now possible, for instance, to lay claim to the properties of a plant remedy as well as its application, even though a local population may have been using the remedy for centuries. Meanwhile, insurers in some countries are already refusing to cover people whose genetic makeup indicates they are likely to develop a serious illness in the future.

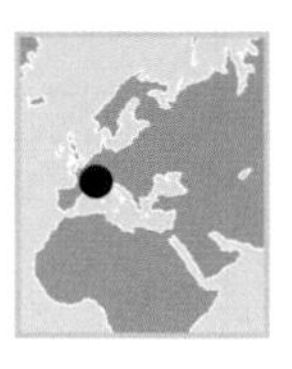

Oceanographic museum, Le Rocher, Monaco (43°44' N, 7°24' E).

Monaco's oceanographic museum, which rises 279 feet (85 m) above the Mediterranean, is built out of freestone and took 11 years to complete. It was inaugurated by Prince Albert I in 1910. It was 20 years ago, at the foot of this mecca for lovers of marine fauna and flora, which the profound ecological transformation that now affects the Mediterranean began. In 1982, Commander Cousteau, who was then still director of the museum, acquired a green alga, *Caulerpa taxifolia*. It was soon to become known as "killer alga." When water was flushed out of the aquariums, the alga began to grow under the museum's very windows, and went on to contaminate almost 25,000 acres (10,000 ha) of shallow sea, from Spain to Croatia. The alga does not kill, but it stifles the biodiversity of marine ecosystems, homogenizing the areas it colonizes and standardizing the fish populations. Measures aimed at combating it—ranging from the mechanical (removal by hand) to the biological (the introduction of alga-loving tropical snails)—have so far proved ineffective.

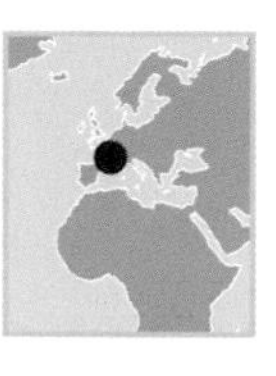

Cherry trees, Bessenay area, Monts du Lyonnais, Rhône, France (45°46' N, 4°33' E).

The Rhône departement is France's third-biggest cherry producer, with an annual yield of 8,000 tons. On the hills around Bessenay, west of Lyon, 998 acres (400 ha) of cherry trees rise in terraces at altitudes of 983 to 2,295 feet (300 to 700 m). Rooted in light, magnesium-rich soil, which these fruit trees love, they contribute about 3,000 tons to the Rhône departement's total output. This relatively young orchard (almost half the trees are less than ten years old) has seen high rates of planting over recent years. The fruit ripens from the end of May until July and gives the village of Bessenay its nickname of "Cherry Capital." A total of 1,000 arboriculturalists cultivate 7,660 acres (3,100 ha) in the departement, growing cherries, pears, apples, apricots, and peaches. These small operations have been reduced by 30 percent in 12 years, but large numbers remain—more than two-thirds of the total. Since 1988, about 250 of the Rhône's agricultural concerns per year have been disappearing. Nationally, the figure is 28,000.

NOVEMBER

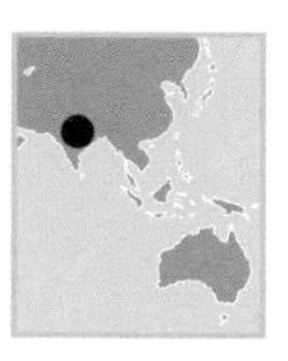

Slaughterhouse near New Delhi, India (28°36' N, 77°12' E).

Over the last 50 years, world meat production has increased from 44 million to 216 million tons a year—an increase twice as great as that of the world's population. Consisting mostly of pork (40 percent), poultry (28 percent), and beef (26 percent), meat production consumes more than a third of the world cereal harvest. In industrial "zero graze" farming (that is, without pasture) and in fattening animals for slaughter, it takes the equivalent of 18.75 pounds (7 kg) of cereals to produce 2.7 pounds (1 kg) of beef, and 5.35 pounds (2 kg) of cereals for 2.7 pounds (1 kg) of poultry. In a world where one in five people is still malnourished, and where cereal production is slowing, this use of cereals as fodder for animals has been criticized. The scandal over bovine spongiform encephalopathy (BSE), which most people know as "mad cow disease," has put the practice of feeding animal remains to farm animals on trial, and legal disputes over the use of growth hormones in livestock have raised growing concerns over how far certain modes of production should be allowed to go.

20

NOVEMBER

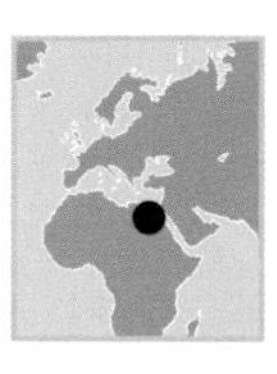

Drying dates in a palm grove south of Cairo, Nile Valley, Egypt (29°43' N, 31°17' E).

Date palm trees are grown only in hot, arid areas with water resources, such as oases. Five million tons of dates are produced each year worldwide. Most of the production from the Near East and North Africa is intended for each country's domestic market and only about 5 percent is exported. Egypt, the world's second-leading producer, after Iran, harvests more than 800,000 tons of dates each year, which are consumed locally at a rate of 22 pounds (10 kg) per person per year. These dates are habitually preserved in traditional ways. Fresh-picked, yellow or red depending on the variety, the dates slowly turn brown as they dry in the sun, protected from the wind and water by a small wall of earth and branches. They are then kept in baskets woven from palms. Although most of the dates produced go on the table, several derivatives (including syrup, flour, dough, vinegar, sugar, alcohol, and pastries) are made from the fruit manually or industrially.

21

NOVEMBER

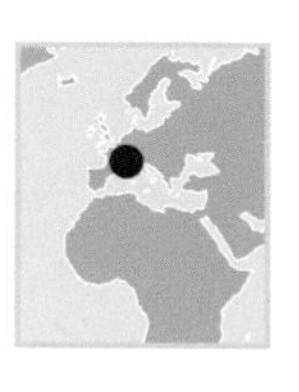

Olive harvest near Les Baux-de-Provence, Côte d'Azur, France (43°44' N, 4°47' E).

From November to February, the Mediterranean's olive harvest is in full swing. Careful harvesting provides work, protects soil from compaction by heavy machinery, and produces extremely high quality oil. Olive oil is popular all over the world for its nutritional and culinary properties, and it plays a part in many cuisines. Consumption has risen by 50 percent since 1990, rising from 1.6 million tons to 2.4 million in 1999. With its 840 million olive trees, olive cultivation in the Mediterranean has a bright future. Nonirrigated olive cultivation, which gets the best out of dry soils in a region where managing fresh water supplies is crucially important and soil degradation is a considerable problem—is an example of sustainable use of soil, preservation of landscape, and support of populations living in economically marginal rural areas. It should retain its role in a region where tourism is devouring space and increasing pressure on land. The Mediterranean receives 30 percent of the world's tourists.

22

NOVEMBER

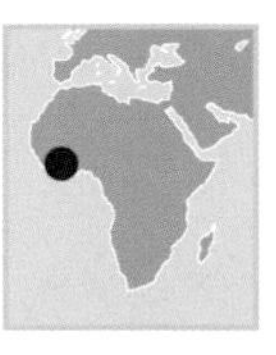

Public waste dump near Abidjan, Republic of Côte d'Ivoire (5°20' N, 4°00' W). Waste dumps near towns and irregular garbage collection are an open sore in Côte d'Ivoire. But the ill health of the "African Manhattan" is but a drop in the ocean on a global scale. The amount of waste generated by industrial countries is a far greater cause for concern: an inhabitant of the United States produces fifty-three times as much as an Ivorian. It should be borne in mind that every pound of finished goods requires a far greater quantity of resources and energy to be taken from the environment. For example, between 8 and 14 tons of nonrenewable raw materials must be used up to manufacture a single personal computer. To maintain their present lifestyle, each individual citizen of the industrialized countries consumes a yearly average of 100 tons of nonrenewable resources, to which must be added 500 tons of fresh water, or between thirty and fifty times the amount available in poor countries.

23

NOVEMBER

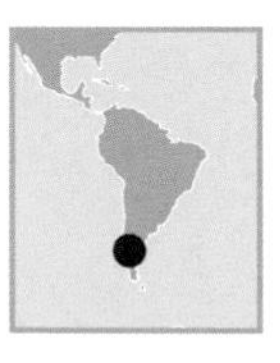

Dairy cows passing between dunes, Maule province, Chile (35°16' S, 73°20' W).

The wind sweeps the volcanic dust before it. Here, an oceanic climate showers the land with abundant moisture, allowing grass to grow rapidly and favoring livestock farming. The country is known for its "crazy geography." It measures 2,608 miles (4,200 km) from north to south, stretching over 35 degrees of latitude, but is only 62 miles wide at its narrowest point—and 250 miles at its widest (100 and 450 km). This means that the north is extremely arid; the Chilean economy here is dominated by copper, iron ore, and sulphur mining. The center has a more Mediterranean climate and contains the biggest cities and associated industry, as well as farming—chiefly fruit and vineyards. In the south, with its oceanic climate, fields give way to pasture, vast forests, and lakes until, gradually, the great glaciers of Patagonia take over. Chile finally comes to an end at the far southern tip of South America, not far from the Antarctic circle.

24

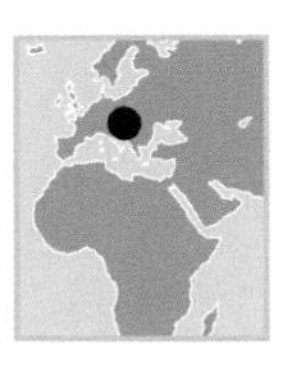

Szentendre, Hungary (47°41' N, 19°03' E).

Szentendre, a town of about 21,000 people, lies about 12.5 miles (20 km) north of Budapest on the Danube. It was founded in the seventeenth century by Serbian colonizers who were fleeing the Turks. Its little winding streets and squares, with their Mediterranean style, as well as many Serbian Orthodox churches, give it its picturesque character. Hungary is experiencing a high rate of economic growth (5.2 percent in 2000) and proved an excellent (and successful) candidate for entry to the European Union. But for many Hungarians, the transition to a market economy after the collapse of the Soviet Union has led to worsening poverty and increased wealth inequality. During the 1990s, the country lost 1.5 million jobs. The death rate (13.3 per thousand in 2001) and a fall in the birth rate have also caused concern. Between 1995 and 2001, the population fell every year by 0.5 percent—or 20,000 people.

NOVEMBER

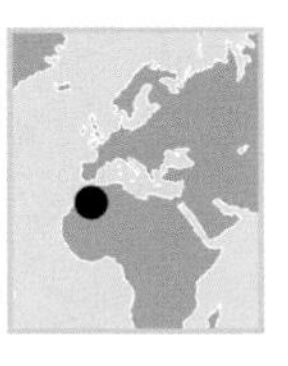

Dades Gorge, Morocco (30°55' N, 06°47' W).

The slender watercourse that flows between the High Atlas and the Anti-Atlas feeds many Berber villages, which huddle against the rocky walls of the Dades river valley. Built out of *pisé* (a mixture of rammed earth and straw), the houses melt into the rocky scenery around them. Yet human presence does not go unnoticed: a multitude of delicate green gardens softens this rocky landscape. Most are tiny fields of grain or potatoes for local consumption. A product typical of this valley is the damask rose, whose flowers are distilled into rose water by a cooperative. By grouping together in this way, farmers can sell their produce at fair prices; if they worked individually, they would have to sell at rock-bottom prices in the local *souk*, or simply to the first middleman they could find. However, this system has remained limited to rose growers, many villagers not feeling the need to change.

26

NOVEMBER

Promotional stand, Lake Balaton, Hungary (46°50' N, 17°45' E).

Lake Balaton is the Hungarians' Mediterranean. This sheet of water 43.5 miles (70 km) long, one of Europe's largest, attracts millions of tourists every year, and is known to Hungarians as "the people's basin." When Europe was divided, Lake Balaton was most of all a meeting place for East and West Germans. Now that the Berlin Wall has fallen, holidaymakers can sometimes be seen gathered around some floating advertisement, drifting on their inflatable mattresses. It is perhaps an image emblematic of a peaceful world, unified by the market economy. Free enterprise, as the forging of the European Union demonstrates, remains the most powerful argument for building peace. Thus, the proliferation of commercial modes of self-expression is defended in the name of the need to maintain the free flow of information. And so the most beautiful places can be transformed into vast advertisement hoardings.

27

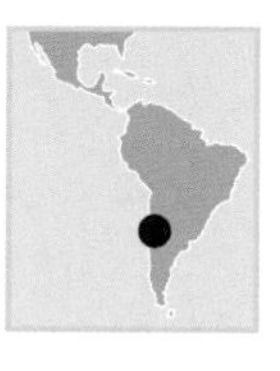

Salar de Atacama, Chile (23°30' S, 68°15' W).

A scattering of pink flamingoes is all that breaks this uniformly mineral scene in the Salar de Atacama. The birds take advantage of the water of the San Pedro river before it vanishes into the landscape, evaporating into the dry air or soaking into the surface of this vast salt plain that covers 1,158 square miles (3,000 km^2). The Salar is part of the world's most arid region, after Antarctica: the Atacama Desert, a great strip of land 1,677 miles (2,700 km) long in Peru and Chile. Its hostile climate is due to the dual action of the cold Humboldt Current, which flows along Chile's Pacific coast and prevents all evaporation, and a warm anticyclone, which traps dry air at ground level. Only a grimy drizzle, the *chamanchaca*, falls here and there. The el Niño phenomenon, which occurs every seven years, may bring more abundant rainfall, but this is often too much, as was the case with the disastrous deluges of 1983 and 1997.

28

NOVEMBER

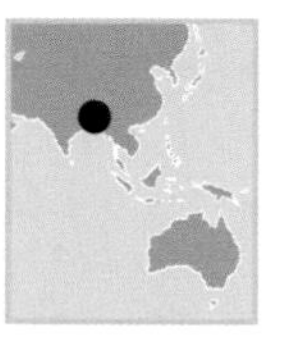

Flooded houses south of Dhaka, Bangladesh (23°21' N, 90°31' E).

Covered by a vast network of 300 waterways, including the Ganges, Brahmaputra, and Meghna rivers, which descend the slopes of the Himalayas to the Bay of Bengal, Bangladesh is a delta plain that is subject to seasonal monsoons. Between June and September huge rains sometimes cause the rivers to overflow their banks and inundate nearly half of the territory. Accustomed to this natural cycle, part of the country's population lives permanently on *char*s, ephemeral river islands made of sand and silt deposited by the rivers. In 1998, however, two-thirds of the country remained under water for several months following the worst flood of the century, which claimed 1,300 lives and left 31 million Bangladeshis homeless. Among the most densely populated territories on Earth, with 360 inhabitants per square mile (922 per km^2), Bangladesh is also one of the poorest countries: 32 percent of the population live on less than one dollar per day. The rising sea level, possibly caused by climatic warming, will only aggravate the difficulties of this country, which might see a considerable portion of its rice fields permanently flooded.

29

NOVEMBER

Dunes bordering the town of Concón, Chile (32°55' S, 71°31' W).

Children like to play with makeshift toboggans on the areas of dunes that border the beach resort of Concón. Judging from the photograph, the sand appears about to engulf the town, but in fact the opposite is true: buildings, roads, and other infrastructure are gaining land at the expense of nature. Begun in 1996, Concón is the latest fashionable district in the metroplex of Viña del Mar, Chile's tourist capital, which sprawls almost continuously over 15.5 miles (25 km) of beaches. Squeezed between the mountains and the sea, and blocked to the south by the large port of Valparaiso, it has had no choice but to expand to the north, at the expense of the dunes. The explosion of tourism, combined with growth in the world's population and the increase in the number of city-dwellers, means that built-up areas are advancing on wild ones. Today, 37 percent of people live less than 40 miles (60 km) from a coast—a number greater than the population of the entire world in 1950.

30

THE PATH TO RESPONSIBLE AND SUSTAINABLE DEVELOPMENT

DECEMBER

01	02	03	04	05

Cheap, safe energy from oil and coal has sustained human progress for more than two centuries.

At some point during the twenty-first century, the world's population is likely to reach 10 billion. More than 80 percent of those people will live in developing countries, where most inhabitants lack the sources of energy they need to improve their quality of life. This population increase, and the desire of the inhabitants of developing countries to attain a standard of living comparable to ours, are expected to increase the world's energy demand by about 66 percent over the next twenty years.

It is vital that we meet this energy challenge in an efficient and responsible way. Growing energy needs will lead to ever higher emissions of greenhouse gases, which will speed up climate change. To hope for a healthier environment while using up more resources is clearly irrational; we need to change our behavior radically, so that the aspirations of all can be met without jeopardizing the prospects and way of life of future generations. This is no less than a challenge to humanity; it is the challenge of sustainable development.

Business has a vital role to play in meeting these challenges, for it possesses the means to meet them. Its great assets are organization, flexibility, creativity, the ability to take risks beyond national borders, and the ability to invest large sums. It is thus able to initiate the changes needed to continue human development in a sustainable fashion.

The Shell Group (properly known as the Royal Dutch/Shell Group of companies) is taking advantage of the structures and means afforded by its global size to contribute to sustainable development and make the future "a better world to live in."

Our recent commitment at the Shell Group to sustainable development is based especially on the lessons to be learned from past failures. The economic, political, and human troubles of the inhabitants of the Niger Delta, where Shell extracts oil, and the outrage over the symbolic desecration of the ocean during the episode of the Brent Spar platform—which Shell planned to sink on the bed of the North Sea, against the advice of environmental non-governmental organizations (NGOs)—revolutionized our view of the world in 1995 and led us to take action.

06 07 08 09 10 11 12 13 14 15

In a world where knowledge is constantly growing and perceptions change, we had a great deal to learn. But we had the will, and the ability, to learn systematically and effect change, with a collective will to move in the right direction.

For us, sustainable development means taking account of economic, social, and environmental factors when formulating our short- and long-term strategies. This new approach has changed the way we work.

We have, indeed, acquired new responsibilities to society: as the past demonstrates, our activities directly affect the environment and society, which also gives us some influence in the choice of what sort of society we want in the future. But we cannot decide this on our own. This is why we endeavor to understand the needs and expectations of citizens and consumers and why we are pledged to transparency in our business.

We hold regular discussion meetings in which governments and NGOs tell us of their concerns; for our part, we take their constructive comments into account in making concrete commitments. We also sound out our employees' opinions through anonymous surveys, and the Internet allows us to listen to a broad sampling of public viewpoints.

Our commitment is translated into practical action, and in the *Shell Report*, an annual report available to all, we report on our achievements and undertakings, including these:

* Faced with energy challenges, we are actively working to find suitable solutions, investing in research, developing renewable energy, and providing development aid for poor countries.
* As part of the battle against the greenhouse effect, we had set ourselves the target of reducing our greenhouse gas emissions by 10 percent in 2002, compared with 1990 levels. This reduction clearly relates to our industrial activities, and not to emissions due to the sale of our products to the consumer. We achieved it in 2002, even though the Kyoto protocol demands it only by 2010.

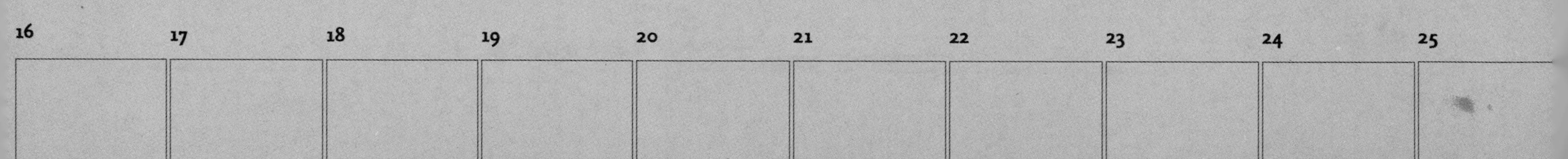

The role and aims of our group have evolved. We are no longer content simply to improve the quality of our production. Today, we also try to influence the way the world can meet its future energy needs by suggesting innovative energy solutions. We firmly believe that working toward sustainable development is good for business because it encourages us to listen. It helps us understand the needs and expectations of citizens and consumers and helps productivity and innovation. For us, therefore, our company's commitment to seeking sustainability is *the* way forward.

If society is to progress toward sustainable development, there must be a commitment from business. But the concrete achievements to contribute to this progress require large investments. Companies must therefore safeguard their profitability if they want to have the resources to meet their commitment. A company's profitability and success depend heavily on its ability to respond to the concerns and expectations of citizens and consumers, who also play an essential part in sustainable development.

Development of the planet can only be sustainable if all are collectively committed to it. So, let's take action!

CHRISTIAN BALMES
General President-Director, Shell Group, France

26	27	28	29	30	31

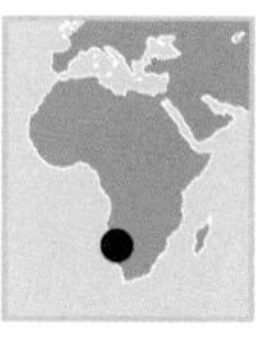

Mountainside in the Zebra Mountains, Kaokoland, Namibia (17°20' S, 13°00' E).

The Zebra Mountains are a strange landscape, where black granite scree is streaked with bands of vegetation. The Kunene river flows through them, and a big dam project is planned near the Epupa waterfalls. It is aimed at contributing to Namibia's development and reducing its dependence on energy from South Africa. Combined with wind power and natural gas, it is expected to reduce the share of imported energy to just 25 percent of the country's consumption. But although it produces no greenhouse gases, the dam has other effects. It is a threat to the Himba people, as it will destroy the pastures where they graze their livestock—their chief resource. The Namibian government, invoking the project's viability and ability to attract foreign investment, refuses to consider alternatives, such as building the dam at Baynes, downstream of Epupa. Although this option would involve flooding only 22 square miles as opposed to 147 (57 km^2, as opposed to 380), it would have the advantage of preserving the Himba culture, as well as the region's tourism value.

DECEMBER

Seals on a rock near Duiker Island, Cape Province, Republic of South Africa (34°05' S, 18°19' E).

Cape fur seals (*Arctocephalus pusillus pusillus*) are highly gregarious. They gather by the hundreds in coastal colonies, chiefly to mate and give birth. Happier in the water than on land, these semiaquatic mammals spend most of their time swimming in coastal waters seeking food: fish, squid, and crustaceans. The species found at the Cape of Good Hope lives only on the coasts of southern Africa, from Cape Cross (Namibia) to Algoa Bay (South Africa), and numbers about 850,000 individuals. Seals belong to the pinniped family, which includes fourteen species of otarid seals (sea lions and eared seals), nineteen species of phocid seals (true seals), and one species of walrus. Pinnipeds live in most seas and total about 50 million individuals, of which 90 percent are phocid seals.

02

DECEMBER

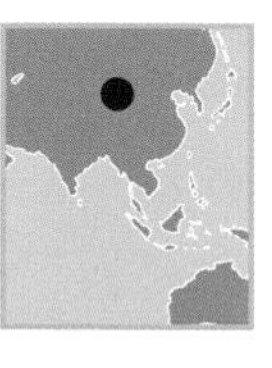

Meanders in the Tuul river, south of Lün, Mongolia (47°52' N, 105°15' E).

The river Tuul rises in the Hentii mountains and travels 508 miles (819 km), passing through Mongolia's capital, Ulaanbaatar (Ulan Bator) before flowing into Lake Baikal, in Russia. In the steppes downstream of Ulaanbaatar, it becomes extremely sluggish, making long meanders and dividing into channels flowing between silty banks and the pale streaks of salt deposits. Nomadic herdsmen pitch their white, circular tents, called yurts, on the islets of close-cropped grass and thickets. Mongolia is unique in having a centuries-old system of nature reserves, which has preserved its environment intact. However, the rapid growth of the capital, whose population has doubled in ten years, is producing alarming levels of pollution. Waste is building up in the Tuul River, which then contaminates the land through which it flows, while air quality in Ulaanbaatar has deteriorated because of increased car ownership, which rose by 30 percent in 2002 alone.

03

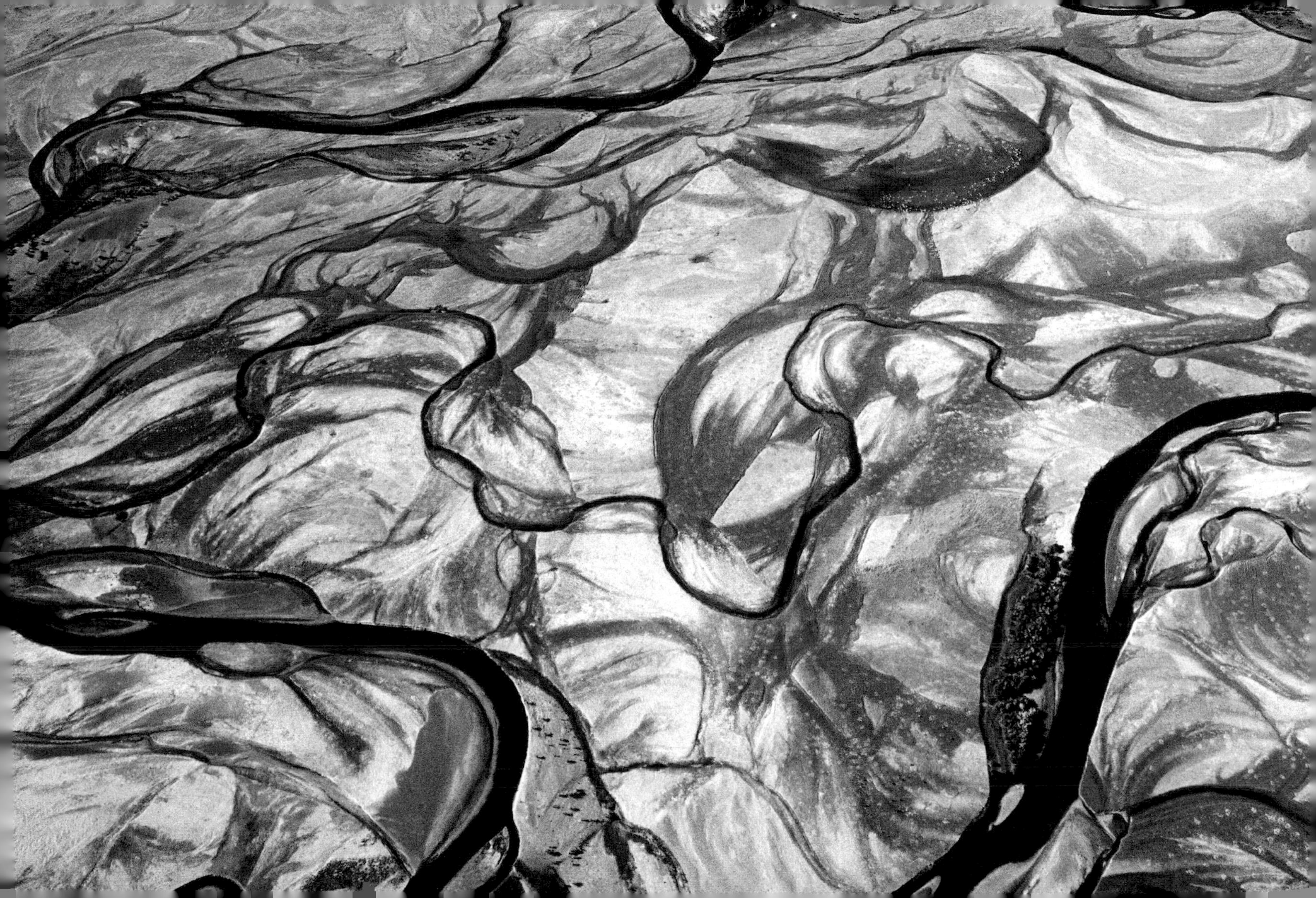

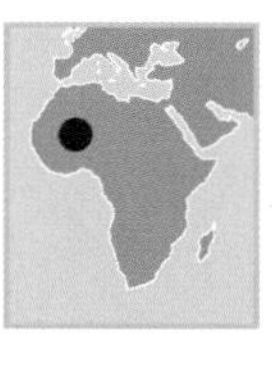

Pink sand dune of Koima, near Gao, Mali (16°15' N, 00°05' W).

On the right bank of the Niger river, this 393-foot (120-m) sand dune "sings." The wind that blows on the sand seems to be celebrating this 2,590-mile (4,170-km) watercourse, which is West Africa's biggest communication artery. Mali prospered until the seventeenth century thanks to the merchants who used the river. But this trade, like other continental traffic, gradually dwindled as it lost out to the Atlantic coast, leading to the general decline of the regions in West Africa's interior. Almost two-thirds of the population of Mali suffers poverty. The results are a high rate of illiteracy, serious malnutrition, reduced life expectancy, poor health, unsanitary living conditions, and exclusion from economic and social life. Although these indicators are generally improving worldwide, they are worsening in most of the countries of sub-Saharan Africa.

04

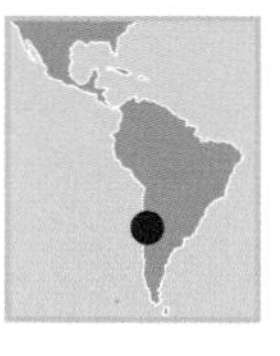

Maricunga salt lake, Chile (26°55' S, 69°05' W).

The salt lake of Maricunga is not completely dry. Its emerald-color pools, linked to the large nearby lagoons of Santa Rosa and Negro Francisco, are still a paradise for pink flamingoes and wild ducks. This wetland area in the heart of the Andes is home to forty-one bird species, including three types of pink flamingo on the red list of threatened species. But mining activity may have an impact on these habitats. Extraction of gold and copper involves not just mining but also drawing off large amounts of the ground water that supplies the lagoons. Since 1996, these lagoons have been designated under the Ramsar Convention, whose mission is to sustainably preserve ecosystems as well as social and economic activity in internationally important wetland areas. It is estimated that half the world's wetlands have been destroyed since 1990.

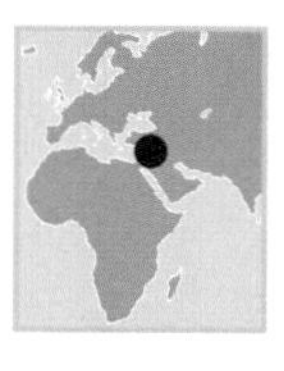

Boat on the Dead Sea near a potash plant, Al Karak region, Jordan (30°50' N, 35°30' E).

The Dead Sea, a landlocked body of water measuring 47 by 9 miles (75 by 15 km), is the lowest point on the planet, at 1,337.8 feet (408 m) below sea level. It greenish color is streaked with white because of its high salt content—nine times greater than the average found in the oceans. Besides common salt (sodium chloride), the waters of the Dead Sea are extremely rich in potash salts, making Jordan the world's eighth-biggest producer. No animal or vegetable life can survive in this part of the valley. Since 1972, the Dead Sea has lost 20 percent of its surface area: its waters, and those of the Jordan, which feeds it, have been diverted to irrigate the Negev Desert. Moreover, excessive use of ground water has led to a severe drop in the water table and deterioration in ground-water quality as a result of salt-water contamination. In the steppes of Syria and Jordan, the salinity of ground water has increased by several grams per liter, threatening many plant and animal species.

DECEMBER

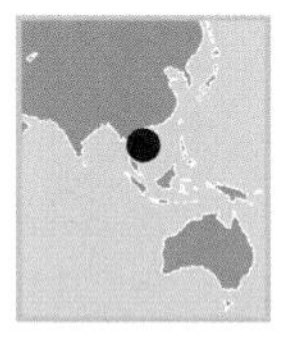

Working in the rice paddies between Chiang Mai and Chiang Rai, Thailand (19°25' N, 98°55' E).

In the mountainous areas beyond Chiang Mai in northern Thailand, these women work in the rice paddies, protected by ample colored fabrics. As everywhere else in the country, rice is harvested by hand. Farms are often family businesses, and the harsh lands of these high plateaus yield little wealth for the peasants, who in this region are mostly Lao. Moreover, large forest fires have led to extensive deforestation, which has been exacerbated by forest clearing. A total of 276,758 acres (112,000 ha) of Thai forest disappeared between 1990 and 2000. This phenomenon, which occurs throughout Southeast Asia, releases a mixture of pollutants into the atmosphere, producing a brown cloud that hangs over the region. By reducing the amount of sunlight that reaches the ground, and by depositing acid particles over fertile land, this artificial fog could seriously affect the climate and agriculture of all Southeast Asia.

07

DECEMBER

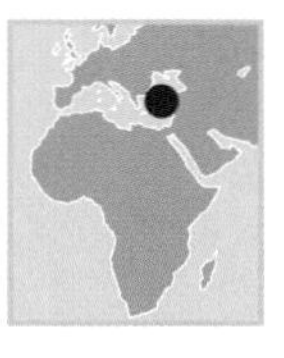

Göreme National Park, Cappadocia, Anatolia, Turkey (38°26' N, 34°54' E).

Produced by volcanic activity and shaped by erosion, the Göreme valley is as notable for its landscapes as for its irreplaceable architectural and artistic monuments. From the fourth century on, hermits in retreat from the world and Christian martyrs hoping to escape from Roman persecution took refuge and then settled here, carving remarkable churches and underground cities in the mountains of Cappadocia. The arrival of the Ottoman Turks led many to convert to Islam, but the churches and the Byzantine frescoes that adorn these mountains were neither destroyed nor pillaged. On UNESCO's World Heritage list since 1985, Göreme National Park and its historic sites are one of the biggest tourist attractions in Turkey, which receives almost 9 million visitors each year.

08

DECEMBER

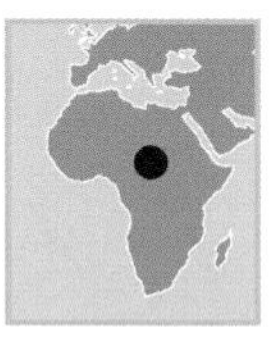

Village on the north shore of Lake Chad, Chad (13°28' N, 14°43' E).

On the islands that dot the northern part of Lake Chad are the huts of the Buduma people, an evocative name that means “reed man” in the Kanembu language. The Buduma are just one of 200 ethnic groups living in Chad, a country torn by conflict between the north, which is predominantly Muslim and “Arab,” and the “black” south, where Christians and animists live side by side. Since the mid-1960s, this divide has fueled a constant struggle for control in this country whose borders are arbitrary, a legacy of the time when it was a French colony. While the brutal dictatorship of Hissene Habré killed 40,000 civilians between 1982 and 1990, the current president, Idriss Déby—who was democratically elected—continues to crush opponents of the government. In 1996, Amnesty International reported that the use of torture was widespread in Chad. At the start of the third millennium, torture using electric shocks was practiced in forty countries, blows on the soles of the feet in thirty countries, partial suffocation in thirty countries, and mock executions in more than fifty countries.

09

DECEMBER

10

Housing plots at Brøndby, on the outskirts of Copenhagen, Seeland, Denmark (55°34' N, 12°23' E).

To combine management of the available space with security and comfort, the building plots at Brøndby, on Copenhagen's southwestern fringe, are arranged in perfect circles, in which each holder has a lot of 4,305 square feet (400 m^2). This type of residential district, which is highly practical, is increasingly common on the outskirts of large cities where there are many jobs. The growth of industry, the attraction exerted by cities, and the expansion of large metroplexes have led to a 13-percent increase in the world's city-dwellers—a trend that is continuing. Almost half the planet's population, 45 percent, is now urban. By 2025, there will be twenty-five giant metroplexes—megalopolises—in the world, each home to between 7 million and 25 million people. Three-quarters of the other billion city dwellers the planet will accommodate in 2025 will live in southern countries.

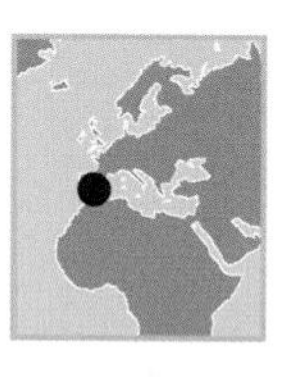

Fishing net off Cartagena, Murcia region, Spain (37°30' N, 0°59' W).

This broad net is floating a few miles offshore of Cartagena, a port in southeast Spain. The countries around the Mediterranean harvest some 1.3 million tons of fish a year from its waters (1.5 percent of the world catch) and consume more than 3 million tons. (Spain is top of the league, with almost 110 pounds [40 kg] per person per year.) Several commercially fished species, such as hake, sole, bass, and monkfish, are already being fished at a higher rate than they can replenish themselves, while the population living around the Mediterranean—which totals 450 million people, with a further 150 million mouths to feed in the tourist high season—could increase by 50 percent over the next 25 years. Not only that—the waters of this almost enclosed sea receive 7,500 tons of heavy metals, 200,000 tons of chemicals, and up to 1 million tons of crude oil every year. Recent international agreements have endeavored to improve management of the Mediterranean's fish resources.

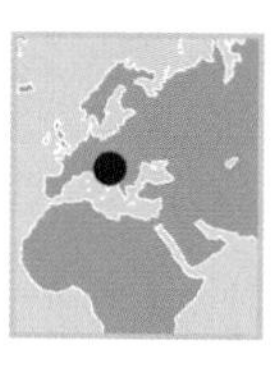

Landscape in the Kukës region, Albania (42°05' N, 20°24' E).

The Kukës region in northeastern Albania contains the country's highest mountain range, the Korab, which rises to 9,027 feet (2,753 m). This mountainous enclave, bordered by the valley of the Drin river, has always been a pocket of resistance in terms of the *kanun*, a form of vendetta based on the old code of "blood vengeance." The northern clans, grouped into vast patriarchal families, have never stopped exercising this traditional right, which is more than 500 years old, despite all the political and judicial reforms that various ruling powers have implemented. Neither the Ottoman decrees, made from the sixteenth century to 1912, nor the laws of King Zog made between World Wars I and II have succeeded in putting an end to this tradition of the mountain people. Today, the *kanun* is still current—indeed, it has even seen something of a revival since the fall of Communism. It is estimated that 10,000 people are exposed to its pernicious influence to some degree. In 2002, between 150 and 800 children were deprived of schooling because their parents kept them confined to their homes, for fear that they might be killed or kidnapped.

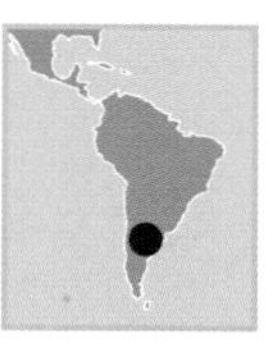

Andean condor in Neuquén province, Argentina (39°00' S, 70°00' W).

An Andean condor, *Vultur gryphus*, spreads its 10-foot (3-m) wingspan and glides effortlessly above the autumn foliage of Patagonia. The world's biggest bird was regarded as sacred by the Incas, and today it remains a symbol of this wild cordillera. Before the arrival of the colonizers, these raptors were common throughout the Andes, from Venezuela to the Peruvian coast and the Tierra del Fuego. The condor—a carrion-eater unfairly plagued by a bad reputation—was persecuted so efficiently that it vanished from many regions, notably from Venezuela and Colombia. Now reintroduction programs have been started in both countries, and there are pockets in Argentina and Chile where the bird remains plentiful. In many parts of the world, birds of prey have benefited from protection and public awareness programs. Nevertheless, it is estimated that 12 percent of all bird species are in serious danger of becoming extinct.

DECEMBER

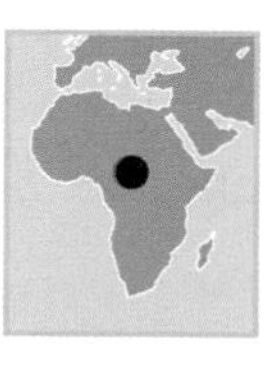

Dugout on Lake Chad, near Bol, Chad (13°28' N, 14°43' E).

Surrounded by a mesh of reeds and papyrus, a narrow dugout follows the indistinct course of one of Lake Chad's countless channels. Africa's fourth-largest lake teems with such craft, which come from Chad, Niger, Nigeria, and Cameroon. These uncontrolled movements worry the authorities because of the illegal fishing in Lake Chad's waters, which are extremely rich in fish. Well-equipped fishermen from Nigeria, Ghana, and Mali compete fiercely with those from Chad, who have rudimentary equipment and pay high taxes. Fishing is, however, an irreplaceable source of wealth and food for this country, which is the fifth poorest in the world. And yet Chad seems to be bouncing back, for its gross domestic product rose by 11 percent in 2002. This increase was closely connected to an oil pipeline that, from 2004, is due to carry 225,000 barrels of oil a day from the oilfields of Doba, in the south, to the port of Douala in Cameroon.

14

Fishermen in Lake Pátzcuaro, Michoacan state, Mexico (19°35' N, 101°35' W).

North of the Sierra Madre del Sur lies Lake Pátzcuaro, on whose shores live the Tarasco and Purepecha Indians. They are celebrated for their traditional crafts, and they also make a living from fishing and farming on the surrounding hills. These large, dragonfly-shape nets are used to catch the *pecito*, a whitefish whose flesh is highly prized but which is now becoming scarce as a result of overfishing but especially because of changes in agricultural methods in the surrounding land. Deforestation and the abandonment of farming on terraces and crop rotation are destabilizing the soil. The earth no longer stays in place, and during the rainy season it washes off the hillsides and into the lake below. Moreover, fertilizers are entering the waters of the lake, where algae and water plants proliferate at the expense of fish. The local inhabitants are now replanting trees on the hills, and building low walls around the fields, in an effort to retain the earth. The livelihood of the communities around the lake depends on this: some have seen half their population migrate to other areas as a result of the exhaustion of local resources.

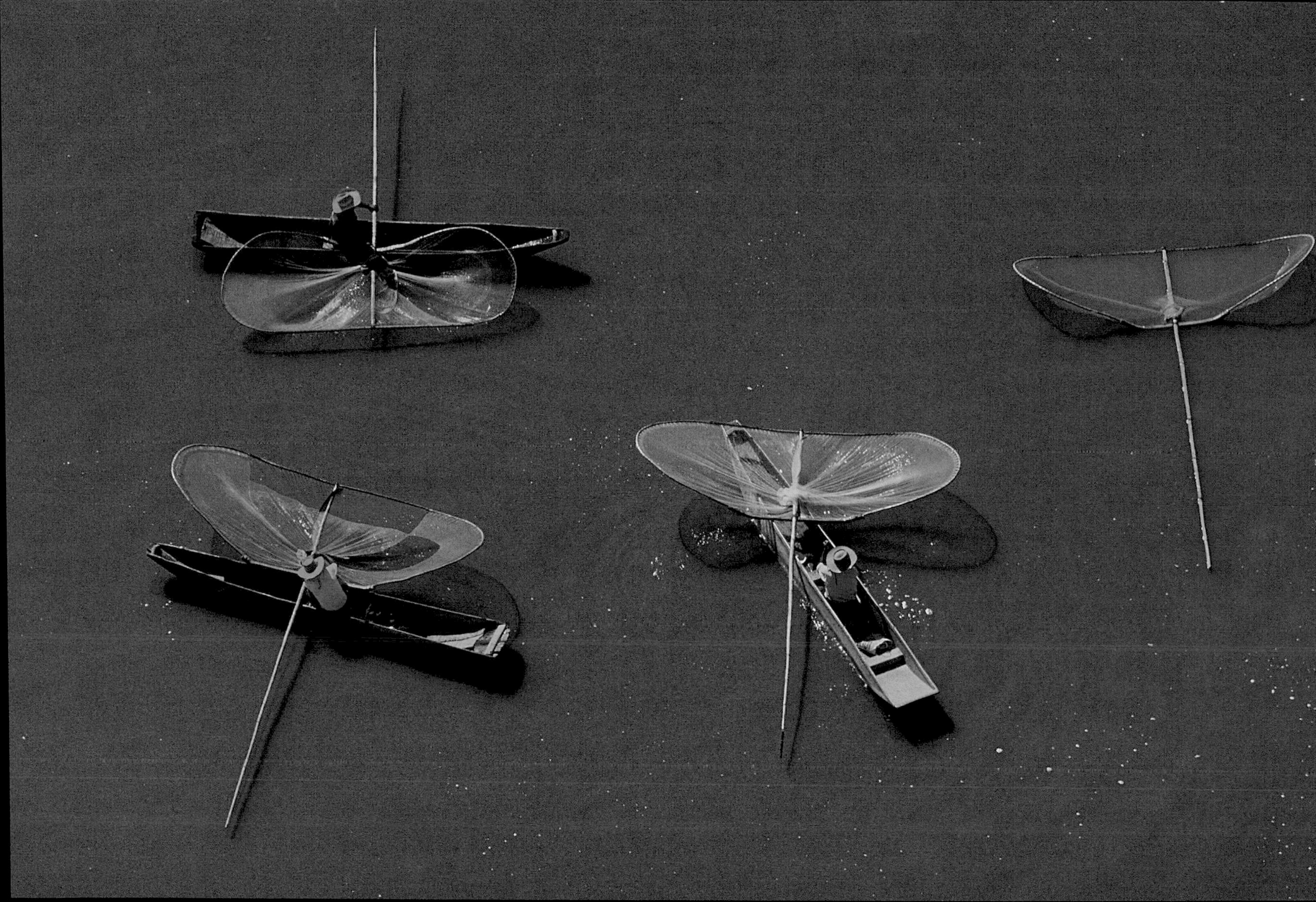

DECEMBER

16

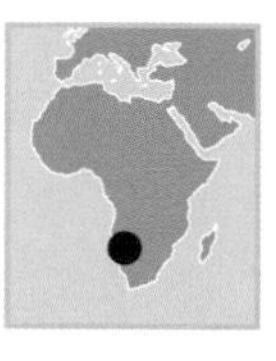

Giraffes in Etosha National Park, Namibia (19°00' S, 15°50' E).

With its 114 mammal species, 340 types of bird, and 16 species of reptiles and amphibians, Etosha National Park is Africa's biggest nature reserve. It covers more than 8,490 square miles (22,000 km^2) on the edge of the salt desert, and consists of semiarid prairies where the savannah's rich animal life can be observed at the rare watering holes. Despite this vast protected area, many of Namibia's animal species, such as the elephant and the desert rhinoceros, are declining because of ruthless poaching. Illegal trafficking in plants and animals is thought to be worth more than $5.175 billion (4.5 billion euros) a year. Worldwide, this puts it in third place behind drug trafficking and arms dealing. Of greatest concern is the situation in Africa's national parks, especially since these former game preserves, which are now defended by armed, and often military, guards, suffer from rampant poaching as well as the disastrous effects of civil war.

DECEMBER

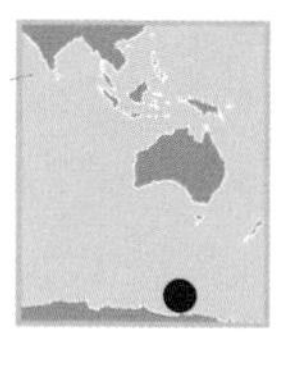

Icebergs off the Adelie coast (South Pole) (67°00' S, 139°00' E).

These drifting icebergs recently detached from the glacial platforms of Antarctica, as can be seen from their flat shape and the ice strata that are still visible on their jagged sides. Like all 480 cubic miles (2,000 km^3) of ice that detach every year from Antarctica, these icebergs will slowly be eroded by the winds and waves before disappearing. Antarctica is a place of extremes: temperatures reach as low as –94° F (–70° C), and winds reach speeds of 200 miles (300 km) an hour. The continent has an area of 5,500 square miles (14 million km^2) and contains 90 percent of the ice and 70 percent of the freshwater reserves of the planet. Antarctica has been governed since 1959 by the Washington Treaty, which gives it international status and restricts its uses to scientific activities. The Russian station at Vostok has extracted, from a depth of 11,800 feet (3,623 m), chunks of ice that have made possible the reconstruction of more than 420,000 years of history of the climate and atmospheric composition. The atmosphere's current content of carbon dioxide—the main gas responsible for global warming—is higher than it has been for 160,000 years.

17

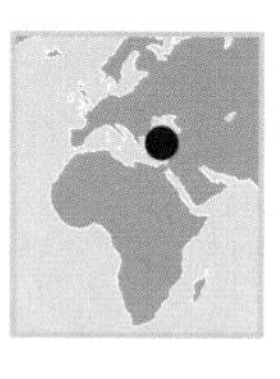

Agricultural landscape between Ankara and Hattousa, Anatolia, Turkey (40°00' N, 33°35' E).
On the Anatolian plateau, north of the line between Ankara and Sivas, the regularity of the agricultural landscape is striking: its orderly, neatly trimmed fields, carefully plowed furrows, and diversified crops of cereals (wheat, barley) and sugar beet. Turkey's agricultural sector still employs 48 percent of the workforce (compared with 17 percent in the United States, or 3.9 percent in France, for example) and, notably, 72 percent of Turkish women who work (compared with 49 percent worldwide). A large proportion of the work that the female workers perform consists of traditional activities and is unpaid; this is the case for women all over the world, including those in developed countries. But in Turkey, the greatest social inequality between the sexes lies in the level of education: 26 percent of women are illiterate, compared with 7.5 percent of men. Mass migration to cities, which offer more jobs and services, leaves young women especially vulnerable, for traditionally they are less educated and therefore less able to adapt.

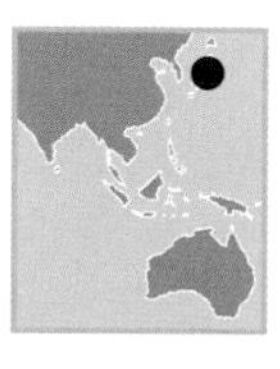

Shinto temple of Meiji-Jingu, Tokyo, Honshu, Japan (35°42' N, 139°46' E). Severely battered by earthquakes and World War II bombing, Tokyo's religious heritage has been reduced to a few temples. One of those is Meiji-Jingu, a Shinto sanctuary that was built on the orders of the emperor Meiji and completed in 1920. This is where Tokyo residents gather to celebrate the New Year. The Shinto religion consists of rituals, ceremonies, and various customs associated with *kami,* spirits that protect communities and people, and which inhabit places in the urban or rural landscape that are considered sacred. Shinto establishes a simple relationship between people and the natural or manmade objects that surround them. It fits very well with Buddhism, which is more metaphysical in character and was brought to Japan in the sixth century. Many Japanese incorporate both religions into their daily lives.

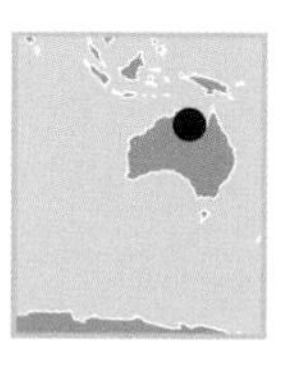

Argyle diamond mine, Western Australia (16°00' S, 128°45' E).

Argyle is an open-cast mine that taps the world's biggest diamond deposit, which on its own produces 20 percent of the world's production. However, most of these gems are of poor quality, destined for drill bits, the teeth of metal saws, or industrial sanders. Only a tiny proportion—5 percent—is cut into precious stones, and these are carefully selected. Pink diamonds, which are the rarest, can be worth tens of thousands of dollars once set into a piece of jewelry. For beautiful, colorless "stones," diamond merchants turn to Africa; since January 2003, however, merchants have been required to make sure that the stones they buy are not "conflict diamonds," whose revenue finances the vicious wars tearing apart countries such as Angola, the Democratic Republic of Congo, and Sierra Leone. Although they account for only a small proportion of the market, an international certification system called the Kimberley process has been put in place to ban them.

Salmon farm near Mechuque, Chauques Islands, Chile (42°17' S, 73°34' W). The cold and unpolluted waters of the Chauques Islands are well suited to salmon farming. This region is the second-biggest producer after Norway, and it has benefited from the world boom in fish farming since the 1970s. Although this industry is an alternative to reducing stocks of wild fish, it still remains poorly regulated. Most producer countries have no controls to limit the impact of fish farms on the environment. The high concentration of fish, and thus of their waste products and the food that is put in their cages, overenriches the surrounding water and deprives it of oxygen—a process known as eutrophication. On top of this, the use of medicines and antibiotics, which are poorly tolerated by the salmon themselves, threatens species living near the cages. Although pressure from ecologists and consumers has led to an improvement in the conditions of animals farmed on land, there has been little research on their marine counterparts.

DECEMBER

22

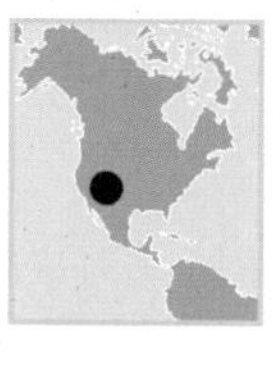

Barringer Crater, near Flagstaff, northern Arizona, United States (35°02' N, 111°01' W).

All that is missing from this lunar landscape are the astronauts. In fact, NASA uses this site for training because the topography closely resembles the that of the moon. The crater interrupts the rocky, desertlike flatness that stretches out around Winslow, Arizona, in the American Southwest. This cosmic scar, discovered in 1871, is 558 feet (170 m) deep and three-quarters of a mile (1.2 km) in diameter. It is the point of impact of a meteorite that collided with the Earth 25,000 years ago at a speed of 39,000 miles per hour (64,000 kph). In the early twentieth century, geologist Daniel Barringer first pronounced this meteorite origin, but his hypothesis was hotly contested because there are also abundant volcanic craters in the area. The examination of meteorite collisions on Earth is assisted by remote-detection satellites, but they still are not able to detect those at the bottom of the ocean. Indeed, we know the surface of Venus or Mars better than the depths of our own planet.

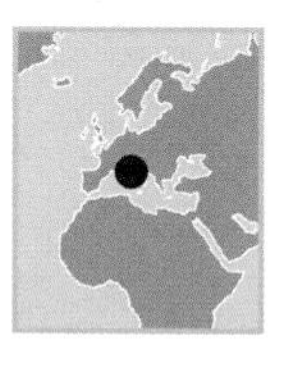

Scheggino, province of Perugia, Umbria, Italy (42°43' N, 12°50' E).

Umbria, the "green heart of Italy," is the sole region on the peninsula that has no seacoast. A world away from the beaches, its hills—dotted with fortified villages—form a gentle landscape with a medieval appearance, celebrated for its historic towns rich in art (including Assisi, Perugia, and Orvieto). Scheggino, perched at an altitude of 1,203 feet (367 m) to the east of Spoleto, stands guard over the valley of the River Nera. Beneath its tower, the village's old center huddles behind its twelfth-century walls around the church of San Niccolò (thirteenth century). All over Europe, villages cluster around their church towers, and wayside crucifixes stand at crossroads. Though secular, Europe bears the stamp of Christianity, which once was the driving force behind strife and conquest. Scheggino is famous for its delicacies: trout and crayfish caught in its clear streams, as well as its prized truffles. The Urbani company, which accounts for 80 percent of Italy's truffle cultivation and 40 percent of the world's production, is based in the district.

DECEMBER

Meanders in the Amazon River near Manaus, Brazil (03°10' S, 60°00' W).

The Amazon is the world's biggest river in terms of the size of its basin, which covers almost 2.6 million square miles (7 million km^2) of land belonging to seven different Latin American countries. The river allows logging companies to penetrate into the heart of Amazonia, but it is the building of new roads that has really allowed them to intensify their activity and, thus, speed up the deforestation that is threatening not just biodiversity and ecosystems but indigenous people, too. Since 1993, wood certification schemes have been in place to ensure the world's forests and their inhabitants are protected. Labels such as that given by the Forest Stewardship Council (FSC) mean that consumers can be sure the wood they buy comes from forests where sustainable logging respects the rights of indigenous people and allows reforesting of felled areas. Well-informed consumers who buy responsibly can thus oblige producer countries to keep a close eye on how their forests are managed.

24

DECEMBER

25

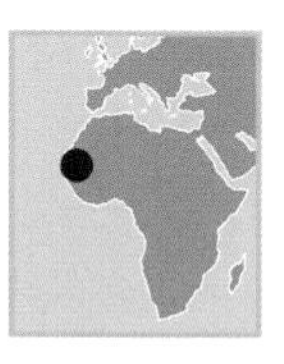

Dromedary caravans near Nouakchott, Mauritania (18°09' N, 15°29' W).

The dromedary, perfectly adapted to the aridity, is an important national livestock in Mauritania and all of the other countries bordering the Sahara. Its domestication several thousand years ago enabled humans to conquer the desert and develop trans-Saharan trade routes. The dromedary eats 25 to 50 pounds (10 to 12 kg) of vegetables a day and can survive without water for many months in the winter. In the summer, because of the heat and expended effort, the dromedary can last only a few days without drinking; by comparison, a human would die of dehydration within twenty-four hours. The reserve fat contained in its single hump helps in thermal regulation, allowing the dromedary to withstand the heating of its body without needing to perspire to cool down. The Maurs, the ethnic majority in Mauritania, raise the dromedary for its milk and meat as well as its skin and wool. In 2001 the country's dromedary livestock numbered about 1 million.

DECEMBER

26

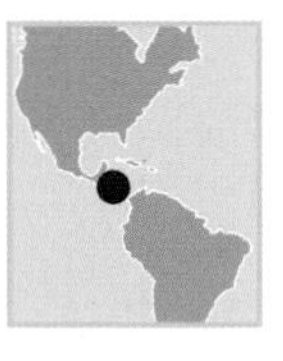

Los Micos lagoon, San Pedro Sula region, Honduras (15°47' N, 87°35' W). The mangrove-fringed lagoon of Los Micos ("the monkeys") is a concentration of tropical luxuriance within the Jeannette Kawas National Park. The park is named after its sadly missed director, the country's leading environmental activist, who was assassinated in 1995 while campaigning against tourism developments that threatened the site. A few recently built hotel complexes coexist with traditional villages. Despite the region's alluring beaches, tourism is still barely developed in Honduras, and the country's economy is dominated by banana plantations owned by U.S. companies. Extreme poverty renders the population highly vulnerable to natural disasters: in 1998, Hurricane Mitch left 11,000 dead and a further 10,000 missing, and claimed 3 million victims in all—out of a total population of 6.6 million.

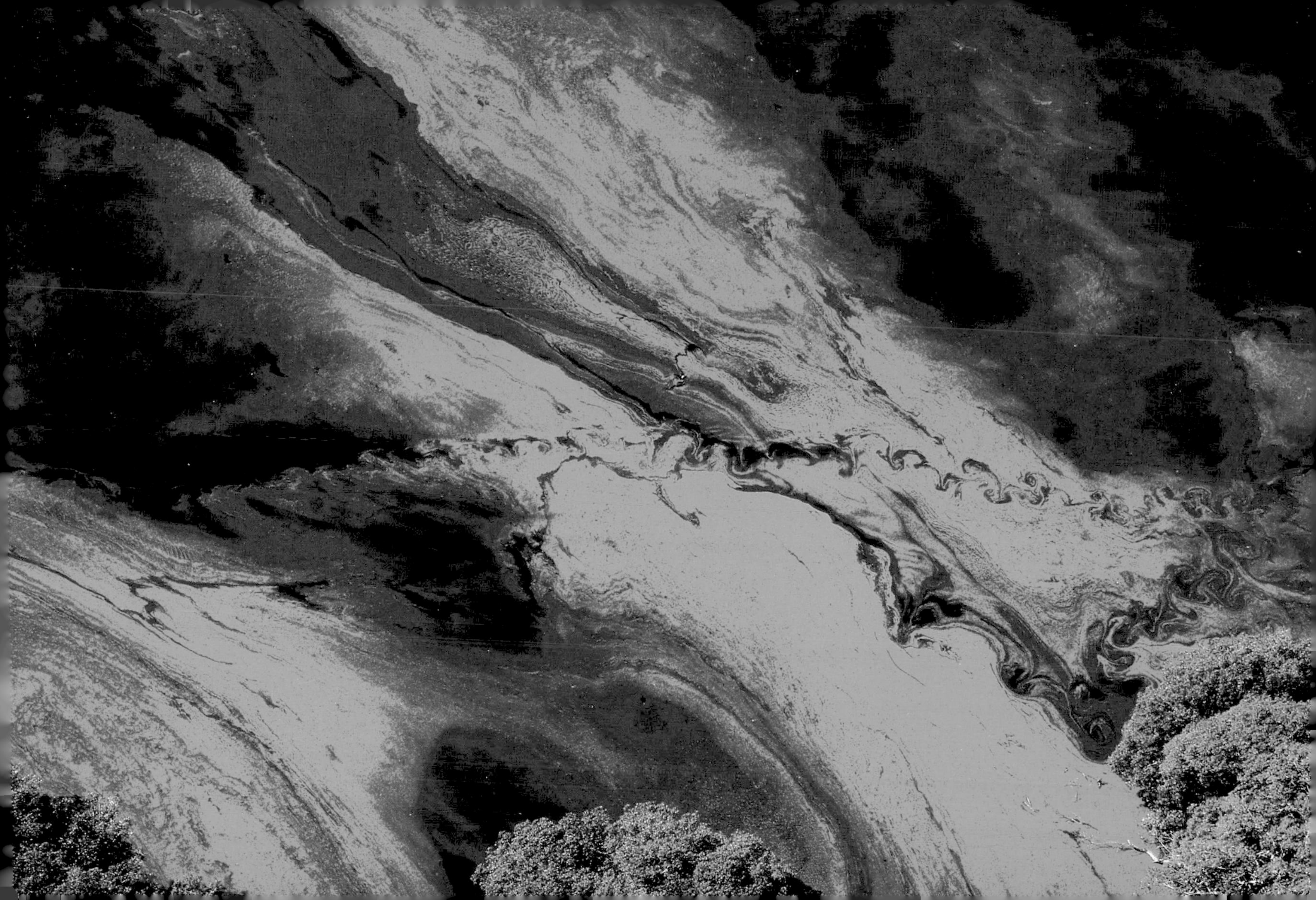

DECEMBER

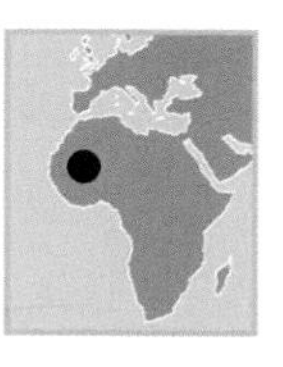

Dwellings on an islet in the river Niger, between Bourem and Gao, Mali (16°30' N, 0°12' W).

The river Niger takes its name from the Tuareg expression "*egerou n-igereou,*v which means "river of rivers." In giving it this name, the nomads meant to emphasize the inestimable value of this river, which flows through Mali's Saharan sands. Tracing a great loop through West Africa, this "Western Nile" floods every year, between July and December. But the local people are not alarmed, for they have adapted their lives to this seasonal fluctuation and are well aware of the floods' importance in fertilizing the soil and helping fish to breed. They do not attempt to build embankments along the river but are content instead to build their houses on land that is not prone to flooding. They prefer to live on *togué*, hillocks whose tops remain above water—like islets—when the river is in flood. What does worry them is drought, which has reduced the river's floods over the last 30 years.

27

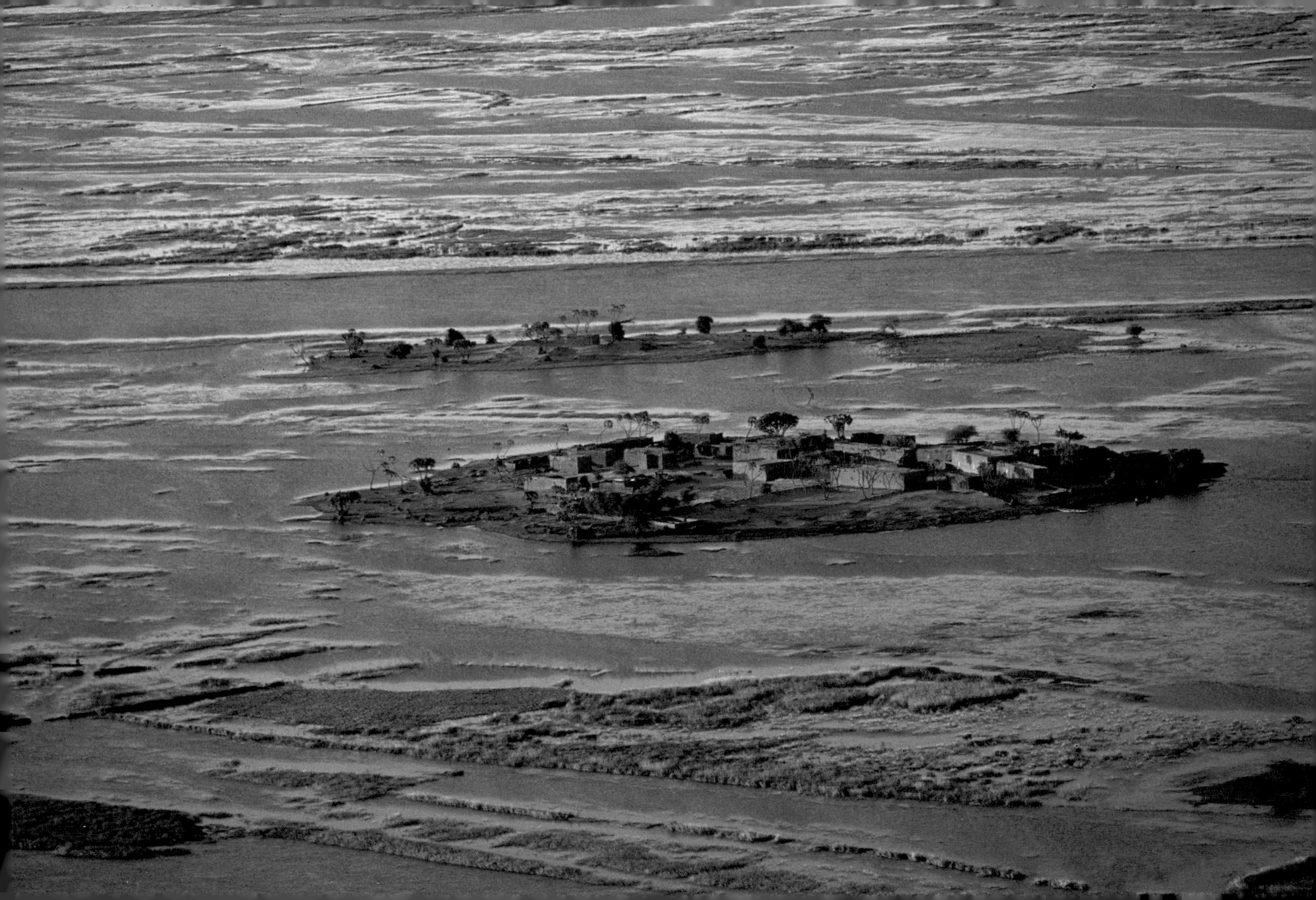

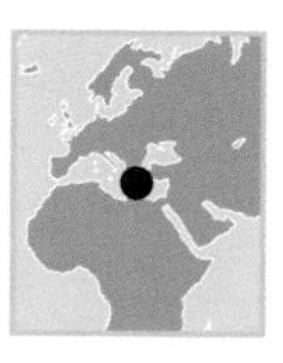

Herd of sheep, Crete, Greece (35°29' N, 24°42' E).

Sheep farming is an important part of Crete's agricultural economy. Every year, shepherds take their flocks on a journey along the stony paths of this rugged, mountainous island in search of grass. This is known as transhumance—the journey to the thin mountain pastures. The situation might be seen as alarming, for the island's 90,000 or so goats devour tree saplings. In fact, transhumance is an important factor in the land's upkeep, for if the forest completely covered the mountains, the landscape would be choked. In eating the undergrowth, the animals also prevent forest fires from spreading—there has even been talk of sheep as firebreaks. Moreover, an ungrazed mountain poses a greater risk of avalanches, for snow slides on long grass, while cropped grass grips it.

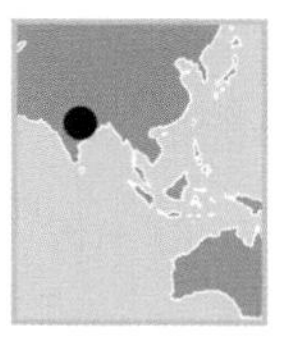

Boat on the river Ganges near Allahabad, Uttar Pradesh state, India (25°27' N, 81°51' E).
Born of the union between two torrents of ice that spring from the highest snows on the planet, and swollen by countless tributaries that also flow from the Himalayan peaks, the Ganges has a basin that occupies a quarter of India. From Rishikesh, north of Uttar Pradesh, where it leaves the Himalayas, to the Bay of Bengal, India's longest river (1,919 miles, or 3,090 kilometers) performs many functions. It is a navigable waterway, a reservoir for irrigation, and a sacred river dotted with places of pilgrimage—Haridwar, Benares, and Allahabad, at the confluence with the river Yamuna, where the Ganges reaches a width of 2.5 miles (2 km)—to which Hindus consign the ashes of their dead. The river also disposes of 3,000 human corpses and 9,000 animal carcasses every year, as well as wastewater from cities, which is responsible for 75 percent of its alarming level of pollution. The quality and quantity of available water is one of the chief concerns of the twenty-first century, and there is a need to reassess certain wasteful practices. Delhi's seventeen luxury hotels use the same amount of water per day (224,400 gallons, or 850,000 liters) as the 1.3 million people who live in the city's poor quarters.

DECEMBER

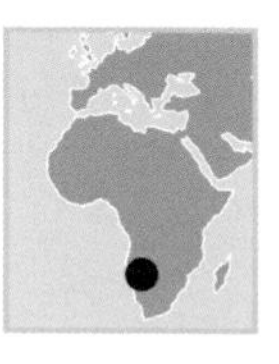

Oryx in the Namib Desert, Swakopmund, Namibia (24°39' S, 15°07' E).

On the Atlantic coast of southern Africa, the Namib Desert covers the entire 800 miles (1,300 km) of the Namibian shoreline and extends inland to a width of 62 miles (100 km), comprising one-fifth of the country's territory. Although its name, in the Nama language, means "place where there is nothing," its biological richness makes it a site unique in the world. The Namib has a secret: the humid air masses coming from the Atlantic condense on contact with the desert surface, which cools at night, enveloping the area in a thick morning fog nearly 100 days each year. This fog adds up to 1.2 inches (30 mm) of annual precipitation and constitutes the desert's main source of water and thus of life. When the orange-red sand is moistened, it allows many vegetal and animal species to subsist in the Namib Desert, such as an insect that specializes in capturing the water vapor. Only species that have evolved the characteristics most adapted to the extreme conditions of desert locations (aridity, harsh temperature, scarce food resources) can survive here, including this gemsbok (*Oryx gazella*), a type of oryx, a large African antelope.

30

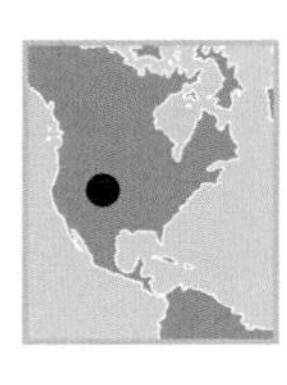

Repairs on a road near Denver, Colorado, United States (39°45' N, 105°00' W).

Like an electronic circuit board, the "bandages" on this old American road weave a fabric on the tarmac, their smooth, dark bands reflecting the light. Applied to repair cracks in the road surface, they bear witness to a long, hard life at the mercy of car tires. In the United States, the expansion of urban areas is not a recent phenomenon—but it is not over yet, either. Expansion is even speeding up as the population grows and the middle classes move to the residential suburbs. Thus cities continue to stretch out their tentacles, eating up almost 3,600 square miles (9,320 km^2) of farmland each year. Worldwide, they gnaw into countryside, but they also wipe out any forests and wetlands that stand in their way. Their growth is a threat to both biodiversity and air quality. The building of higher-density neighborhoods, however, along with downtown revitalization of city centers that are often decaying and inhabited by the poor, can help limit the impact of urban growth on the environment while also encouraging a more diverse social mix.

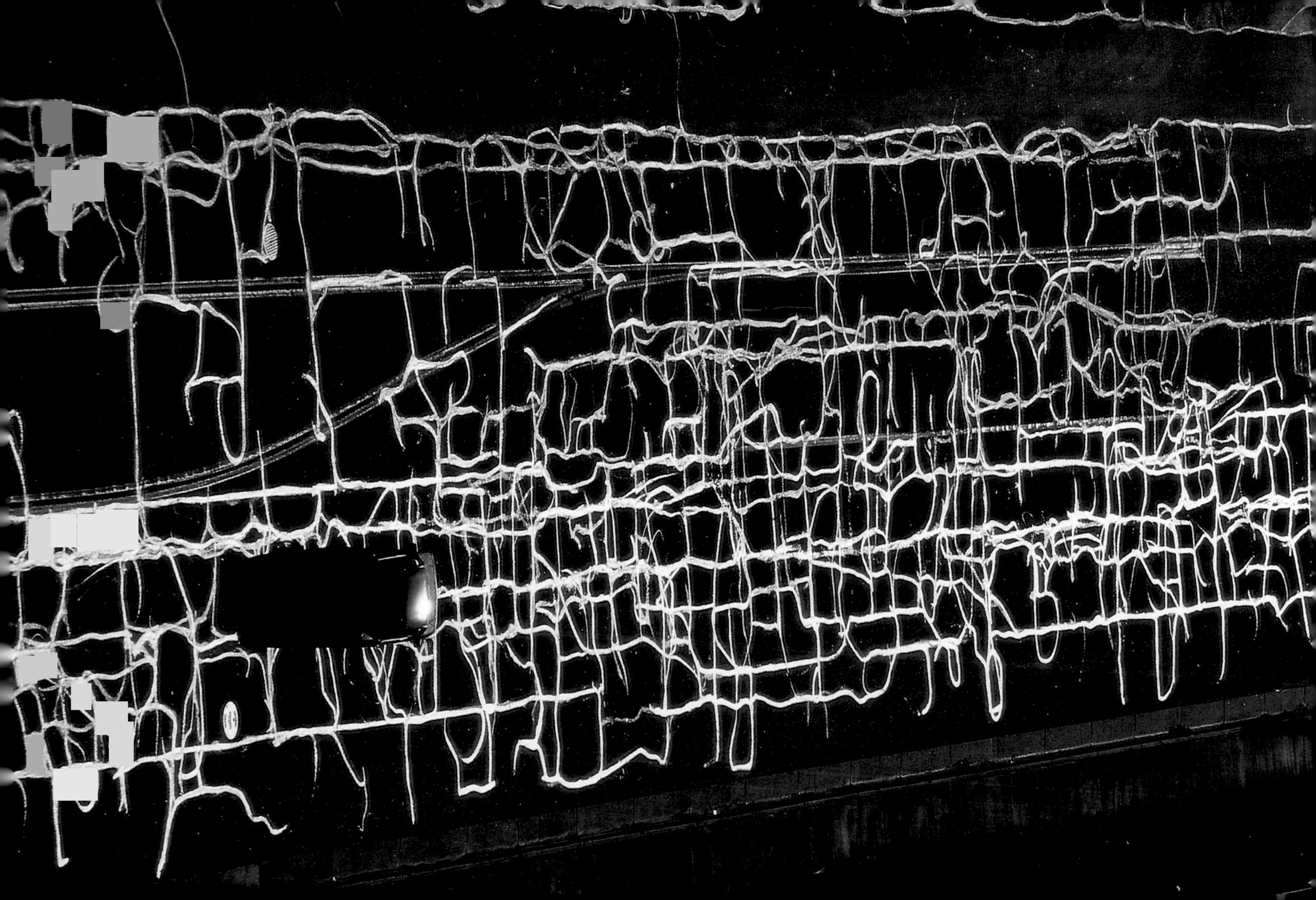

INDEX

ACKNOWLEDGMENTS

UNESCO: Mr. Federico Mayor, director-general, Mr. Pierre Lasserre, director of the ecological sciences division, Ms. Mireille Jardin, Ms. Jane Robertson, Ms. Josette Gainche and Mr. Malcolm Hadley, Ms. Hélène Gosselin, Mr. Carlos Marquès, Mr. Oudat-chine, of the public information office, Mr. Francesco di Castri and Ms. Jeanne Barbière, of the environmental coordination, as well as Mr. Gérard Huber, who was kind enough to support our project. When we go to press, with many happy memories from the four corners (!) of the Earth, we fear we may have forgotten some of you who helped us realize this project. We are truly sorry for this, and thank you all most warmly. Nor have we forgotten the many "anonymous" individuals who, out of the limelight, contributed to our incredible undertaking.
FUJIFILM: Mr. Masayuki Muneyuki, president, Mr. Toshiyuki "Todd" Hirai, Mr. Minoru, "Mick" Uranaka, of Fujifilm Tokyo, Mr. Peter Samwell, of Fujifilm Europe, and Ms. Doris Goertz, Ms. Develey, Mr. Marc Héraud, Mr. François Rychelewski, Mr. Bruno Baudry, Mr. Hervé Chanaud, Mr. Franck Portelance, Mr. Piotr Fedorowicz, and Ms. Françoise Moumaneix and Ms. Anissa Auger, of Fujifilm France.
CORBIS (1996-99): Mr. Stephen B. Davis, Mr. Peter Howe, Mr. Malcolm Cross, Mr. Charles Mauzy, Mr. Marc Walsh, Ms. Vanessa Kramer, Ms. Tana Wollen, and Ms. Vicky Whiley.
AIR FRANCE: Mr. François Brousse and Ms. Christine Micouleau, as well as Ms. Dominique Gimet, Ms. Mireille Queillé, and Ms. Bodo Ravoninjatovo.
EUROCOPTER: Mr. Jean-François Bigay, Mr. Xavier Poupardin, Mr. Serge Durand, and Ms. Guislaine Cambournac.
ALBANIA: ECPA, LtCol. Aussavy, DICOD, Col. Baptiste, Capt. Maranzana and Capt. Saint Léger, SIRPA, Mr. Charles-Philippe d'Orléans, DETALAT, Capt. Ludovic Janot; crews of the French air force, Mr. Etienne Hoff, Mr. Cyril Vasquèz, Mr. Olivier Ouakel, Mr. José Trouille, Mr. Frédéric Le Mouillour, Mr. François Dughin, Mr. Christian Abgral, Mr. Patrice Comerier, Mr. Guillaume Maury, Mr. Franck Novak, pilots.
ANTARCTICA: French Institute for Polar Research and Technology; Mr. Gérard Jugie; L'Astrolabe, Capt. Gérard R. Daudon, Sd. Capt., Alain Gaston; Heli Union France, Mr. Bruno Fiorese, pilot; Mr. Augusto Leri and Mr. Mario Zucchelli, Projetto Antartida, Italie Terra Nova.
ARGENTINA: Mr. Jean-Louis Larivière, Ediciones Larivière; Ms. Mémé et Ms. Marina Larivière; Mr. Felipe C. Larivière; Ms. Dudú von Thielman; Ms. Virginia Taylor de Fernández Beschtedt; Cdt. Sergio Copertari, pilot, Emilio Yañez and Pedro Diamante, co-pilots, Eduardo Benítez, mechanic; squadron of the federal air police, Capt. Norberto Edgardo; Gaudiero Capt. Roberto A. Ulloa, former governor of Salta province; police station of Orán, Salta province, Cdt. Daniel D. Pérez; Military Geographic Institute; Capt. Rodolfo E. Pantanali; Aerolineas Argentinas.
AUSTRALIA: Ms. Helen Hiscocks; Australian Tourism Commission, Ms. Kate Kenward and Ms. Gemma Tisdell and M. Paul Gauger; Jairow Helicopters; Heliwork, Mr. Simon Eders; Thai Airways, Ms. Pascale Baret; Club Med at Lindeman Island and Byron Bay Beach.
AUSTRIA: Mr. Hans Ostler, pilot.
BAHAMAS, THE: Club Med Eleuthera, Paradise Island, and Columbus Isle.
BANGLADESH: Mr. Hossain Kommol and Mr. Salahuddin Akbar, external publicity wing of the ministry of foreign affairs, His Eminence Tufail K. Haider, Bangladeshi ambassador in Paris and Mr. Chowdhury Ikthiar, first secretary, Her Excellency Ms. Renée Veyret, French ambassador in Dhaka, Mr. Mohamed Ali and Mr. Amjad Hussain, of Biman Bangladesh Airlines, as well as Vishawjeet, Mr. Nakada, Fujifilm Singapore, Mr. Ezaher of the Fujifilm laboratory in Dhaka, Mr. Mizanur Rahman, director, Rune Karlsson, pilot and J. Eldon. Gamble, technician, MAF Air Support, Ms. Muhiuddin Rashida, Sheraton Hotel in Dhaka, Mr. Minto.
BELGIUM: Mr. Thierry Soumagne, Mr. Wim Robberechts, Mr. Daniel Maniquet, Mr. Bernard Séguy, pilot.
BOTSWANA: Mr. Maas Müller, Chobe Helicopter.
BRAZIL: Government of Mato Grosso do Norte e do Sul; Fundação Pantanal, Mr. Erasmo Machado Filho and the French Regional Parks, Mr. Emmanuel Thévenin and Mr. Jean-Luc Sadorge; Mr. Fernando Lemos; His Eminence Pedreira, Brazilian ambassador to UNESCO; Dr. Iracema Alencar de Queiros, Instituto de Proteção Ambiental do Amazonas and his son Alexandro; Brasilia tourist office; Mr. Luis Carlos Burti, Burti Publishers; Mr. Carlos Marquès, of the OPI division of UNESCO; Ms. Ethel Leon, Anthea Communication; TV Globo; Golden Cross, Mr. José Augusto Wanderley and Ms. Juliana Marquès, Hotel Tropical in Manaus, VARIG.
CAMEROON: His Eminence Jacques Courbin, French ambassador in Chad, Mr. Yann Apert, cultural counsellor, Ms. Sandra Chevalier-Lecadre and the French Embassy in Chad, Mr. Lael Weyenberg and *A Day in the Life of Africa*, Mr. Thierry Miaillier, of RJM aviation, Mr. Jean-Marie Six and Aviation Sans Frontières, Mr. Bruno Callabat and Mr. Guy Bardet, pilots, Mr. Gérard Roso.
CANADA: Ms. Anne Zobenbuhler, Canadian Embassy in Paris and Canadian tourist office, Ms. Barbara di Stefano and Mr. Laurent Beunier, Destination Québec; Ms. Cherry Kemp Kinnear, Nunavut tourist office; Ms. Huguette Parent and Ms. Chrystiane Galland, Air Canada; First Air; Vacances Air Transat; Mr. André Buteau, pilot, Essor Helicopters; Mr. Louis Drapeau, Canadian Helicopters; Canadian Airlines.
CHAD: His Eminence Jacques Courbin, French ambassador in Chad, Mr. Yann Apert, cultural counsellor, Ms. Sandra Chevalier-Lecadre and the French embassy in Chad, Mr. Lael Weyenberg and *A Day in the Life of Africa*, Mr. Thierry Miaillier, of RJM aviation, Mr. Jean-Marie Six and Aviation Sans Frontières, Mr. Bruno Callabat and Mr. Guy Bardet, pilots, Mr. Gérard Roso.
CHILE: Ms. Véronica Besnier, Mr. Luis Weinstein, Mr. Jean-Edouard Drouault, of Eurocopter Chile, Capt. Fernando Perez of "Aviación del Ejército de Chile," Capt. Carlos Lopez, pilot, Capt. Patricio Gallo, director of operations for Eurocopter Chile, Capt. Carlos Ruiz, pilot, Capt. Gonzalo Maturana, co-pilot, Capt. Yerko Woldarski, pilot, Capt. Hernán Soruco, co-pilot, and A1C David Espinoza.
CHINA: Hong Kong tourist office, Mr. Iskaros; Chinese embassy in Paris, His Eminence Caifangbo, Ms. Li Beifen; French embassy in Beijing, His Eminence Pierre Morel, French ambassador in Beijing; Mr. Shi Guangeng, of the ministry of foreign affairs, Mr. Serge Nègre, kite flyer, Mr. Yann Layma.
CÔTE D'IVOIRE: Vitrail & Architecture; Mr. Pierre Fakhoury; Mr. Hugues Moreau and the pilots, Mr. Jean-Pierre Artifoni and Mr. Philippe Nallet, Ivoire Hélicoptères; Ms. Patricia Kriton and Mr. Kesada, Air Afrique.
CROATIA: Mr. Franck Arrestier, pilot.
DENMARK: Weldon Owen Publishing, the whole production team at "Over Europe"; Stine Norden.
DJIBOUTI, REPUBLIC OF: Mr. Ismaïl Omar Guelleh, president of the Republic, Mr. Osman Ahmed Moussa, minister of presidential affairs, Mr. Fathi Ahmed Houssein, division general, head of state, major general of the army, Mr. Hassan Said Khaireh, head of the military cabinet, Ms. Mouna Musong, counsellor to the president, National tourist office of Djibouti.
ECUADOR: Mr. Loup Langton and Mr. Pablo Corral Vega, Descubriendo Ecuador; Mr. Claude Lara, Ecuadorian ministry of foreign affairs; Mr. Galarza, Ecuadorian consulate in France; Mr. Eliecer Cruz, Mr. Diego Bouilla, Mr. Robert Bensted-Smith, Galapagos national park; Ms. Patrizia Schrank, Ms. Jennifer Stone, "European Friends of Galapagos"; Mr. Danilo Matamoros, Jaime and Cesar, Taxi Aero Inter Islas M.T.B.; Mr. Etienne Moine, Latitude 0°; Mr. Abdon Guerrero, San Cristobal airport.
EGYPT: Rally of the Pharaohs, "Fenouil," organizer, Mr. Bernard Seguy, Mr. Michel Beaujard, and Mr. Christian Thévenet, pilots; the staff of Paris-Dakar 2003 and Mr. Etienne Lavigne, of ASO.
FINLAND: Mr. Dick Lindholm, pilot.
FRANCE: Ms. Dominique Voynet, planning and environment minister; Ministry of Defense/SIRPA Paris police headquarters, Mr. Philippe Massoni and Ms. Seltzer; Montblanc Hélicoptères, Mr. Franck Arrestier and Mr. Alexandre Antunes, pilots; Corsica

tourist office, Mr. Xavier Olivieri; Auvergne tourist committee, Ms. Cécile da Costa; Côtes d'Armor general council, Mr. Charles Josselin and Mr. Gilles Pellan; Savoie general council, Mr. Jean-Marc Eysserick; Haute-Savoie general council, Mr. Georges Pacquetet and Mr. Laurent Guette; Alpes-Maritimes general council, Ms. Sylvie Grosgojeat and Ms. Cécile Alziary; Yvelines general council, Mr. Franck Borotra, president, Ms. Christine Boutin, Mr. Pascal Angenault and Ms. Odile Roussillon; Loire tourist board; Rémy Martin, Ms. Dominique Hériard-Dubreuil, Ms. Nicole Bru, Ms. Jacqueline Alexandre; Éditions du Chêne, Mr. Philippe Pierrelee, artistic director; Hachette, Mr. Jean Arcache; Moët et Chandon/Rallye GTO, Mr. Jean Berchon and Mr. Philippe des Roys du Roure; Printemps de Cahors, Ms. Marie-Thérèse Perrin; Mr. Philippe Van Montagu and Willy Gouere, pilot, SAF Hélicoptères, Mr. Christophe Rosset, Hélifrance, Héli-Union, Europe Helicoptère Bretagne, Héli Bretagne, Héli-Océan, Héli Rhône-Alpes, Hélicos Légers Services, Figari Aviation, Aéro service, Héli Air Monaco, Héli Perpignan, Ponair, Héli-inter, Héli Est; La Réunion: La Réunion tourist office, Mr. René Barrieu and Ms. Michèle Bernard; Mr. Jean-Marie Lavèvre, pilot, Hélicoptères Helilagon; New Caledonia: Mr. Charles de Montesquieu, Mr. Daniel Pelleau d'Hélicocéan and Mr. Bruno Civet d'Héli Tourisme; Antilles: Club Med Boucaniers and La Caravelle; Mr. Alain Fanchette, pilot; Polynesia: Club Med Moorea; Haute-Garonne: Ms. Carole Schiff, Mr. Alexandre Antunès, pilot; Lyon and the area: Ms. Béatrice Shawannn, Mr. Christophe Schereich, Mr. Daniel Pujol (Flood pilots of the Saône, Taponas); Pyrénées-atlantiques: DICOD and SIRPA.
GERMANY: Mr. Peter Becker, pilot, Ms. Ruth Eichhorn, Ms. Geneviève Teegler and the entire staff at *GEO* Germany, Mr. Wolfgang Mueller-Pietralla, of Autostadt, Mr. Frank Müller-May and Mr. Tom Jacobi, of *Stern* magazine.
GIBRALTAR: Mr. David Durie, governor of Gibraltar, Mr. John Woodruffe, of the governor's office, Col. Purdom, Lt. Brian Phillips, Ms. Béatrice Quentin, Ms. Peggy Pere, Mr. Franck Arrestier, pilot, Mr. Jérôme MARX, mechanic.
GREECE: Ministry of culture in Athens, Ms. Eleni Méthodiou, Greek delegation to UNESCP; Greek tourist office; Club Med Corfou Ipsos, Gregolimano, Helios Corfou, Kos, and Olympia; Olympic Airways; Interjet, Mr. Dimitrios Prokopis and Mr. Konstantinos Tsigkas, pilots, and Kimon Daniilidis; Athens weather center.
GUATEMALA AND HONDURAS: Mr. Giovanni Herrera, director, and Carlos Llarena, pilot, Aerofoto in Guatemala City; Mr. Rafael Sagastume, STP villas in Guatemala City.
HUNGARY: The staff of the French embassy in Budapest, the Mayor of Budapest, L'Institut Français de Budapest.
ICELAND: Mr. Bergur Gislasson and Mr. Gisli Guestsson, Icephoto Thyrluthjonustan Helicopters; Mr. Peter Samwell; National tourist office in Paris.
INDIA: Indian embassy in Paris, His Eminence Kanwal Sibal, ambassador, Mr. Rahul Chhabra, first secretary, Mr. S.K. Sofat, air brigade genereal, Mr. Lal, Mr. Kadyan and Ms. Vivianne Tourtet; ministry of foreign affairs, Mr. Teki E. Prasad and Mr. Manjish Grover; Mr. N.K. Singh, of the prime minister's office; Mr. Chidambaram, member of parliament; Air Headquarters, S.I. Kumaran, Mr. Pande; Mandoza Air Charters, Mr. Atul Jaidka, Indian International Airways, Capt. Sangha Pritvipalh; French embassy in New Delhi, His Eminence Claude Blanchemaison, French ambassador in New Delhi, Mr. François Xavier Reymond, first secretary.
INDONESIA: Total Balikpapan, Mr. Ananda Idris and Ms. Ilha Sutrisno; Mr. and Mrs. Didier Millet.
IRELAND: Aer Lingus; Irish national tourist office; Capt. David Courtney, Irish Rescue Helicopters; Mr. David Hayes, Westair Aviation Ltd.
ITALY: French embassy in Rome, Mr. Michel Benard, press office; Heli Frioula, Mr. Greco Gianfranco, Mr. Fanzin Stefano, and Mr. Godicio Pierino.
JAPAN: Eu Japan Festival, Mr. Shuji Kogi and Mr. Robert Delpire; Masako Sakata, IPJ; NHK TV; Japan Broadcasting Corp.; Asahi Shimbun newspaper group, Me. Teizo Umezu.
JORDAN: Ms. Sharaf, Mr. Anis Mouasher, Mr. Khaled Irani, and Mr. Khaldoun Kiwan, Royal Society for Conservation of Nature; Royal Airforces; Mr. Riad Sawalha, Royal Jordanian Regency Palace Hotel.
KAZAKHSTAN: His Eminence Nourlan Danenov, Kazakhstan ambassador in Paris; His Eminence Alain Richard, French ambassador in Almaty, and Ms. Josette Floch; Prof. René Letolle; Heli Asia Air and its pilot Mr. Anouar.
KENYA: Universal Safari Tours of Nairobi, Mr. Patrix Duffar; Transsafari, Mr. Irvin Rozental.
KUWAIT: Kuwait Centre for Research & Studies, Prof. Abdullah Al Ghunaim, Dr. Youssef; Kuwait National Commission for UNESCO, Sulaiman Al Onaizi; Kuwait delegation to UNESCO, His Excellency Dr. Al Salem, and Mr. Al Baghly; Kuwait Airforces, Squadron 32, Maj. Hussein Al-Mane, Capt. Emad Al-Momen; Kuwait Airways, Mr. Al Nafisy.
LEBANON: Mr. Lucien George, Mr. Georges Salem, Lebanese military.
LUXEMBURG: Mr. Bernard Séguy, pilot.
MADAGASCAR: Mr. Riaz Barday and Mr. Normand Dicaire, pilots, Aéromarine; Sonja and Thierry Ranarivelo, Mr. Yersin Racerlyn, pilot, Madagascar Hélicoptère; Mr. Jeff Guidez and Lisbeth.
MALAYSIA: Club Med Cherating.
MALDIVES: Club Med Faru.
MALI: TSO, Paris-Dakar Rally, Mr. Hubert Auriol; Mr. Daniel Legrand, Mr. Arpèges Conseil, and Mr. Daniel Bouet, Cessna pilot.
MAURITANIA: TSO, Paris-Dakar Rally, Mr. Hubert Auriol; Mr. Daniel Legrand, Mr. Arpèges Conseil, and Mr. Daniel Bouet, Cessna pilot; Mr. Sidi Ould Kleib.
MEXICO: Club Med Cancun, Sonora Bay, Huatulco, and Ixtapa.
MONACO: His Royal Highness Prince Albert of Monaco, Col. Lambelin, Col. Jouan, Ms. Catherine Alestchenkoff, of Grimaldi Forum, Mr. Patrick Lainé, pilot.
MONGOLIA: His Eminence Jacques-Olivier Manent, French ambassador in Mongolia, His Eminence Louzan Gotovddorjiin, Mongolian ambassador in France, Tuya of Mongolia Voyages, Mongolian military.
MOROCCO: Royal Moroccan police headquarters, Gen. El Kadiri and Col. Hamid Laanigri; Mr. François de Grossouvre.
NAMIBIA: Ministry of Fisheries; French cooperation mission, Mr. Jean-Pierre Lahaye, Ms. Nicole Weill, Mr. Laurent Billet, and Mr. Jean Paul Namibian; Tourist Friend, Mr. Almut Steinmester.
NEPAL: Nepal embassy in Paris; Terres d'Aventure, Mr. Patrick Oudin; Great Himalayan Adventures, Mr. Ashok Basnyet; Royal Nepal Airways, Mr. JB Rana; Mandala Trekking, Mr. Jérôme Edou, Bhuda Air; Maison de la Chine, Ms. Patricia Tartour-Jonathan, director, Ms. Colette Vaquier, and Ms. Fabienne Leriche; Ms. Marina Tymen and Ms. Miranda Ford, Cathay Pacific.
NETHERLANDS, THE: Paris-Match; Mr. Franck Arrestier, pilot.
NIGERIA: TSO, Paris-Dakar Rally, Mr. Hubert Auriol; Mr. Daniel Legrand, Mr. Arpèges Conseil, and Mr. Daniel Bouet, Cessna pilot.
NORWAY: Airlift A.S., Mr. Ted Juliussen, pilot, Mr. Henry Hogi, Mr. Arvid Auganaes, and Mr. Nils Myklebust.
OMAN: His Majesty Sultan Quabous ben Saïd al-Saïd; ministry of defense, Mr. John Miller; Villa d'Alésia, Mr. William Perkins and Ms. Isabelle de Larrocha.
PERU: Dr. Maria Reiche and Ms. Ana Maria Cogorno-Reiche; ministry of foreign affairs, Mr. Juan Manuel Tirado; Peruvian national police; Faucett Airline, Ms. Cecilia Raffo and Mr. Alfredo Barnechea; Mr. Eduardo Corrales, Aero Condor.
PHILIPPINES, THE: Filipino Airforces; *Seven Days in the Philippines* by Millet Publishers, Ms. Jill Laidlaw.
PORTUGAL: Club Med Da Balaia, Ms. Ana Pessoa and ICEP, HeliPortugal, and Ms. Margarida Simplício, IPPAR.
RUSSIA: Mr. Yuri Vorobiov, vice-minister, and Mr. Brachnikov, Emerkom; Mr. Nicolaï Alexiy Timochenko, Emerkom in Kamtchatka; Mr. Valery Blatov, Russian delegation to UNESCO.
ST. VINCENT AND THE GRENADINES: Mr. Paul Gravel, SVG Air; Ms. Jeanette Cadet, The Mustique Company; Mr. David Linley; Mr. Ali Medjahed, baker; Mr. Alain Fanchette.
SENEGAL: TSO, Paris-Dakar Rally, Mr. Hubert Auriol; Mr. Daniel Legrand, Mr. Arpèges Conseil, and Mr. Daniel Bouet, Cessna pilot; Club Med Almadies and Cap Skirring.
SOMALIA: His Royal Highness Sheikh Saud Al-Thani of Qatar; Mr. Majdi Bustami, Mr. E. A. Paulson and Osama, office of His Royal Highness Sheikh Saud Al-Thani; Mr. Fred Viljoen, pilot; Mr. Rachid J. Hussein, UNESCO-Peer Hargeisa, Somalia; Mr. Nureldin Satti, UNESCO-Peer, Nairobi, Kenya; Ms. Shadia Clot, representative of the Sheikh in France; Waheed,

Al Sadd travel agency, Qatar; Cécile et Karl, Emirates Airlines, Paris.
SOUTH AFRICA: SATOUR, Mrs. Salomone, South African Airways, Jean-Philippe de Ravel, Victoria Junction, Victoria Junction Hotel.
SPAIN: His Eminence Jesus Ezquerra, Spanish ambassador to UNESCO; Club Med Don Miguel, Cadaquès, Porto Petro, and Ibiza; Canary Islands: Mr. Tomás Azcárate y Bang, Viceconsejería de Medio Ambiente Fernando Clavijo, Protección Civil de las Islas Canarias; Mr. Jean-Pierre Sauvage and Mr. Gérard de Bercegol, Iberia; Ms. Elena Valdés and Ms. Marie Mar, Spanish tourist office; Basque country: office of the Basque government presidency; Mr. Zuperia Bingen, director, Ms. Concha Dorronsoro and Ms. Nerea Antia, press office of the Basque government presidency, Mr. Juan Carlos Aguirre Bilbao, head of the helicopter unit of the Basque police (Ertzaintza) *la corrida*.
SWEDEN: Stine Norden.
TAIWAN: Ms. Helene Lai, civil aviation office of the ministry of transport of Taiwan.
THAILAND: Royal forestry department, Mr. Viroj Pimanrojnagool, Mr. Pramote Kasemsap, Mr. Tawee Nootong, Mr. Amon Achapet; NTC Intergroup Ltd, Mr. Ruhn Phiama; Ms. Pascale Baret, Thai Airways; Thai national tourist office, Ms. Juthaporn Rerngronasa and Watcharee, Mr. Lucien Blacher, Mr. Satit Nilwong, and Mr. Busatit Palacheewa; Fujifilm Bangkok, Mr Supoj; Club Med Phuket.
TUNISIA: Mr. Zine Abdine Ben Ali, president of the Republic; presidency of the Republic, Mr. Abdelwahad Abdallah and Mr. Haj Ali; Tunisian air force, Laouina base, Col. Mustafa Hermi; Tunisian embassy in Paris, His Eminence Bousnina, ambassador and Mr. Mohamed Fendri; Tunisian national tourist office, Mr. Raouf Jomni and Mr. Mamoud Khaznadar; Editions Cérès, Mr. Mohamed and Mr. Karim Ben Smail; Hotel The Residence, Mr. Jean-Pierre Auriol; Basma-Hôtel Club Paladien, Mr. Laurent Chauvin; Tunisian weather center, Mr. Mohammed Allouche.
TURKEY: Turkish Airlines, Mr. Bulent Demirçi and Ms. Nasan Erol; Mach'Air Helicopters, Mr. Ali Izmet, Öztürk and Seçal Sahin, Ms. Karatas Gulsah; General Aviation, Mr. Vedat Seyhan and Faruk, pilot; Club Med Bodrum, Kusadasi, Palmiye, Kemer, Foça.
UKRAINE: Mr. Alexandre Demianyuk, secretary-general of UNESCO; Mr. A. V. Grebenyuk, director of the Chernobyl exclusion zone; Ms. Rima Kiselitza, of Chernobylinterinform, Ms. Marie-Renée Tisné, office for the protection against ionizing radiation.
UNITED KINGDOM: *England:* Aeromega and Mike Burns, pilot; Mr. David Linley; Mr. Philippe Achache; Environment Agency, Mr. Bob Davidson and Mr. David Palmer; Press Office of Buckingham Palace; *Scotland:* Ms. Paula O'Farrel and Mr. Doug Allsop of Total oil marine in Aberdeen; Mr. Iain Grindlay and Mr. Rod de Lothian, Helicopters Ltd in Edinburgh.
UNITED STATES: *Wyoming:* Yellowstone National Park, Ms. Marsha Karle and Ms. Stacey Churchwell; *Utah:* Classic Helicopters; *Montana:* Carisch Helicopters, Mr. Mike Carisch; *California:* Ms. Robin Petgrave, of Bravo Helicopters in Los Angeles and the pilots Ms. Akiko K. Jones and Mr. Dennis Smith; Mr. Fred London, Cornerstone Elementary School; *Nevada:* Mr. John Sullivan and the pilots Mr. Aaron Wainman et Mr. Matt Evans, Sundance Helicopters, Las Vegas; *Louisiana:* Suwest Helicopters and Mr. Steve Eckhardt; *Arizona:* Southwest Helicopters and Mr. Jim McPhail; *New York:* Liberty Helicopters and Mr. Daniel Veranazza; Mr. Mike Renz, Analar helicopters, Mr. John Tauranac; *Florida:* Mr. Rick Cook, Everglades National Park, Rick and Todd, Bulldog Helicopters in Orlando, Chuck and Diana, Biscayne Helicopters, Miami, Club Med Sand Piper; *Alaska:* Mr. Philippe Bourseiller, Mr. Yves Carmagnole, pilot; *Colorado:* Ms. Elaine Hood of Raytheon Polar Services Company and Ms. Karen Wattenmaker, Denver.
UZBEKISTAN: (not flown over) Uzbekistan embassy in Paris, His Eminence Mamagonov, ambassador, and Mr. Djoura Dekhanov, first secretary; His Eminence Jean Claude Richard, French ambassador in Uzbekistan, and Mr. Jean Pierre Messiant, first secretary; Mr. René Cagnat and Natacha; Mr. Vincent Fourniau and Mr. Bruno Chauvel, Institut Français d'Etudes sur l'Asie Centrale (IFEAC).
VENEZUELA: Centro de Estudios y Desarrollo, Mr. Nelson Prato Barbosa; Hoteles Intercontinental; Ultramar Express; Lagoven; Imparques; Icaro, Mr. Luis Gonzales.

We would also like to thank the companies that have allowed us to work thanks to contracts or exchanges:
AÉROSPATIALE, Mr. Patrice Kreis, Mr. Roger Benguigui, and Cotinaud.
AOM, Ms. Françoise Dubois-Siegmund and Ms. Felicia Boisne-Noc, Mr. Christophe Cachera.
CANON, Mr. Guy Bourreau, Mr. Pascal Briard, Service Pro, Mr. Jean-Pierre Colly, Mr. Guy d'Assonville, Mr. Jean-Claude Brouard, Mr. Philippe Joachim, Mr. Raphaël Rimoux, Mr. Bernard Thomas, and of course Mr. Daniel Quint and Ms. Annie Rémy who helped us so often throughout the project.
CLUB MED, Mr. Philippe Bourguignon, Mr. Henri de Bodinat, Ms. Sylvie Bourgeois, Mr. Preben Vestdam, Mr. Christian Thévenet.
CRIE, world express mail, Mr. Jérôme Lepert and his whole staff.
DIA SERVICES, Mr. Bernard Crepin.
FONDATION TOTAL, Mr. Yves le Goff and his assistant Ms. Nathalie Guillerme.
JANJAC, Mr. Jacques and Olivier Bigot, Mr. Jean-François Bardy and Mr. Eric Massé.
KONICA, Mr. Dominique Bruguière.
MÉTÉO FRANCE, Mr. Foidart, Ms. Marie-Claire Rullière, Mr. Alain Mazoyer and all the forecasters.
RUSH LABO, Mr. Denis Cuisy and all our friends at the lab.
WORLD ECONOMIC FORUM of Davos, Dr. Klaus Schwab, Ms. Maryse Zwick, and Ms. Agnès Stüder.

The team of "La Terre vue du Ciel," Altitude photo agency:
Photography assistants: Franck Charel, Françoise Jacquot, Ambre Mayen, and Erwan Sourget, who followed the entire project, not forgetting Sibylle d'Orgeval and Arnaud Prade, who rejoined us these past two years, and all those who were involved at different times during these years of flying: Denis Lardat, Tristan Carné, Christophe Daguet, Stefan Christiansen, Pierre Cornevin, Olivier Jardon, Marc Lavaud, Franck Lechenet, Olivier Looren, and Antonio López Palazuelo.

Pilot of the Colibri EC 120 Eurocopter: Wilfrid Gouère aka "Willy".

COORDINATING OFFICE:
Production coordinator: Hélène de Bonis (1994-99) and Françoise Le Roch'-Briquet.
Publication and editing of captions coordinator: Isabelle Delannoy, with assistance from Emilie Tran-Phong, Nicolas Cennac, Julien Nennault, and Audrey Salas, as well as Nadia Auriat of UNESCO.
Exhibition coordinator: Catherine Arthus-Bertrand, Tiphanie Babinet, and Jean Poderos.
Production assistants: Antoine Verdet, Catherine Quilichini, Gloria-Céleste Raad (Russia).
Editing staff: Danielle Laruelle, Judith Klein, Hugues Demeude, Sophie Hurel and PRODIG, geographical laboratories, Ms. Marie-Françoise Courel and Ms. Lydie Goeldner, Mr. Frédéric Bertrand.
Research: Isabelle Lechenet, Florence Frutoso, Claire Portaluppi.

PHOTOGRAPH CREDITS

All photographs in *The Earth From The Air: 366 Days* are by Yann Arthus-Bertrand, except:
January 21, February 1, February 2, February 27, April 14, May 10, May 22, July 19, September 13, October 4, October 7, November 14: © Helen Hiscocks
March 21: © Max PPP
May 17: © Jim Wark
June 16, June 20, October 16, December 11: © Philippe Bourseiller
August 12: © François Jourdan
August 19: © Renaud Van der Meeren

All the images in this book are distributed through Altitude photo agency
(except March 24):
altitude@club-internet.fr
www.yannarthusbertrand.org

All photographs in this book were taken on Fuji Velvia (50 ASA) film. Yann Arthus-Bertrand worked primarily with CANON EOS 1N camera bodies and CANON L series lenses. Some pictures were taken with a PENTAX 645N camera and a FUJI GX 617 panoramic camera.

The printing process used in the publication of this book respects the environment.

Translated from the French by Simon Jones

First published in the United Kingdom in 2003 by Thames & Hudson Ltd, 181A High Holborn, London WC1V 7QX

www.thameshudson.com

British Library Cataloguing-in-Publication Data
A catalogue record for this book is available from the British Library

ISBN 0-500-54278-3

Printed and bound in Italy